MONSOONAL AUSTRALIA
LANDSCAPE, ECOLOGY AND MAN IN THE NORTHERN LOWLANDS

Bark painting: Ochre on stringbark, artist: Bargudubu, tribe: Gunwinggu, Western Arnhemland.

'Borlung' is one of the names given to the mythical rainbow snake who is the most revered totemic ancestor of Western Arnhemland Aborigines. She is believed to be the mother creator controlling human fertility, seasonal change and the regeneration of nature and is therefore the most revered totemic ancestor. Borlung's spirit lives in water holes on earth, often appearing as a rainbow arched across the sky or as rainbow hues in drops of water. Thunder is her voice raised in anger and lightning her forked tongue flickering across the sky.

Photography: Jennifer Steele.

MONSOONAL AUSTRALIA

Landscape, ecology and man
in the northern lowlands

Edited by
C.D.HAYNES
Department of Conservation and Land Management, Perth
M.G.RIDPATH
Tropical Ecosystems Research Centre, CSIRO, Darwin
M.A.J.WILLIAMS
Department of Geography and Environmental Science, Monash University, Melbourne

A.A.BALKEMA / ROTTERDAM / BROOKFIELD / 1991

Published by
A.A. Balkema, P.O. Box 1675, 3000 BR Rotterdam, Netherlands
A.A. Balkema Publishers, Old Post Road, Brookfield, VT 05036, USA

ISBN 90 6191 638 0
© 1991 A.A. Balkema, Rotterdam
Printed in the Netherlands

In memory of Harry Frith

Table of contents

Foreword

Dreams and schemes have been held for Australia's tropical lowlands, and most have foundered. Cropping has been tried and failed on the plains of the Daly, Ord, and Adelaide Rivers, for example. Summer rainfall is high and natural productivity of much of the lowland is high, so it seems odd that the human population is really very small. When we note that Dr. Holze had over 60 crop plants in his botanic gardens at Darwin, at the turn of the century, it also seems odd that most cereals, fruit, and vegetables eaten today are imported from southern parts of Australia. These and other peculiarities of white peoples's adaptation – or lack of it – in the Australian tropical lowlands, arise because for them this is a very unfamiliar ecologic theatre. In contrast, the Aboriginal people, who had a culture integrated with the seasons and biology of the monsoonal tropics for hundreds of generations, became victims of the white society which floundered in these torrid lands despite its achievements elsewhere.

Eventually civilisation in the north will be enriched if adaptation – in the fullest sense – is realised through knowledge of the environment and its natural ways. Education of many people in the results of scientific enquiry is one of our important methods of progress. It therefore gives me pleasure to introduce this book, produced by three men whose work in the north has been extensive, sympathetic, and careful. In keeping with the progression from explorers to scientific teams, contributors to this book have done their research in the north under the auspices of various organisations. Of the editors (who also are authors of some chapters herein), Chris Haynes was responsible for development of Kakadu National Park during his time with Australian National Parks and Wildlife Service, Mike Ridpath was in charge of CSIRO's Darwin laboratories and developed the research programme at its Kapalga field station in the park, and Martin Williams enlarged upon his PhD research near Adelaide River by spending a period with the CSIRO Land Research team in the region, and has since maintained his northern interest. The entire group of nineteen scientists who write in this book provide, through direct experience as well as scholarship, a volume of natural history which should be widely assimilated.

The theme is ecological and the focus is on the profound wet-dry seasonal contrast to which flora, fauna, and humans adapted. Scientific questions about evolution of the biota over the geologically long period of Australia's desiccation are sharpened by recognition of large changes of climate and sea level over the last 20,000 years; knowledge gained about processes on these time scales improves our understanding of today's landscape and ecology. Plant and animal communities migrated and regrouped as coasts and the northern monsoon advanced upon an ancient land with poor soils, while sea level rose across the broad Sahul shelf between 20,000 and 6000 years ago. The relative proportions of sclerophyll forests, woodlands, rainforest, and wetlands doubtless changed during this march of vegetation.

The annual cycle of rains alternating with drought has drawn different responses from different organisms; the monsoon forests include deciduous trees, for example. As water availability changes, and the climaxes of new growth and of fruit or seed production move amongst plants with the march of seasons, so too do birds and animals move around the region. Aborigines lived by these interlocking annual cycles, and skilfully used fire to promote selected plants and animals to their advantage, thus adding their imprint to change and adaptation of the biologic communities. Termites, man, and Magpie Geese equally were components of the ecosystems, before Europeans arrived with their attempts at temperate styles of agriculture, with plants which have become weeds, and livestock which have gone feral. Although experiments with farming methods and crops more suited to the north will continue, successful moulding of our monsoonal lowlands to needs of the future ultimately depends on our learning the vocabulary and ways of nature in the region. This book provides a very suitable foundation.

John Chappell
Department of Biogeography & Geomorphology
Australian National University, Canberra

Preface

The natural gateway into Australia is by its monsoonal North. The first Aboriginal colonization probably came across now submerged land to enter the north of the continent more than 40,000 years ago (Nix 1981). In historic times the north coast appears to have been long known to the Macassans, conceivably even to a 15th century Chinese navigator (McKnight 1969). It now seems likely that the first Europeans to see Australia were Portuguese sailors who visited its northern regions in the early 16th century (Wallis 1982), followed 100 years later by the Dutch.

This book is about the natural features of the Australian monsoonal lowlands, generally extending 100-200 km inland from the tropical coasts, and their biota. Together with a small area of humid tropics in northeast Queensland, this region forms the most fertile and biologically productive part of the Australian tropics. When Australia was colonized the monsoonal lowlands supported one of the most dense Aboriginal populations on the continent. The first European colonists largely failed to understand this indigenous culture. In particular they were unaware of how the Aborigines had come to terms with the tropical environment and developed appropriate and often sophisticated responses to it. The book also pays some attention to these aspects.

The monsoonal lowlands and life in them are dominated by four characteristics of the tropical environment. Of these, the constantly high temperatures and 12 hours of daily light virtually year round are shared with the rest of the tropics. However the monsoonal lowlands fall into the zone of the wet-dry (or seasonal) tropics, where annual rainfall is heavy and markedly seasonal but of variable onset and duration (Ridpath 1985). We are just beginning to appreciate the multifarious consequences for land and life of this unique regime. The effects of the savage alternation each year between the climatic conditions of a jungle and those of a desert are the major and recurrent theme throughout the book (Plates 1-3).

Scientific research in the monsoonal lowlands gathered momentum after the end of the 1939-1945 war. The last 25 years in particular have seen a dramatic upsurge in systematically organized and increasingly integrated field studies of this environment. At a workshop on tropical wetlands held in Darwin in August 1980 in honour of his pioneering research, the late Dr Harry Frith, Chief of CSIRO Division of Wildlife Research, commented on these advances. He suggested that the accumulation of published and unpublished results of research had reached the stage at which an attempt should be made to put them together. At his prompting we undertook the task of producing such a volume. The authors have written their respective chapters with all those in mind who have an interest in this environment, be they other scientists or members of the wider public (for whom a glossary of technical terms has been included). It is our hope, as it was that of Harry Frith, that the contributions printed here will also be useful in establishing some principles upon which the wise management of the monsoonal lowlands can be based.

This book deals with the area of the monsoonal lowlands between the Victoria and Roper Rivers shown in Figure 1, especially that between the Daly and Blyth Rivers. Nonetheless much of the book may also be relevant to monsoonal lowlands beyond this immediate context. The book comprises four parts each introduced by a short commentary. The first part deals with the landscape and the physical factors acting on it; the second concerns vegetation and the impact upon it of fire; the third concerns the ecology of the vertebrate animals; and the fourth describes the ecology of humans with some discussion of their present and future roles. No book on the monsoonal lowlands can yet be definitive. The editors and authors will be well rewarded if this volume encourages the pursuit of that goal.

Our grateful thanks go to John Cleasby and Gary Swinton for skilful draughting of the figures, and to Alan Fried, Helen Quilligan, Nicola Smith and Tonia Stokes for help with the proofs and index.

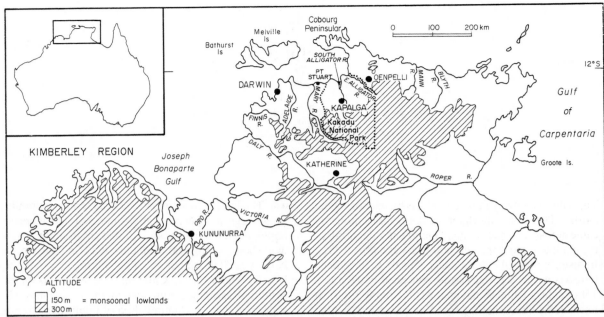

Figure 1. Map of the 'Top End' of the Northern Territory and adjacent regions. The monsoonal lowlands are defined as those areas from 0 to 150 m above sea level, generally extending 100-200 km inland from the coast. The scope of this book covers the monsoonal lowlands from the Victoria River in the west to the Roper River in the east.

REFERENCES

McKnight, C.C. 1969. Introduction. In C.C. McKnight (ed.), *The Farthest Coast*: 1-35. Melbourne: Melbourne Univ. Press.

Nix, H.A. 1981. The environment of Terra Australis. In A. Keast (ed.), *Ecological Biogeography of Australia*: 103-133. The Hague: Junk.

Ridpath, M.G. 1985. Ecology in the Wet-Dry Tropics – How different? In M.G. Ridpath & L.K. Corbett (eds.), *The Ecology of the Wet-Dry Tropics, Proc. Ecol. Soc. Aust.* 13: 3-20.

Wallis, H. 1982. The Rotz Atlas. A royal presentation. *The Map Collector* 20: 39-42.

Plate 1. Sedgelands (mainly *Eleocharis* spp.) at Kapalga (Northern Territory) in the late part of the dry season. Compare with Plates 2 and 3 which show the same point in the wet season. This is the habitat of the dusky rat *Rattus colletti* (Photograph: C.K.Williams).

Plate 2. Sedgelands at Kapalga in the early part of the wet season, at the same point as Plates 1 and 3 (Photograph: C.K.Williams).

Plate 3. Sedgelands at Kapalga in the late part of the wet season, at the same point as Plates 1 and 2 (Photograph: C.K.Williams).

Part 1:
The physical environment

Introduction

Monsoonal Australia is a land of stark relief and sharp climatic contrasts. Dry and dusty for seven months of the year, it is rapidly transformed into a land of floods and downpours during the ensuing five months of the wet season, from November until March. Over many generations, plants, animals and human societies have striven to adapt themselves to the constraints imposed by the prevailing highly seasonal hydrological regime. Apart from being able to cope with monsoon floods as well as the day to day vagaries of the weather, the biota of northern Australia must also be able to survive in what is effectively a desert for more than half the year. How they manage to do all this is the subject of the last ten chapters of this book. In these first three chapters on landscape, climate and hydrology, our aim is to set the scene for the later detailed discussions of how the plant, animal and human communities of this part of Australia have responded to the challenge and possibilities inherent in living in the seasonally-wet tropics.

The northern landscape is a mosaic of elements that are exceedingly ancient and others that are geologically very young. In the opening chapter to this volume (Chapter 1), Williams describes the progressive development of the landscape from earliest Precambrian times until the last few thousand years. Although geographically restricted in area to some 50,000 km², the region between Darwin, Arnhem Land and Pine Creek contains a generous slice of the geological history of northern Australia.

Until about 1,800 million years ago, much of this region was a broad subsiding basin in which sediments eroded from the adjacent uplands accumulated to a depth of 14,000 metres. Deposition was arrested by earth movements of sub-continental extent, and the sedimentary and volcanic rocks within the basin were altered by heat and pressure and deformed by folding. Prolonged erosion ensued. The resulting landscape of low relief was buried beneath several hundred metres of what are now the Arnhem Land sandstones. Henceforth the area remained geologically stable.

A brief incursion of the sea 110 million years ago submerged much of the region, so that when the sea retreated the land was covered with a thin blanket of marine sands and muds. For long the land remained low lying and exposed to deep weathering. Slow uplift began some 15 million years ago, resulting in enhanced erosion and accentuated relief.

During the past few million years world sea levels have risen and fallen through 150 metres as the great continental ice sheets of the Northern Hemisphere have waxed and waned, at roughly 100,000 year intervals, but with many shorter term fluctuations. The last time of very low sea level was 20,000 years ago, when much of Australia was colder, drier and windier than today. As the ice sheets melted and sea levels rose, the prehistoric inhabitants of northern Australia were forced relentlessly inland. The land bridge between Australia and New Guinea was submerged by 9,000 years ago, and by 6,000 years ago the sea had reached its present level. Extensive mangrove swamps developed on what are now the coastal plains. These swamps were soon buried beneath several metres of alluvial clay laid down by meandering tidal rivers. As the tidal rivers changed shape and became less sinuous, the zone of saltwater incursion extended further upstream. The alluvial and coastal elements of the landscape are still evolving rapidly, in contrast to the rocky foothills and rugged sandstone plateau, where the tempo of change is much slower.

Running water is of course the dynamic agent of landscape change, and in Chapter 2, McDonald and McAlpine review the climatic causes of the floods and droughts that afflict the North in alternating seasons. Although highly seasonal, the rainfall is remarkably reliable in terms of incidence and quantity. Nowhere else in Australia is precipitation so seasonal; nowhere else is it so reliable.

Nevertheless, in certain years characterised by unusually low sea surface temperatures and unusually high atmospheric pressures in the general region of Indonesia and northern Australia, the monsoonal rains begin late and end early. These are often years of drought in the eastern half of Australia, and are linked to changes in global atmospheric circulation associated with El Nino events. Conversely, some years are wetter than average, and in those years air pressures remain conspicuously lower than normal and sea surface temperatures offshore are also anomalously high.

Given the low year to year wet season variability in rainfall, it comes as no surprise to discover that although the growing season for plants is restricted to six months of the year, it never fails. Furthermore, despite the fact that false starts to the growing season are common, false ends are very rare.

The drainage characteristics of the northern plains are considered by Kingston in Chapter 3. Until quite recently, very little was known about the hydrological behaviour of the coastal rivers in the northern wetlands. The first task of the water engineers was to install stream gauges for monitoring changes in river level. Analysis of such data makes it possible to define the number of days when floods of a given magnitude are likely to occur within the various coastal rivers in the area.

In addition, Kingston summarises twenty years of his own professional observations in the form of a map showing drainage status on different parts of the alluvial and coastal plains. In some sectors of the plains, seasonal flooding is largely a function of direct inputs from rainfall; elsewhere, tidal influence may dominate; elsewhere again, overflow from river channels may provide the main source of floodwater.

It is clear from Kingston's analysis that it is unwise to generalise from one river basin to another. The outlet of the Wildman River is silting up, that of the Mary River is virtually blocked, but the East Alligator has adequate runoff and tidal discharge to scour and maintain its channel.

An excellent model for future hydrological and geomorphological research in this region is the recently completed intensive three-year study of the South Alligator tidal river and plains by a group of scientists attached to the North Australian Research Unit of the Australian National University. Their work is summarised very briefly in Chapter 1 of this book, and has important implications for future management of the northern wetlands.

CHAPTER 1

Evolution of the landscape

M.A.J.WILLIAMS

INTRODUCTION

It is interesting to compare certain Aboriginal percep-
tions of the northern landscape with the more prosaic
accounts offered by some of the first non-Aboriginal
explorers to see this country after their gruelling overland expeditions from the south. In their respective
journals both Leichhardt (1847) and Stuart (1863) give
brief but accurate descriptions of the geology and
scenery of the country they were traversing, and the
otherwise laconic Stuart, no doubt much influenced by
the dry and barren country he had previously crossed,
was moved to enthusiastic speculation about the economic future of this region: 'From Newcastle Water to
the sea-beach, the main body of the horses have been
only one night without water ... If this country is
settled, it will be one of the finest colonies under the
Crown, suitable for the growth of any and everything –
what a splendid country for producing cotton!' (Stuart
1863, p.57).

Consider now an Aboriginal viewpoint. The clan
elders at Milingimbi on the north coast of Arnhem
Land speak of a time when many of their sacred sites
were accessible on foot; today they lie submerged
beneath the sea.

In a letter to the author dated the 3rd of September
1980, Mr Ian Morris, Aboriginal Training Officer at
Kakadu National Park with the Australian National
Parks and Wildlife Services, wrote as follows: 'A
friend of mine, Steve Davis, who lives at Milingimbi,
has been attempting to map sites of significance offshore with the clan elders of the region. The interesting
thing is that, when mapped, these sites form tranverse
bands which seem to advance into the Arafura Sea
roughly like contours. Whether or not this has any
significance in relation to previous sea levels I do not
know ... The old Aboriginal people of the Arnhem
Land coast certainly have an extensive knowledge of
the sea and the features of the sea floor. This knowledge is all tied up in stories, songs, legends, etc. and it
is hard to say how far back it goes.'

It is tempting to dismiss such folk memories as
fanciful, until one recalls that the rising of the sea from
18,000 to 6,000 years ago was a world-wide phenomenon, linked to the melting of the great continental ice
sheets of the Northern Hemisphere. Reliably dated
archaeological evidence of an active Aboriginal presence in and around Arnhem Land during the past
25,000 years and more is not wanting (White 1967,
Kamminga & Allen 1973, Schrire 1982, Jones 1985),
so that an oral tradition consistent with a prehistoric
marine flooding of the land rings strangely true. The
remarkable point is that the sea reached its present
level some 6,000 years ago; the memory enshrined in
song and dance would therefore seem to be even
older.

And so we are confronted with two contrasting
views of the landscape: one firmly rooted in an accumulated knowledge of the past; the other, anticipating
change, looking to the future. Some of the tensions
inherent in these two approaches to the land are epito-
mised in the conflicting values associated on the one
hand with the desire to extract and develop, on the
other with the desire to maintain and conserve –
matters which are the subject of the last two chapters
of this book.

The aim of this first chapter is to set the stage for
later chapters on the climate, hydrology, plants,
animals and human communities of the northern wetlands by describing how the present landscape came
into being. It is worth emphasising at the outset that the
ancient sandstones that make up the Arnhem Land
plateau mantle a very much older landscape, which
became buried when the Arnhem Land sandstones
were laid down some 1,700 million years ago. Subsequent erosion and removal of these sandstones has
exposed extensive portions of the Early Precambrian
landscape, and these have persisted with surprisingly
little modification to form significant elements of the
landscape we see today. For this reason, we begin with
the tectonic events of Archaean and Precambrian
times, over 2,000 million years ago, before moving to
the fluctuations of more recent years (see Table 1).

Table 1. Geological time scale.

Million years	Period	Era	Eon
0			
	Quaternary*		
1.8			
	Pliocene		
5.0			
	Miocene	CAINOZOIC	
22.5			
	Oligocene		
38			
	Eocene		
54			
	Palaeocene		
65			PHANEROZOIC
	Cretaceous		
141			
	Jurassic	MESOZOIC	
195			
	Triassic		
230			
	Permian		
280			
	Carboniferous		
345			
	Devonian	PALAEOZOIC	
395			
	Silurian		
435			
	Ordovician		
500			
	Cambrian		
570		PRE-CAMBRIAN	PROTEROZOIC
2500			ARCHAEAN

* The Quaternary comprises the Pleistocene (1.8 million to 10,000 years) and the Holocene or Recent (10,000 to present).

TECTONIC SETTING

The Precambrian geology of the area shown on Figure 2 has been analysed in considerable if sometimes conflicting detail by contributors to the international symposium on the Pine Creek Geosyncline. The account which follows draws selectively upon the published proceedings of this symposium, edited by Ferguson & Goleby (1980), and upon earlier reviews by Dunn (1962), Malone (1962a, 1962b), Walpole (1962), and Williams (1969a).

The tectonic history of this region is, in essence, very simple. A granitic basement of Archaean rocks occupied the area between what are now Darwin, Pine Creek and the Arnhem Land plateau until about 2,500 million years ago (Ferguson et al. 1980, Page et al.

1980). A trough developed as a result of faulting and subsidence of the Archaean basement rocks, possibly in response to incipient rifting (Rossiter & Ferguson 1980, Stuart-Smith et al. 1980, Needham & Stuart-Smith 1980). As it became progressively deeper and broader this trough functioned as a very effective regional sediment trap or depocentre. Within this depression (conventionally known as the Pine Creek Geosyncline) over 14,000 metres of interbedded sediments and volcanic rocks accumulated over an interval of several hundred million years. The sediments consisted of non-marine gravels, sands and muds, and of marine sands, clays and carbonates (Needham et al. 1980).

About 1,900 million years ago a localised episode of orogenic and metamorphic activity heralded, or maybe even triggered, the major regional metamorphism of 1,800 million years ago (Page et al 1980). During this widespread event the geosynclinal sedimentary and volcanic formations, together with some of the adjacent basement rocks, were folded, faulted and strongly altered by heat and pressure to form the metasediments and other metamorphic rocks of the Pine Creek Geosyncline (Ferguson 1980, Ferguson et al. 1980, Needham et al. 1980). One outcome of the regional metamorphism was a resetting of the radiometric clocks of some Archaean rocks back to zero, so that these now have apparent ages of 1,800 million years, when in fact they first originated towards 2,500 million years ago (Ferguson et al. 1980, Page et al. 1980).

Close on the heels of the widespread regional metamorphism, uplift and erosion of 1,870-1,800 million years ago came several subsidiary episodes of igneous activity, with sporadic orogenic activity until about 1,700 million years ago (Riley 1980, Page et al. 1980). (In Figure 2, for the sake of simplicity, these younger intrusive rocks are referred to, rather arbitrarily, as Younger Granites). During and for several tens of millions of years after these localized intrusive and volcanic events, the land was faulted, uplifted and eventually worn back by erosion to form a gently undulating Precambrian surface studded with scattered hills and scored with shallow valleys – but for the absence of trees and grasses, a landscape not unlike that of the present day.

Henceforth the land remained tectonically fairly stable. From about 1,690 to 1,650 million years ago several hundred metres of more or less flat-bedded sands, intercalated basaltic lavas, and minor basal gravels were laid down across the early Precambrian erosion surface (Needham & Stuart-Smith 1980). These sediments, which crop out today in the cliffs and gorges of the Arnhem Land plateau, comprise the oldest cover rocks in the region. Lithologically similar flat-bedded Precambrian sandstones crop out in Mount

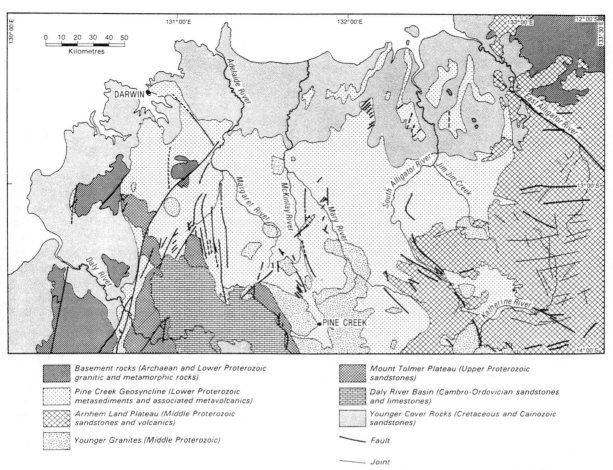

Figure 2. Geological map of the Darwin – Pine Creek – East Alligator region, generalised from Walpole et al. (1968).

Tolmer plateau and other plateaux south of Darwin and west of Pine Creek. On Figure 2 they are shown as Upper Proterozoic in age, and the Arnhem Land sandstones as Middle Proterozoic, following Walpole et al. (1968). In fact, they are not particularly well dated and the Precambrian plateau sandstones which crop out on either flank of the roughly wedge shaped area of exposed Pine Creek Geosyncline rocks may well prove to be in large measure coeval.

Palaeozoic sediments, and in particular sandstones and limestones of Cambro-Ordovician age, occupy extensive areas of the Daly River valley and adjacent areas (Figure 2). The summits of the Precambrian plateaux alluded to earlier appear to be bereft of any Palaeozoic cover rocks, but they are mantled in numerous areas by up to 200 metres of nearly horizontal Mesozoic sandstones and mudstones. A possible inference to be drawn from the differential distribution of Palaeozoic and Mesozoic cover rocks depicted on Figure 2 is that until the marine transgressions of Mesozoic times, the Precambrian plateaux were always relatively elevated areas of land. As a result of prolonged denudation during the remarkably long stable interval from about 1,650 to 150 million years ago, they would have been net exporters of sediment, particularly along their steeper margins. Away from their eroding edges, they acted as extraordinarily efficient caprocks, preserving the underlying Precambrian rocks and landforms from erosion and in the process also protecting the uranium-bearing formations.

Many geologists have noted that the uranium deposits of western Arnhem Land and the East Alligator area are 'stratabound' in the sense that they coincide with the uppermost Lower Proterozoic metasediments and are invariably adjacent to or capped by the unmetamorphosed Middle Proterozoic sandstones of the Arnhem Land plateau (Hegge et al. 1980, Needham & Stuart-Smith 1980). However, as Needham et al. (1980) have pointed out, the geographical association between plateau margin and ore bodies may ultimately prove to be more apparent than real, for younger sediments are often very thin near the Arnhem Land scarp making it easier to discover the uranium in such a geomorphic setting. Nevertheless, it seems undeniable that the plateau sandstones have long operated as very effective caprocks, protecting whatever lay beneath (including ore bodies) from later erosion.

By about 1,800-1,700 million years ago, much of northern Australia had become tectonically stable, forming the geological feature known as the North Australian Craton (Clarke & Cook 1983, p.433). We noted earlier that during Middle and Upper Proterozoic times (1,690 to 575 million years ago) gravels, sands, basalts and dolerites accumulated more or less horizontally within shallow depressions on this otherwise stable landscape. The sandstones and intercalated volcanic rocks of Arnhem Land were laid down in one such sedimentary basin.

The end of the Precambrian (575 million years ago) coincided with intense orogenic activity in central Australia (Veevers et al. 1984) and with gradual breakup of the very much bigger continent of which Australia was then a part. Throughout the ensuing 200 million years, Australia lay close to the equator and enjoyed a tropical climate, often wet, sometimes dry, but almost invariably warm (Embleton 1984, Quilty 1984, Talent 1984).

During much of the Palaeozoic era, from Cambrian to Carboniferous times, Australia, Africa, South America, India and Antarctica seem to have formed part of a relatively stable continental mass comprising a crystalline basement of mainly igneous Precambrian rocks on which were scattered occasional sedimentary basins in which thick sequences of Palaeozoic rocks accumulated. This mega-continent we term Gondwanaland, following Suess (1885) and later works. Of modern import is the fact that the distribution of many of our economically important mineral ores coincides with that of the Precambrian Shield in northern and western Australia and of the folded and metamorphosed Archaean to Palaeozoic sediments of the Hammersley, Davenport, Macdonnell and Musgrave Ranges in the west and centre of Australia.

Continental arching in Late Carboniferous times (some 300 million years ago) ended Palaeozoic marine deposition in western Australia and proved the forerunner of the widespread Jurassic to Early Cretaceous rifting between the western margin of Australia and the eastern border of India on the one hand and between Antarctica and southern Australia on the other (Veevers 1971, Veevers et al. 1971 and 1984). Such rifting was the precursor to the drifting apart of India and Australia, which began in Early Cretaceous times, some 130 million years ago. The fragmentation of Gondwanaland had in fact started before that time, in the Middle Jurassic, 160 million years ago, with the separation of western Gondwanaland (South America

and Africa) from its eastern province (Australia, Antarctica and India).

A major marine transgression in Early Cretaceous times accompanied the collapse of the continental margins consequent upon the separation of India from Australia. During the Aptian (110-107 million years ago), the sea flooded much of northern Australia, including the Daly, Pine Creek and McArthur Basins (Skwarko 1966, Veevers 1984). The marine sands and muds laid down by the Aptian Sea overlie Mesozoic freshwater sandstones and together form the Mullaman Beds. First described by Tenison Woods (1886), and later identified by Brown (1895, 1908) as being at least in part of Lower Cretaceous age, the Mullaman Beds crop out in low sea cliffs between Darwin and the Arnhem Land coast. Deeply weathered tabular remnants of the Mullaman Beds are also scattered across the undissected portions of the surface of the Arnhem Land plateau and of the other Middle to Upper Proterozoic plateaux shown on Figure 2, indicating that these localities have been uplifted by several hundred metres since the Aptian.

During the long interval from 600 to 100 million years ago, Australia's climate was essentially tropical for the first 250 million years, episodically glacial during the ensuing 100 million years, and subtropical to temperate to glacial during the 150 million years that followed. As a very rough generalisation, it seems that times of globally widespread glaciation may coincide with episodes of widespread mountain-building in middle and high latitudes and times of more equable climate with prolonged intervals of tectonic stability. Uplift in Australia and South America certainly appears to have triggered the Permo-Carboniferous glaciations of Gondwana (Powell & Veevers 1987).

The separation of Australia from Antarctica began 95 million years ago (Johnson & Veevers 1984). Thereafter, Australia moved northwards into warmer sub-tropical latitudes dominated by semi-permanent high pressure zones over land and sea. Latitudinally static for much of the Mesozoic, Australia moved northwards through 40° of latitude during the ensuing 65 million years of the Cainozoic.

The sea had largely withdrawn from Australia by the Late Cretaceous (Brown et al. 1968) and prolonged deep weathering of the exposed land surface ensued. A brief marine incursion took place again in Early Miocene times in parts of Northern Territory and was followed by gentle uplift with localized folding and faulting in Middle to Late Miocene times (Veevers 1971, 1984). If the ages of Tertiary basalts in the Eastern Highlands of Australia are plotted against latitude (Wellman & McDougall 1974), there is an apparent southward migration of the locus of volcanism at a rate of 66 ± 5 mm / year. Totally independent evidence from palaeomagnetism and sea-floor spreading indicates that since the Mid-Miocene, Australia has moved about 800 km northwards, at a mean rate of 50-70 mm/year (Veevers 1984). What was the overall impact of these tectonic events upon the northern landscape? We consider this question in the following sections.

TERTIARY EROSION AND DEPOSITION

Although it is true that the northern landscape contains features of very great antiquity which reflect its long geological history and the cumulative impact of ancient tectonic events, it was not until after the retreat of the Lower Cretaceous sea that the erosional and depositional events of the Tertiary and Quaternary created the major elements of the present-day landscape depicted in Figure 3.

Tertiary and Quaternary together make up the Cainozoic, which began 65 million years ago. The Quaternary spans only the last 2 million years, but is significant for two main reasons. One concerns climate, the other humans. The Quaternary is remarkable alike for the speed and amplitude of its climatic fluctuations, and for the fact that it coincides with the

emergence from Africa and the spread to all parts of the world of prehistoric humans who walked upright, lit fires, made stone tools and were capable of complex speech (Williams 1985a). In the next section we discuss the influence of Quaternary climatic fluctuations upon the landscape; here we deal with Tertiary events which were responsible for fashioning much of the present landscape.

Even prior to the invasion of the Aptian Sea 110 million years ago, the northern landscape was already one of low relief. Where the sub-Cretaceous surface has been recently exhumed by erosion, it appears as a gentle undulating and often deeply weathered landscape which bevels Archaean and Proterozoic rocks alike and falls gradually away in height towards the north (Hays 1967, Williams 1969c).

For a long time after the Aptian Sea had withdrawn the land remained tectonically stable and close to sea level. Prolonged deep weathering was accompanied by minor erosional bevelling of the Cretaceous Mullaman Beds and older rocks. The resulting land surface with its intensely leached bedrock capped by pisolithic ironstone has been termed the Bradshaw surface by Wright (1963) and today coincides with the summits

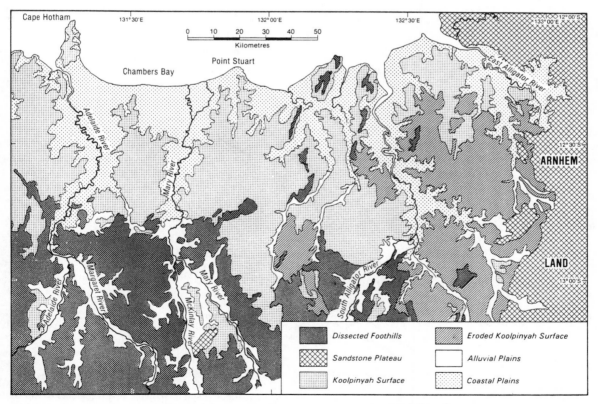

Figure 3. Geomorphic map of the Adelaide – Alligator Rivers region, generalised from Williams (1969b) and Story et al. (1976).

of the dissected foothills and of Arnhem Land plateau in the southwest and southeast of the area shown in Figure 3.

It is difficult to specify an exact age for this surface. Certainly it must be younger than the Lower Cretaceous rocks across which it is cut, but if close to sea level and of very low relief the latter would not have been prone to much denudation anyway. Elsewhere in Northern Territory and Queensland similar deeply weathered Lower Cretaceous rocks are overlain by siliceous limestones of Mid-Miocene age and younger (see Williams 1969c). An erosional unconformity or a depositional gap separates the deeply weathered Cretaceous formations from the overlying unweathered limestones. It is therefore probable that the Bradshaw surface is older than Mid-Miocene.

Noakes (1949) has suggested that slow uplift during the Miocene initiated a process of drainage incision and scarp retreat. The geomorphic evidence is certainly consistent with river rejuvenation triggered by very gentle epeirogenic warping. Erosional stripping of the Bradshaw (and older) weathering profiles has etched out a stark relief of rocky strike ridges, rugged granite hills, and steep-sided sandstone tablelands (Figures 3 and 4). Narrow valleys traverse the plateau and the dissected foothills; included among them are the headwaters of all the major rivers in the area.

Sustained scarp retreat during the past 15 million years or so was accompanied by slow uplift. The impressive scarps bordering the Arnhem Land plateau and the rugged relief of the dissected foothills were a legacy of this dual process of entrenchment and erosional backwearing. Galloway (1976) considers that the Arnhem Land cliffs have been retreating at a rate amounting to 1 m per 1000 years over the past 15-20 million years. If the present elevation of 200-300 m above sea level is indeed a product of Middle to Late Miocene uplift, then the corresponding rate of erosional downcutting would be about 20 mm per 1000 years.

As the Late Tertiary rivers debouched from the rocky confinements of the plateau and dissected foothills, they spread out across the northern erosional plains and deposited a series of low-angle alluvial fans. The sediments comprising these fans were rather more sandy in the east (understandable, given their proximity to the sandstone plateau) and more clay-rich in the west. In due course the valleys became aggraded and blocked with alluvial debris from their elevated hinterland, rocky strike ridges became bevelled and progressively buried in their own detritus, and once discrete fans coalesced to form a broad depositional apron which sloped gently northwards from its abrupt contact with the steep and rocky upland borders.

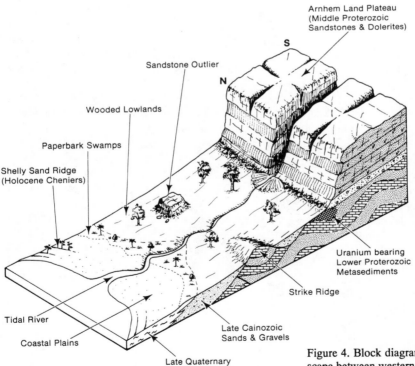

Figure 4. Block diagram showing major features of the landscape between western Arnhem Land and the coast, based on author's fieldwork.

Nearly half the area mapped in Figure 3 was formed in this way. Following Christian & Stewart (1953), Hays (1967) and Williams (1969c), it is here termed the Koolpinyah surface.

Although it is primarily a late Tertiary depositional surface, the Koolpinyah surface is in fact complex and polygenic. Borehole data show that the sediment mantle is up to 30-40 m thick, sometimes more, but usually much less (Williams 1969a). The sediments consist of gravels, sands, silts and clays which have been repeatedly weathered, eroded, redeposited and reweathered. Parent rocks include Cretaceous, Upper Proterozoic and Lower Proterozoic formations which range from siltstone through sandstone and limestone to granites and basalts. The end product of the late Tertiary – early Quaternary weathering has been removal of bases and alkali earths and the formation of a residual concentration of partially hydrated oxides of iron and aluminium which geologists and soil scientists refer to as laterite (see Paton & Williams 1972 for a critical discussion of this term). The Koolpinyah surface is littered with relict blocks of pisolithic laterite which are being destroyed by current erosional and weathering processes (Williams 1969c).

Nearly 40% of the area mapped on Figure 3 consists of Koolpinyah surface sediments, but in the east of the area, and in particular between the East Alligator river and Arnhem Land, very little of the original level surface remains intact, for erosion has exposed much of the underlying bedrock. A series of erosional stages ranging from minimal modification to complete stripping of the Koolpinyah deep weathering profile is reflected in the soils and vegetation, and gives rise to a repetitive sequence of minor landforms mapped in some detail during the 1965 and 1972-73 CSIRO land system surveys of this region (Story et al. 1969, 1976).

One interesting feature of the Koolpinyah landscape which is of considerable importance for animal life during the long dry season is the presence of numerous closed depressions which frequently contain freshwater billabongs throughout the year. Some are scattered haphazardly across the surface, and their origin remains obscure. Others, the majority, seem to be aligned and partially segmented former channels (Williams 1969c). They are floored with alluvial sands which may overlie gravelly alluvial clays. A detailed study of this now defunct Late Tertiary or Early Quaternary drainage system has never yet been carried out, but would be of considerable interest. In contrast to the numerous small streams that are now dissecting the Koolpinyah surface east of the South Alligator, the ancestral Koolpinyah surface rivers to the west are now long dead and buried. Whether this reflects uplift to the east and subsidence to the west remains unknown.

The gently rolling landscape with its seemingly monotonous cover of mostly *Eucalyptus tetrodonta* and *E. miniata* tall open forest rising above the yellow dry season straw of the annual sorghums and spiky spear-grass (*Heteropogon contortus*) will be familiar to travellers who have made the journey from the Stuart Highway to Kakadu. Concealed beneath this apparent monotony is a geomorphic history and associated subtle variety of soils and plants as complex and interesting as anywhere in Australia.

MAJOR FEATURES OF THE LANDSCAPE

By the end of the Tertiary, five of the major landscape elements mapped in Figure 3 and depicted as a block diagram in Figure 4 had evolved into roughly their present form.

In the east is the Arnhem Land plateau, a rugged *sandstone plateau* which rises 200-300 m above the adjacent plains. Criss-crossed by tensional joints which have weathered into deep vertical clefts, the plateau surface is pitted and fretted by corrosion, and is in parts largely devoid of soil and trees. The narrow valleys within the plateau are wooded and usually well watered. Occasional steep outliers of the plateau, such as Cannon Hill and Nourlangie Rock, testify to former scarp retreat. Spectacular features in an otherwise featureless plain, their walls are adorned with Aboriginal rock paintings and at their feet lie prehistoric occupation sites with a known antiquity of at least 25,000 years. One of the great tragedies of our time is the fact that the very ground that has long sustained the Aboriginal people of the northern wetlands – furnishing them with rock shelters and galleries, living floors and sacred sites – is also host to the stratabound uranium deposits which lie close beneath the unconformable contact between Lower Proterozoic metasediments and the younger Proterozoic sandstones of the Arnhem Land plateau (Fig. 4).

To the south lie the *dissected foothills* – rocky hills, boulder-covered strike-ridges, stony hillocks and occasional granite tors which rise up to 200-300 m above the Koolpinyah surface.

Extending north from the steeply sloping margins of the dissected foothills and the sandstone plateau is the *Koolpinyah surface*. This very widespread feature consists of level to gently sloping wooded or forested lowlands which descend imperceptibly from an elevation of 100-200 m in the south to a few tens of metres above sea level in the north.

The *eroded Koolpinyah surface* in the east is rolling to undulating wooded country with occasional inliers of Precambrian rocks (Fig. 4). The summits of the strike ridges which protrude through the Koolpinyah lateritic weathering mantle are bevelled remnants of yet older surfaces. Some of these steep, boulder-mantled hills extend as far north as the coast, and reflect structural lineaments which were determined 1,800 million years ago.

Incised into the Koolpinyah surface is a series of rivers which rise in the uplands to the south and flow north to the Arafura Sea. From west to east, the major river basins are the Adelaide-Margaret, the Mary-McKinlay, the South-Alligator – Jim Jim and the East Alligator (Figs. 2 and 3). During the height of the wet season between January and March these rivers are deep and fast-flowing. By the end of the dry season they may dry out into a string of billabongs or water-holes, separated by exposed rock-bars or by sandy to gravelly channel bars. The total thickness of alluvium is of the order of 10-12 m, but it is quite variable. The *alluvial plains* bordering the main river channels may be grassy or wooded, and are sandy north of the Arnhem Land plateau and more silt- and clay-rich alongside streams which issue from the dissected foot-hills, with their scattered outcrops of greywacke siltstones.

Certain landforms on the alluvial plains are not in equilibrium with the geomorphic and hydrological processes that are operating today. Some of the levees are now inactive and strongly weathered, and certain portions of the flood-plain have long ceased aggrad-ing, and form slightly elevated and rather patchy al-luvial terraces (Williams 1969c; Galloway 1976). The soils and vegetation on these old river terraces are easily distinguished from those on the active flood plains. Although poorly studied so far, they represent the impact of former changes in river load and dis-charge which themselves reflect the influence of Quaternary climatic fluctuations. Since a proper un-derstanding of the sixth and final landscape element shown in Figure 3 – the *coastal plains* – is not possible without some appreciation of the repercussions of Quaternary climatic changes we will now consider these in relevant detail.

QUATERNARY CLIMATIC FLUCTUATIONS

With the separation of Australia from Antarctica 95 million years ago, and the eventual creation of a cold circumpolar current around Antarctica, the stage was set for a 10-15°C drop in surface temperatures of the Southern Ocean over the past 40 million years since the Late Eocene. Thermally isolated from warmer northern waters, ice began to accumulate on Antarc-tica, first as mountain glaciers, later as ice-caps, so that by the Mid-Miocene, 10-15 million years ago, even West Antarctica was covered in ice. Since that time, as we noted previously, Australia has advanced a further 800 km north into more tropical latitudes, accentuating the temperature contrast between north and south, and bringing northern Australia under the influence of the summer monsoon. Central Australia, dominated by an easterly sub-tropical jet stream aloft and by tropical anticyclones in the lower atmosphere, is subject to the parching influence of dry subsiding air as part of the normal Hadley cell circulation over the tropic. It was not always so, and the gradual disintegration of major drainage systems throughout inland Australia was in part related to Australia's northward drift into drier latitudes, and in part to Miocene earth movements.

With closure of the Panama Isthmus late in the Pliocene and reinforcement of the North Atlantic Gulf Stream, ice began to accumulate over North America, and by 2.5 million years ago the volume of ice was such that thereafter the waxing and waning of the huge North American ice sheets had a major influence not only upon global sea level fluctuations but also upon world climates.

During glacial maxima, when the ice sheets covered much of North America and significant portions of Europe, world sea levels were lowered by 100-150 m, New Guinea – Australia – Tasmania formed a single land mass, and the land area of Australia was increased by 20 per cent. Glacial maxima were also times of enhanced aridity in the intertropical zone: lakes and rivers dwindled, forests shrank, formerly vegetated sand dunes became active once more, and the deserts and tropical savannas encroached upon the former domain of woodland or forest.

With a return to interglacial conditions there was a rapid rise in sea level as the northern ice caps melted. New Guinea and Tasmania again became islands. The inland dunes were colonised by plants and stabilised. Dust storms became less frequent. Lake levels rose, salt lakes became fresh, and ephemeral or highly sea-sonal rivers began to flow more regularly and for most years.

Because we live in the present we tend to view it as the norm. In fact, the present climate of Australia is quite atypical of the past 100,000 years, as a glance at the sea level curve in Figure 5 will show. The last time world sea level, and world climate, was comparable to that of today was during the last interglacial 125,000 years ago. In the 1.8 million years of Quaternary time, there have been at least 17 recognisable glacial-interglacial cycles, with an average duration of roughly 100,000 years (Fink & Kukla 1977). The interglacials were all relatively brief events of perhaps 5,000 – 10,000 years; the remainder of the cycle was characterised by a saw-toothed build-up towards full glacial conditions (Broecker & van Donk 1970), again well illustrated in the sea level curve (Fig. 5). In other words, the norm for Australia over the past 2 million years was a climate tending towards glacial: certainly colder, probably windier, and no doubt much drier than

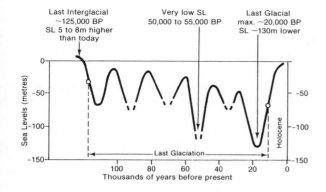

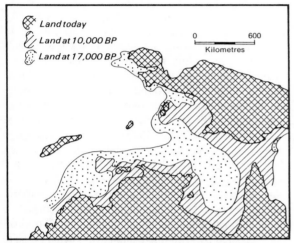

Figure 5. Sea level fluctuations during the past 125,000 years, and some associated changes in the coastline of northern Australia and Papua New Guinea, slightly modified from Chappell (1976, Figures 1 and 3).

today (Williams 1984b, 1985b).

At least as far as northern Australia is concerned, we do not have data on the sixteen or so Quaternary glacial-interglacial cycles which preceded the last high sea level stage of 125,000 years ago (Fig. 5). For that reason we will focus upon the environmental changes which took place in the north from the time of the last glacial maximum (18,000 ± 2,000 years ago) until the present. Furthermore, the range of climatic variability within that time was about as great as at any time in the Quaternary.

The discussion which follows is based upon information culled from a variety of sources, including some excellent general reviews (Jones & Bowler 1980; Chappell & Grindrod 1983), some interesting new work in the Gulf of Carpentaria (Torgersen et al. 1985; De Deckker et al. 1988) and some finely detailed stratigraphic work on the Holocene sediments of the coastal plains (Bardsley et al. 1985, Woodroffe et al. 1985, 1986).

As Figure 5 shows very clearly, a key feature of northern Australia's geography at the height of the last glaciation is the land bridge between Australia and New Guinea: with sea level nearly 150 m lower, the Torres Strait was dry land (Jennings 1972). The warm ocean current which now flows through the Torres Straits westwards between Australia and the Indonesian archipelago, was deflected by the newly exposed land, thereby depriving the northern coast of a source of moist maritime air, and hence of potential rainfall. Sea surface temperatures were also somewhat lower in that part of the Indian Ocean adjacent to Indonesia and northern Australia (McIntyre et al. 1976), perhaps below the threshold limit of 26-27°C necessary for tropical cyclones to form. Since the northern wetlands receive a very significant proportion of their present-day rainfall from tropical cyclones (the incidence of which is one per year, as discussed in Chapter 2), annual precipitation then may have been half the present quantity at this time because of this cause alone. In addition, the northern coastline lay some 400 km further north, a factor of great importance in a region where rainfall decreases rapidly away from the coast, from 1,600 mm today at Darwin to 750 mm some 400 km inland.

During this time of low sea level a very large body of fresh to slightly saline water – Lake Carpentaria – covered much of the otherwise dry floor of what is now the Gulf of Carpentaria (Torgersen et al. 1985). Whether this lake was fed mainly from the New Guinea uplands, or from northern Australian rivers, is still unclear; but there is a problem here, not yet resolved. Why was it possible for such a large lake to persist for so long (≥ 40,000 to 13,000 years ago), until submerged beneath the rising sea, when pollen spectra from Lynch's Crater on the Atherton Tableland of northern Queensland indicate a drastic reduction in rainforest and an effective precipitation perhaps 2-3 times lower than today (Kershaw 1978)?

Also of relevance here is the thermoluminescence age of 20,000 (± 3,000) years for a large foredune built when Lake Woods was 6 m deeper than its present wet season level (Hutton et al. 1984). Lake Woods lies immediately south of Newcastle waters, and 600 km inland from Darwin. The field evidence suggests that Pleistocene Lake Woods was not formed by a brief but exceptional flood; the size of its coastal foredune argues rather for a lake of some duration.

Dulhunty (1983) has argued that the perennial lake ancestral to modern Lake Eyre – Lake Dieri – was full and fresh in its surface layers between perhaps 45,000 to 20,000 years ago. The catchment area of Lake Dieri would have been similar to that of modern Lake Eyre, namely the Eastern Highlands of central Queensland, but it is thought to have flowed on south to the coast. However, the precise age of Pleistocene Lake Dieri is far from well established, as is the age of Lake Woods.

What does now appear to be reasonably certain is that between 20,000 and 12,000-11,000 years ago much of Australia, India, tropical South America and large parts of Indonesia were very much drier than they are today. The equatorward migration of Australia's concentric inland vegetation zones postulated by Nix and Kalma (1972) is thus entirely consistent with world-wide evidence of very late Pleistocene inter-tropical aridity (Bowler 1978, Williams 1985b).

Temperature and pressure gradients between equator and pole were steeper than today towards 18,000 years ago. Trade winds were stronger and considerable quantities of dust were blown from inland Australia out to sea, particularly in the southeast and northwest of the continent (Thiede 1979, see also Chapter 13). It is quite likely that many of the finer-textured northern soils contain a high proportion of wind-blown dust, and it is equally likely that some of the extensive sand-sheets on the Koolpinyah surface may be alluvium that has been resorted by wind during drier episodes in the Quaternary. However, many of the sand-sheets are known to be created by termite activity, and in particular by termites bringing sand particles to the surface and by the later disintegration of their mounds (Williams 1968, 1978).

With the melting of the great Northern Hemisphere ice sheets soon after 18,000 years ago, world sea level rose, rapidly at first, then more slowly (Fig. 5). The mean horizontal loss of land to the sea amounted to nearly one metre per week until about 10,000 years ago, when it began to slow down. The rate of vertical encroachment of the sea upon the land amounted to about 15 mm per year. To the early Aboriginal coastal dwellers of the north such changes would have been visible and potentially traumatic, forcing the people relentlessly inland towards the once barren sandstone plateau of Arnhem Land, which they reached at least 25,000 years ago, or possibly more than 15,000 years after their arrival in Australia from the north by sea.

Despite the loss of land there were several bonuses linked indirectly to the post-glacial rising of the sea. One was the renewed influx of warm moist maritime air, another was an increase in the duration of the summer monsoon, and a greater penetration inland of the wet season rains. The plant cover soon adapted to these changes, and the previously much expanded arid zone shrank back to its heartland in the interior. Desert gave way to savanna, savanna yielded to woodland or forest, isolated patches of monsoon rainforest expanded and sometimes coalesced. By 14,000 the sea had begun to flood Lake Carpentaria, and soon after 9,000 Torres Strait was submerged, enabling a warm current to lap the northern coast. Between about 9,000 and 7,000 years ago Australia was somewhat warmer and very much wetter than today, and rainforests attained their maximum extent at about this time (Williams 1984a).

The post-glacial rise in sea level was accomplished by 6,000-7,000 years ago (Thom & Chappell 1975). Thereafter the coastal plains began to form. The coastal valleys of the larger rivers became estuaries, mangroves colonised the tidal mud flats, and were at their most extensive between 7,500 and 6,000 years ago. This 'big swamp' phase is particularly well dated in the lower South Alligator plains, but is clearly evident at a number of other localities scattered across the northern coast of Australia (Woodroffe et al. 1986). The big mangrove swamp acted as a very efficient sediment trap, so that fine mud washed in from the rivers during the summer wet season was deposited among the mangroves, which eventually became buried. By about 4,000 years ago, the coastal plain was traversed by large sinuous tidal rivers and the mangroves became confined to the coastal fringe and to the muddy banks of the tidal streams, much as today (Woodroffe et al. 1986). By 2,000 years ago many of the sinuous channels had been cut off from their parent rivers, which acquired a very character-istic cuspate form in the lower valleys of the South and East Alligator rivers. The geomorphic and hydro-logical reasons for these changes are discussed in detail and with great clarity by Chappell and Woodrof-fe (1985) and by Woodroffe and co-workers (1986) in their recent monograph. It is clear from their work that the coastal plains are a very dynamic feature, and that the influx of salt water into the plains is a result of long term geomorphic processes, although no doubt locally aggravated by the activities of feral water buffaloes (see Chapters 11 and 12 for a further discussion of the impact of the buffalo). To quote from Chappell and Woodroffe (1985, p.85): 'The transition from sinuous to cuspate river phases appears to involve considerable channel instability, with increased salt-water incursion to upstream tidal reaches.'

One other intriguing feature of the coastal plains deserves mention. Roughly parallel to the shoreline but converging towards the various headlands is a series of low, narrow ridges built of shelly sand and studded with Pandanus palms. (These ridges are che-niers, so called because the first French settlers in coastal Louisiana discovered similar landforms there, the chief difference being that there they support nat-ive American oaks (Fr., chêne) rather than Pandanus palms). In his diary for Thursday July 24, 1892, Stuart noted that the beach (at Point Stuart) was covered with a soft blue mud, 'which had nearly covered all the shells' (Stuart 1863, p.56). The following day, after erecting a flag on one of the tallest trees in the patch of monsoon vine forest at the headland (Fig. 6), he reported that the ground was 'very soft where the salt water had covered it, in others not so bad. Judging from the number of shells banked up in different places, the sea must occasionally come over this' (Stuart 1863, p.58). He was alluding to the shelly

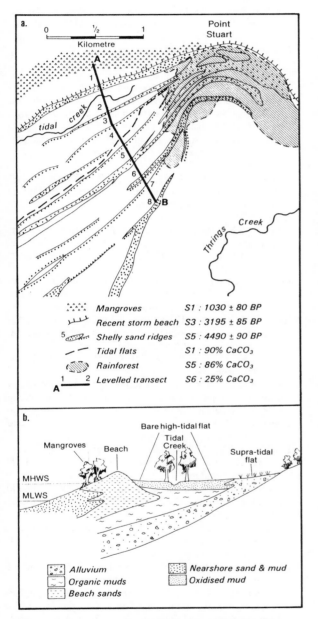

Figure 6. Point Stuart chenier plain (a), showing preliminary radiocarbon ages of some of the chenier ridges, after Clarke et al. (1979). Dated shell samples (S1 ... S6) came from numbered chenier ridges (S1 from ridge 1, etc.) along line of transect. Lower diagram (b) shows generalised stratigraphy of a mangrove coast in the late Holocene, following the postglacial rise of sea level to its present elevation some 6,000 years ago, after Chappell & Thom (1977).

chenier ridges, some of which are quite recent features, but most are in fact several thousand years old (Fig. 6a).

Cheniers may develop in several different ways, one being as storm beaches during infrequent catastrophic storms which severely damage the mangroves, smashing the mangrove shells and hurling them above the normal mean high water spring tide level (MHWS, in Fig. 6b) to form a beach. In due course muds accumulate anew in the intertidal zone, mangrove seedlings re-establish themselves, and trap the muds among their roots and pneumatophores. The mangrove belt advances slowly seawards, and the coastline progrades accordingly. Sedges gradually colonise the tidal mud-flats inland, muds accumulate, until the flats emerge above tidal limits. Figure 6b shows an idealised version of the coastal plains stratigraphy, encompassing these events. At Point Stuart, Clarke et al. (1979) obtained radiocarbon ages on shells from some of the main chenier ridges (Fig. 6a). Later workers will need to obtain more dates, including several dates from each ridge. Meanwhile, and accepting the first dates at face value, it would seem that progradation of the shoreline amounted to about 45 cm per 1000 years along the levelled line of transect (Fig. 6b). The relative absence of carbonate (i.e., shells) in chenier ridge S6 is interesting, and may reflect its greater age and more prolonged leaching, or may simply be because it was mainly composed of quartz sand from its inception.

Today the coastal plains are extensively flooded during the monsoonal wet season (see Chapter 3 for details). Parts of the plains remain under several metres of water for up to half the year, some for all the year. Large perennially flooded paperbark swamps are a feature of the plains along their inland contact with the Koolpinyah surface. The seasonally flooded coastal plains of sedges and grasses form a slightly elevated buffer zone between paperbark swamp and bare tidal flats with their sporadic chenier ridges. The coastal plains are the youngest feature in an otherwise ancient landscape. Such a recent history may be reflected in Aboriginal fire practices (see Chapter 5), and is certainly reflected in the evolutionary histories, diversity and adaptive strategies of the vertebrate animals described in Chapters 6 to 11 of this book.

The links between plants and their environment are equally subtle and complex (Chapter 4). They, too, reflect the evolutionary history of the landscape, as well as the geomorphic processes operating upon it today. The next two chapters in this book discuss the climate and hydrology of the northern wetlands, and the remaining ten chapters deal with the manifold ways in which plants, animals and human societies have striven to adapt themselves to this highly seasonal tropical environment.

REFERENCES

Bardsley, K. N., Davie, J. D. S. & Woodroffe, C. D. (eds.) 1985. *Coasts and Tidal Wetlands of the Australian Monsoon Region.* Australian National University North Australia Research Unit, Mangrove Monograph No.1, Darwin.

Bowler, J. M. 1978. Glacial age aeolian events in high and low latitudes: a Southern Hemisphere perspective. In: E. M. van Zinderen Bakker (ed.), *Antarctic Glacial History and World Palaeoenvironments,* Balkema, Rotterdam: 149-172.

Broecker, W.S. & van Donk, J. 1970. Insolation changes, ice volumes, and the 18$_O$ record in deep-sea cores. *Rev. Geophys. Space Physics* 8: 169-198.

Brown, D.A., Campbell, K.W.W. & Crook, K.A.W. 1968. *The Geological Evolution of Australia and New Zealand,* Pergamon.

Brown, H.Y.L. 1895. Government geologist's report on explorations in the Northern Territory. *South Aust. Parliamentary Pap.* No.82.

Brown, H.Y.L. 1908. Report on recent mineral discoveries and further record of Northern Territory boring operations. *South Aust. Parliamentary Pap.* No. 85.

Chappell, J. 1976. Aspects of late Quaternary palaeogeography of the Australian – East Indonesian region. In: R. L. Kirk & A.G.Thorne (eds.), *The Origin of the Australians,* Australian Inst. Aboriginal Studies Human Biology Series No.6, Canberra: 11-22.

Chappell, J.M.A. & Grindrod, A. 1983. *CLIMANZ. A Symposium of Results and Discussions concerned with Late Quaternary Climatic History of Australia, New Zealand and Surrounding Seas.* 2 Vols. Dept. Biogeography & Geomorphology, Australian National University, Canberra.

Chappell, J. & Thom, B.G. 1977. Sea levels and coasts. In: J.Allen, J.Golson & R.Jones (eds.), *Sunda and Sahul. Prehistoric Studies in Southeast Asia, Melanesia and Australia,* Academic Press, London: 275-291.

Chappell, J. & Woodroffe, C. 1985. Morphodynamics of Northern Territory tidal rivers and floodplains. In: K. N. Bardsley et al., *op. cit.,* 85-108.

Christian, C.S. & Stewart, G.A. 1953. *General report on survey of the Katherine-Darwin region, 1946.* CSIRO Aust. Land Res. Ser. No.1.

Clarke, I. F. & Cook, B.J. 1983. *Geological Science, Perspectives of the Earth,* Australian Academy of Science, Canberra.

Clarke, M. F., Wasson, R.J. & Williams, M.A.J. 1979. Point Stuart chenier plain and Holocene sea levels in northern Australia. *Search* 10: 90-92.

De Deckker, P., Chivas, A. R. & Shelley, J.M.G. 1988. Ostracod shell chemistry: a new palaeoenvironmental indicator applied to a regressive/transgressive record from the Gulf of Carpentaria, Australia. *Palaeogeogr., Palaeoclimatol., Palaeoecol.* 66: 231-241.

Dulhunty, J.A. 1983. Lake Dieri and its Pleistocene environment of sedimentation. South Australia. *J. Proc. R. Soc. N.S.W.* 116: 11-15.

Dunn, P.R. 1962. Alligator River, Northern Territory. 1:250,000 Geological series. *Explan. Notes Bur. Miner. Resour. Geol. Geophys. Aust.*

Embleton, B.J.J. 1984. Continental palaeomagnetism. In: J.J.Veevers (ed.), *Phanerozoic Earth History of Australia,* Clarendon Press, Oxford: 11-16.

Ferguson, J. 1980. Metamorphism in the Pine Creek Geosyncline and its bearing on stratigraphic correlations. In: J.Ferguson & A.B.Goleby, *op. cit.,* 91-100.

Ferguson, J., Chappell, B.W. & Goleby, A.B. 1980. Granitoids in the Pine Creek Geosyncline. In: J.Ferguson & A.B.Goleby, *op. cit.,* 73-90.

Ferguson, J. & Goleby, A.B. (eds.), 1980. *Uranium in the Pine Creek Geosyncline.* Proc. Int. Symp., Sydney 1979. International Atomic Energy Agency, Vienna.

Fink, J. & Kukla, G.J. 1977. Pleistocene climates in Central Europe: at least 17 interglacials after the Olduvai event. *Quaternary Research* 7: 363-371.

Galloway, R.W. 1976. Geomorphology of the Alligator Rivers area. In: R.Story et al., *op. cit.,* 52-70.

Hays, J. 1967. Land surfaces and laterites in the north of the Northern Territory. In: J.N.Jennings & J.A.Mabbutt (eds.), *Landform Studies from Australia and New Guinea,* Australian National University Press, Canberra: 182-210.

Hegge, M.R., Mosher, D.V., Eupene, G.S. & Anthony, P.J. 1980. Geologic setting of the East Alligator uranium deposits and prospects. In: J.Ferguson & A.B.Goleby, *op. cit.,* 259-272.

Hutton, J.T., Prescott, J.R. & Twidale, C.R. 1984. Thermoluminescence dating of coastal dune sand related to a higher stand of Lake Woods, Northern Territory. *Aust. J. Soil Res.* 22: 15-21.

Jennings, J.N. 1972. Some attributes of Torres Strait. In: D.Walker, *op. cit.,* 29-38.

Johnson, B.D. & Veevers, J.J. 1984. Oceanic palaeomagnetism. In: J.J.Veevers, *op. cit.,* 17-38.

Jones, R. (ed.) 1985. *Archaeological Research in Kakadu National Park.* Australian National Parks and Wildlife Service, Special Publications 13: 317 pp.

Jones, R. & Bowler, J. 1980. Struggle for the savanna: northern Australia in ecological and prehistoric perspective. In: R.Jones (ed.), *northern Australia: Options and Implications,* Res. School Pac. Studies, Australian National University, Canberra: 3-31.

Kamminga, J. & Allen, H. 1973. *Report of the Archaeological Survey.* Alligator Rivers Environmental Fact-Finding Study, Canberra: 119 pp.

Kershaw, A.P. 1978. Record of last interglacial-glacial cycle from northeastern Queensland. *Nature* 272: 159-161.

Leichhardt, L. 1847. *Journal of an Overland Expedition in Australia from Moreton Bay to Port Essington.* T. & W.Boone, London.

Malone, E.J. 1962a. Darwin, Northern Territory. 1:250,000 Geological Series. *Explan. Notes Bur. Miner. Resour. Geol. Geophys. Aust.*

Malone, E.J. 1962b. Pine Creek, Northern Territory. 1:250,000 Geological Series. *Explan. Notes Bur. Miner. Resour. Geol. Geophys. Aust.*

McIntyre, A. & CLIMAP project members 1976. The surface of the Ice-age earth. *Science* 191: 1131-1144.

Needham, R.S., Crick, I.H. & Stuart-Smith, P.G. 1980. Regional geology of the Pine Creek Geosyncline. In: J. Ferguson & A.B. Goleby, *op. cit.,* 1-22.

Needham, R.S. & Stuart-Smith, P.G. 1980. Uranium in the Pine Creek Geosyncline. In: J.Ferguson & A.B.Goleby, *op. cit.,* 233-257.

Nix, H.A. & Kalma, J.D. 1972. Climate as a dominant control in the biogeography of northern Australia and New Guinea. In: D.Walker, *op. cit.,* 61-91.

Noakes, L.C. 1949. *Geological Reconnaissance of the*

Katherine-Darwin Region, Northern Territory, with Notes on the Mineral Deposits. Bur. Miner. Resour. Geol. Geophys. Aust. Bull. No.16.

Page, R.W., Compston, W. & Needham, R.S. 1980. Geochronology and evolution of the Late-Archaean Basement and Proterozoic rocks in the Alligator Rivers uranium field, Northern Territory, Australia. In: J.Ferguson & A.B.Goleby, *op. cit.*, 39-68.

Paton, T.R. & Williams, M.A.J. 1972. The concept of laterite. *Ann. Ass. Amer. Geogr.* 62: 42-56.

Powell, C.McA. & Veevers, J.J. 1987. Namurian uplift in Australia and South America triggered the main Gondwana glaciation. *Nature* 326: 177-179.

Quilty, P.G. 1984. Phanerozoic climates and environments of Australia. In: J.J.Veevers, *op. cit.*, 48-57.

Riley, G.H. 1980. Granite ages in the Pine Creek Geosyncline. In: J.Ferguson & A.B.Goleby, *op. cit.*, 69-72.

Rossiter, A.G. & Ferguson, J. 1980. A Proterozoic tectonic model for northern Australia and its economic implications. In: J.Ferguson & A.B.Goleby, *op. cit.*, 209-232.

Schrire, C. 1982. *The Alligator Rivers: Prehistory and Ecology in Western Arnhem Land.* Terra Australis No.7, Dept. of Prehistory Research School of Pacific Studies. Australian National University: 277 pp.

Skwarko, W.K. 1966. Cretaceous stratigraphy and palaeontology of the Northern Territory. *Bur. Miner. Resour. Geol. Geophys. Aust. Bull.* No.73.

Story, R., Galloway, R.W., McAlpine, J.R., Aldrick, J.M. & Williams, M.A.J. 1976. *Lands of the Alligator Rivers area, Northern Territory.* CSIRO Land Research Series No.38, Melbourne.

Story, R., Williams, M.A.J., Hooper, A.D.L., O'Ferrall, R.E. & McAlpine, J.R. 1969. *Lands of the Adelaide-Alligator area, Northern Territory.* CSIRO Land Research Series No.25, Melbourne.

Stuart, J.M'Douall 1863. *Explorations across the Continent of Australia, with Charts, 1861-62.* F.M.Baillière, Melbourne.

Stuart-Smith, P.G., Wills, K., Crick, I.H. & Needham, R.S. 1980. Evolution of the Pine Creek Geosyncline. In: J.Ferguson & A.B.Goleby, *op. cit.*, 23-37.

Suess, E. 1885. *Das Antlitz der Erde,* Vol.I. F.Tempsky, Vienna.

Talent, J.A. 1984. Australian biogeography past and present: determinants and implications. In: J.J.Veevers, *op. cit.*, 57-93.

Tenison Woods, J.E. 1886. Report on the geology and mineralogy of the Northern Territory of Australia. *South Aust. Parliamentary Pap.* No.122.

Thiede, J. 1979. Wind regimes over the late Quaternary southwest Pacific Ocean. *Geology* 7: 259-262.

Thom, B.G. & Chappell, J. 1975. Holocene sea levels relative to Australia. *Search* 6: 90-93.

Torgersen, T., Jones, M.R., Stephens, A.W., Searle, D.E. & Ullman, W.J. 1985. Late Quaternary hydrological changes in the Gulf of Carpentaria. *Nature* 313: 785-787.

Veevers, J.J. 1971. Phanerozoic history of Western Australia related to continental drift. *Geol. Soc. Aust. J.* 18: 87-96.

Veevers, J.J. (ed.) 1984. *Phanerozoic Earth History of Australia,* Clarendon Press, Oxford.

Veevers, J.J., Jones, J.G. & Talent, J.A. 1971. Indo-Australian stratigraphy and the configuration and dispersal of Gondwanaland. *Nature* 229: 383-388.

Veevers, J.J., Jones, J.G., Powell, C.McA. & Talent, J.A. 1984. Synopsis. In: J.J.Veevers, *op. cit.*, 351-364.

Walker, D. (ed.) 1972. *Bridge and Barrier: the Natural and Cultural History of Torres Strait.* Res. School Pac. Studies, Australian National University, Canberra.

Walpole, B.P. 1962. Mount Evelyn, Northern Territory. 1:250:000 Geological Series. *Explan. Notes Bur. Miner. Resour. Geol. Geophys. Aust.*

Walpole, B.P., Crohn, P.W., Dunn, P.R. & Randall, M.A. 1968. *Geology of the Katherine-Darwin region, Northern Territory.* Bureau of Mineral Resources, Australia, Bulletin 82.

Wellman, P. & McDougall, I. 1974. Cainozoic igneous activity in eastern Australia. *Tectonophysics* 23: 49-65.

White, C. 1967. Early stone axes in Arnhem Land. *Antiquity* 41: 147-152.

Williams, M.A.J. 1968. Termites and soil development near Brocks Creek, N.T. *Aust. J. Sci.* 31: 153-154.

Williams, M.A.J. 1969a. Geology of the Adelaide-Alligator area. In: R.Story et al., *op. cit.*, 56-70.

Williams, M.A.J. 1969b. 1:500,000 geomorphology map of Adelaide-Alligator area, Northern Territory. In: R.Story et al., *op. cit.*, map folder.

Williams, M.A.J. 1969c. Geomorphology of the Adelaide-Alligator area. In: R.Story, *et al.*, *op. cit.*, 71-94.

Williams, M.A.J. 1978. Termites, soils and landscape equilibrium in the Northern Territory of Australia. In: J.L.Davies & M.A.J.Williams (eds.), *Landform Evolution in Australasia,* Australian National University Press, Canberra: 128-141.

Williams, M.A.J. 1984a. Quaternary environments. In: J.J.Veevers, *op. cit.*, 42-47.

Williams, M.A.J. 1984b. Cenozoic evolution of arid Australia. In: H.G.Cogger & E.E.Cameron (eds.), *Arid Australia,* Australian Museum, Sydney: 59-78.

Williams, M.A.J. 1985a. On becoming human: geographical background to cultural evolution. *Australian Geographer* 17: 175-185.

Williams, M.A.J. 1985b. Pleistocene aridity in tropical Africa, Australia and Asia. In: I.Douglas & T.Spencer (eds.), *Environmental Change and Tropical Geomorphology,* George Allen & Unwin, London: 219-233.

Woodroffe, C.D., Chappell, J.M.A., Thom, B.G. & Wallensky, E. 1986. *Geomorphological Dynamics and Evolution of the South Alligator Tidal River and Plains, Northern Territory.* Australian National University North Australia Research Unit, Mangrove Monograph No.3, Darwin.

Woodroffe, C.D., Thom, B.G. & Chappell, J. 1985. Development of widespread mangrove swamps in mid-Holocene times in northern Australia. *Nature* 317: 711-713.

Wright, R.L. 1963. Deep weathering and erosion surfaces in the Daly River basin, Northern Territory. *J. Geol. Soc. Aust.* 10: 151-164.

CHAPTER 2

Floods and droughts: The northern climate

N.S. McDONALD & J. McALPINE

INTRODUCTION

The climate of the coastal plains area of the Northern Territory is distinct in that it is highly reliable in its occurrence from year to year. It is this reliability in combination with the amount of rainfall that distinguishes the climatic pattern present in the region from that of the surrounding areas, particularly to the south. Within the coastal plains region there is only a relatively low degree of spatial variability in climate.

The seasonal alternation of flood and drought is the dominant feature of the climate. While flood and drought elsewhere in Australia are regarded as natural hazards and lack predictability, here they are merely expressions of the regular yearly alternation of excess of surface water and lack of soil water. This cycle is so regular that the physical and biological environments are closely adjusted to it.

In this chapter we describe and analyse the chief characteristics of the climate of the region and the atmospheric conditions that give rise to them. Particular emphasis is placed on variability and interactions among meteorological components that can be masked by condensed tabular statistical summary. We conclude with a water balance analysis to translate the effects of these atmospheric conditions into the ground surface phenomena that play a direct role in biological activity.

PRINCIPAL CLIMATIC ELEMENTS

The climatic elements and statistics for the area have been discussed and presented previously by a number of authors (Christian & Stewart 1953, Anon. 1961, and McAlpine 1969). Only a brief summary of the available climatic data is presented here.

Mean annual rainfall is 1600 mm at Darwin and decreases to the east and south to 1250 mm (Fig. 7). The rainfall distribution is highly seasonal, with the rainy season occurring between late October and early April. An outstanding feature of the rainfall pattern is the reliability of its incidence and the low variability of

the range of values within each month. For instance, at Oenpelli (Table 2) in January, 90% of all values over a 60 year period were between 200 and 450 mm and the lowest and highest monthly rainfalls for the same period were 135 and 777 mm respectively. Similar patterns of reliability are exhibited for each month at other stations. The characteristics of the climate on a seasonal basis are dealt with later. Mean annual minimum and maximum temperatures on the coast at Darwin are 32° and 23°C respectively and inland these change to 34° and 22°C. Mean monthly temperatures throughout the year vary only slightly from the annual means. The greatest variation in mean temperatures is that associated with the diurnal range which in January is 7°C and in July is 11°C. Data for temperature and a range of other climatic elements are presented in Table 3. The main features are higher relative humidities during the rainy season and higher rates of evaporation in the dry season. A comparison of Tables 2 and 3 reveals that although mean annual evaporation exceeds rainfall in most years there is an excess of rain over evaporation during the wet season.

Global radiation over the area shows a small annual amplitude with mean daily values of approximately 20 mJ/m^2 (478 1y/day). During the summer, when day length and solar elevation are greatest, the percentage of clear sky is at a minimum. In the dry season, the lower cloud cover compensates for the shorter day length and lower solar elevation (11.2 hours and 55° at solar noon on the winter solstice). The actual reduction caused by reduced solar elevation amounts to only 20% of the incident radiation on 17 February and 26 October when the sun is directly overhead at solar noon (1250 hours CST at Darwin). There is considerable daily variation dependent upon cloud cover as shown by the one standard deviation range in Figure 8.

GENERAL CLIMATIC CONTROLS

The large scale atmospheric circulation over northern Australia which controls the regional climates is an integral part of the monsoonal circulation extending

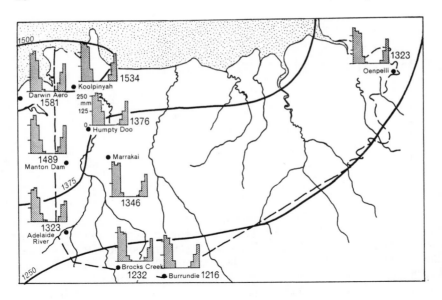

Figure 7. Isohyets of mean annual rainfall (mm) with distribution of mean monthly rainfall (mm) shown by histograms for a number of stations.

Table 2. Rainfall variability at Oenpelli (mm).

	Jan.	Feb.	Mar.	Apr.	May	Jun.	Jul.	Aug.	Sept.	Oct.	Nov.	Dec.
Lowest	135	113	49	0	0	0	0	0	0	0	3	55
10%	213	117	101	3	0	0	0	0	0	0	40	122
50%	310	272	274	37	1	0	0	0	0	0	92	202
90%	449	498	473	143	24	3	2	1	17	92	200	362
Highest	777	655	594	414	194	35	62	15	36	168	276	583

Table 3. Mean monthly data for various climatic elements at Darwin.

	Jan.	Feb.	Mar.	Apr.	May	Jun.	Jul.	Aug.	Sept.	Oct.	Nov.	Dec.	Ann.
Maximum temperature (°C)	32	32	32	33	32	31	30	31	32	33	33	33	32
Minimum temperature (°C)	25	25	24	24	22	20	19	20	23	25	25	25	23
Rh 09.00 hours (%)	79	81	81	75	66	62	60	65	68	68	70	74	71
Rh 15.00 hours (%)	68	69	65	51	42	39	35	40	45	50	55	63	52
Evaporation (mm)	180	145	150	170	180	175	200	220	240	250	220	200	2330

across tropical latitudes from Africa to the Coral Sea. The high annual range of temperature over the continents in this sector permits a noticeable seasonal latitudinal migration of the tropical atmospheric circulation: the Hadley regime. In the Southern Hemisphere winter, low level currents emanating from the subtropical anticyclones near 25°S – the Southeast Trades – pass over northern Australia moving towards the Intertropical Convergence Zone (ITCZ) located over the Asian mainland. At the height of this exchange of air, the weather in northern Australia is essentially warm and dry during the 'annual drought' or dry season. In summer, when the circulation is reversed, the Trades over the area are replaced by an inflow of moist air to the ITCZ or the monsoon trough. The combination of moist air and the ascending limb of the Hadley cell produces widespread heavy rainfall during the wet season or the season of 'floods'.

The strength and position of the ITCZ are influenced by the overall strength of the Hadley regime and as well by a longitudinal oscillation termed the Walker Circulation. Bjerknes (1969) recognized that there is a quasi-periodic circulation between the atmospheric pressures over the Eastern Pacific Ocean and the area to the north of Darwin. For the high index situation there is higher than normal pressure and cold ocean water in the Eastern Pacific and lower than normal pressures in the Australian sector. The reverse occurs for the low index situation. Tanaka (1980) demonstrated that the interannual fluctuations in the monsoonal characteristics are correlated with this Walker Circulation. Composite maps of seasons when an index of the Walker Circulation is alternatively high and low show markedly different circulation features.

Strongly developed Walker Circulation

The areas of high rainfall anomalies are located near the two active ITCZs, one in the Northern Hemisphere and the other well to the south of the north Australian coast (Fig. 9). A tongue of cold water extends across the equator almost to 180°W. Precipitation deficiencies, though not shown on Figure 9, occur over Eastern New Guinea. Only slight convergence associated with perturbations or surges in the moist inflow is needed to produce widespread rainfall over the monsoonal lowlands of northern Australia. 1963, 1968, 1971 and 1974 are recent examples of this situation.

Weakly developed Walker Circulation

The Northern Hemisphere ITCZ is not present and the Australian ITCZ barely crosses the mainland (Fig. 10). The area of maximum precipitation anomaly is to the north and east of New Guinea and the cold water in the Eastern Pacific is replaced with warm water. As the mean position of the monsoon trough is adjacent to the monsoon lowlands, rainfall is still abundant but it is reduced due in part to the more frequent influx at mid-levels of the atmosphere of drier air from the Australian subtropical ridge. Recent examples of this case are the years of 1964, 1965, 1966, 1969, 1970 and 1972.

Apart from its two or three year quasi-periodicity, the Walker Circulation has increased in amplitude in recent times from a minimum in the decades from 1920 to 1950. This change in amplitude is noticeable in the time sequence of annual rainfall for the Darwin-Daly River region where the annual rainfall in recent years has become decidedly more variable (Fig. 11). The mean rainfall for this division in January, February and March for the years used by Tanaka (1980) in establishing his map is 1066 mm and 866 mm for the strong and weak cases respectively, a statistically significant difference.

While a similar relationship exists between the Northern Hemisphere monsoon and fluctuations in the

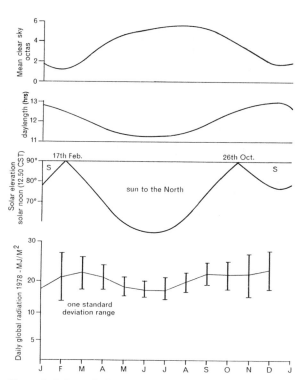

Figure 8. Solar radiation parameters, Darwin.

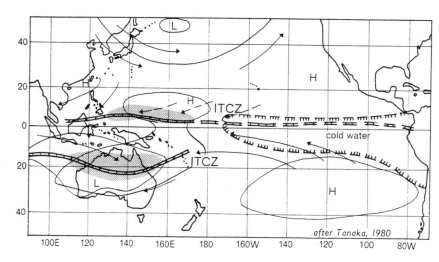

Figure 9. Schematic map of strong summer monsoon in northern Australia. Thin solid lines show strong circulation at the 850 mb level; shaded areas are the regions of heavy precipitation.

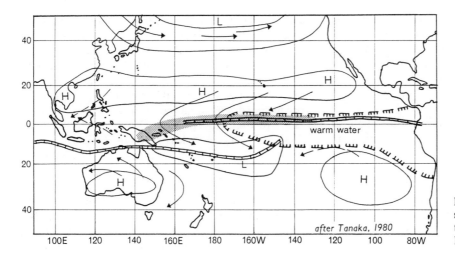

Figure 10. Schematic map of weak summer monsoon in northern Australia. Thin lines show weak circulation at the 850 mb level.

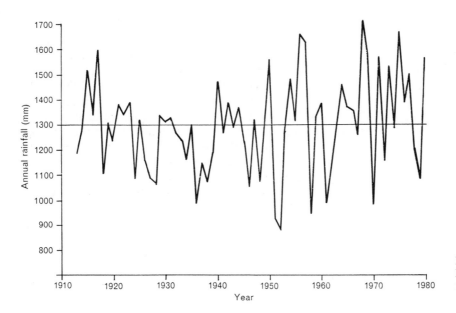

Figure 11. Annual rainfall Darwin-Daly River rainfall districts, 1913-1980.

Walker Circulation, since there is little or no precipitation at this time of the year over the monsoonal lowlands of Australia, the effects of such fluctuations are imperceptible.

THE SEASONS

The approach taken here will be to select weather sequences which are representative of the diurnal variation of weather elements and which also highlight particular meteorological events central to our understanding of the seasonal climate. For each season, half hourly observations taken at Darwin Airport are graphed allowing a comparison of the mean values and the variation of individual elements (over a 24 hour period, between days in a season, and between seasons), and the variation or correlation between these elements. The elements considered are temperature, dew point temperature, wind speed and direction, cloud cover by type (cumulus, cumulonimbus and other low or mid atmosphere clouds) and, finally, current weather. From these elements other parameters or meteorological processes can be deduced. For example, relative humidity can be estimated from the spacing between the air and dew point temperature. Although the graphs are representative of Darwin and other coastal localities, inland mesoscale processes such as convection, and land and sea breeze effect, may well be different.

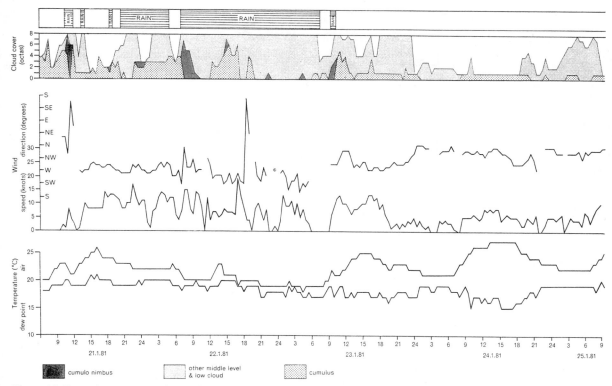

Figure 12. Synoptic wet season observations at Darwin aerodrome, 21-25 January 1981.

The wet season

The onset of the 'wet' season occurs towards the end of December or early in January as the monsoon trough approaches the Australian mainland. Wet conditions last until late March, though the actual year by year timing depends on the large scale parameters discussed earlier. During the wet season, Southern (1966) recognized that there can be periods of excessive rainfall, periods of occasional showers and lengthy dry spells. The periods of sustained rainfall are associated with macroscale rain-producing systems such as tropical cyclones, monsoon depressions and surges. There are usually three to four such sequences lasting up to ten days in any wet season. The periods of occasional showers are associated with local convection and with the sea breeze convergence. Dry spells can be lengthy, for example during February 1979, there were 21 days without rainfall in Darwin owing to the presence in the mid atmosphere of dry southern air which restricted cumulus development.

The sequence chosen here (21-25 January, 1981) illustrates the macroscale influence of a surge in the monsoonal flow with resultant widespread and continuous rainfall (Fig. 12). In this instance a total of 438 mm of rain fell in Darwin in the first three days. The latter half of the sequence is a two day period of no rainfall even though surface temperatures and humidities are not appreciably different from those of the preceding two days. At the commencement of the sequence a deep moist north-west airstream penetrated well inland to the monsoon trough located as far south as Alice Springs. For the 36 hours from 21.00 hrs CST on 21st January there was almost continuous rainfall from an overcast sky of altostratus clouds with the occasional embedded convective cloud. Winds were from the west to southwest and were reasonably strong, averaging 10 knots. Humidities were high and there was only minimal diurnal variation in the temperatures. For the latter half of this sequence, organized convergence was not present, the wind direction turned gradually to the north and the wind strength was reduced. Surface dew point temperatures remained high and, given the lower cloud cover, the diurnal temperature range increased slightly. A land and sea breeze circulation is not conspicuous. The daily rainfalls for this period were 86 mm on the 21st, 147 mm on the 22nd, 205 mm on the 23rd, 2 mm on the 24th and no rainfall on the 25th. Note the rainfall is measured at 09.00 hrs and refers to the rainfall of the preceding 24 hours, i.e. 147 mm fell between 09.00 hrs on the 21st and 09.00 hrs on the 22nd. Rainfall amounts of similar magnitude were recorded at other stations in the region.

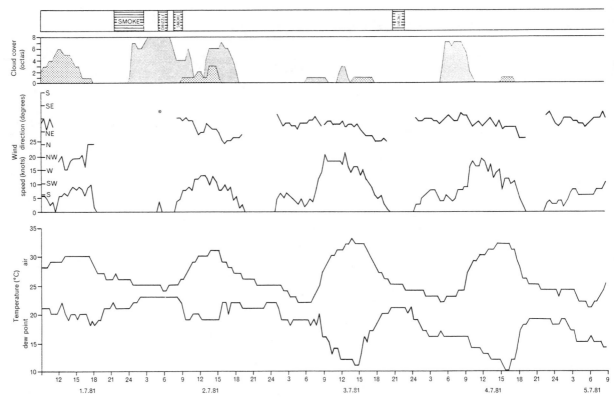

Figure 13. Synoptic dry season observations at Darwin aerodrome, 1-5 July 1981.

One feature not shown in this example is the effect of tropical cyclones. As a general rule tropical cyclones develop in the monsoon trough or in a deep moist easterly flow located over the warm ocean water to the north. The positioning of this trough will decide when and where the hazardous consequences and widespread rainfall will occur. Tropical cyclones are more likely when the trough is located just to the north of the continent, i.e. early or late in a strong wet season. One cyclone per year on average affects the Northern Territory coast (Lee & Neal 1981).

The dry season

Lasting from May to September, the dry season is a period of little or no rainfall. Over this period Darwin has a median rainfall of less than 10 mm. There are, however, some changes in the day to day weather caused by the passage of the anticyclones across the southern half of the continent. When an anticyclone is centred over South Australia, there is normally a strengthening or a surge in the general southeasterly flow over the monsoon wetlands. As the anticyclone moves gradually eastward the local winds tend to lighten and become more easterly. Land and sea breeze

circulation is well developed. Prior to the next surge there is a period of relatively calm conditions and it is at this time that infrequent showers may occur. The duration of each stage in this cycle is dependent upon the strength, orientation and rate of movement of the southern systems. Persistent strong southeasterly winds associated with slow moving systems will allow drier, dusty air to replace the moist coastal air over the local region which enhances nocturnal radiative cooling. Periods in the dry season, notably June and early July when this cooling occurs, are recognized by traditional societies as the 'cold' season.

The sequence selected (1-5 July, 1981) illustrates the surge in the southeasterly flow and the calm period before the surge (Fig. 13). On 1st July, 1981, a ridge extended eastwards across central Australia to about Alice Springs, and by the 3rd an anticyclone and its associated ridge were established across the continent. On the first day in this sequence, there was a noticeable build up in the early afternoon of convective clouds and a pronounced change in the wind direction with the onset of the sea breeze. Atmospheric humidities were high, dew point temperatures approached summer values, and the diurnal temperature range was low. During the early hours of 2nd July

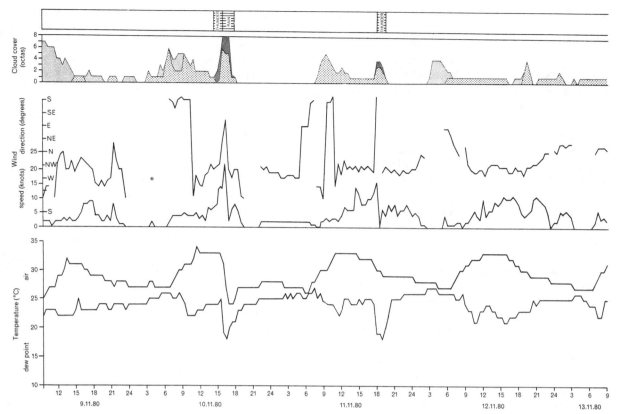

Figure 14. Synoptic observations (transition period) at Darwin aerodrome, 9-13 November 1980.

drizzle was observed at Darwin associated with an upper level trough and an overcast altostratus sky. On the 3rd there was a significant strengthening of the winds from the southeast. For the next five days, strong east to southeast winds gusting to 30 knots, low dew point temperatures and a diurnal temperature range of up to 15°C were observed. In the early afternoon, the sea breeze circulation opposed the offshore motion resulting in light to no winds, winds more northerly and an increase in humidity.

The transition season

Before the monsoon trough affects the mainland and after the influence of the southern anticyclones, there is a gradual build-up in humidity in October and November before the onset of the wet season. The reverse occurs in April. The most significant meteorological phenomena of this period are the occasional thunderstorms resulting from mesoscale activity. These storms provide the only relief from the unpleasant hot and humid weather.

The sequence selected, from 9th to 13th November 1980, illustrates the uniform hot and humid conditions and the development of a thunderstorm (Fig. 14). Over the four day period light north to westerly winds were directed towards the Australian heat low and cumulonimbus development occurred in the mid-afternoon on two occasions. On the 10th, from a storm centred southeast of Darwin airport, 40 mm of rain were recorded at Noonamah some 25 km away but none at the recorded site. This storm, however, was accompanied by an increase in wind speed and the downdraft caused both the air and dew point temperatures to fall suddenly. Apart from a similar drop in the dew point temperatures on the 11th, other interesting features of this period include the light and variable winds and the high maximum and minimum temperatures.

These examples demonstrate the variability of atmospheric conditions and changes that occur in what is commonly perceived as a uniform seasonal climate. These atmospheric processes can exert subtle influences on many aspects of the regional ecology that would not be apparent from an analysis restricted to the summary information provided in surface climatic statistics.

WATER BALANCE

The analysis of the water balance of the area presented here is based on estimates derived from a model developed initially by Slatyer (1960) to study the agroclimatology of the Katherine region to the south. The model was subsequently modified and applied in regional and environmental investigations of northern Australia by Fitzpatrick (1965) and Slatyer (1970), and more specifically to the coastal plains area east of Darwin by McAlpine (1969, 1976). These applications dealt chiefly with the water balance in relation to pasture productivity and such specific matters as water levels in tailing dams. The purpose here is to deal with two major components of the water balance, soil moisture and water surplus or runoff, that are directly related to biological productivity.

A detailed description of the model and its computer implementation has been published elsewhere (Keig & McAlpine 1969). The model is based on a weekly time interval and applied on a point basis, usually for the location of the climate station for which data are directly available. In this analysis a standard period of 60 years of rainfall data from 1912 to 1972 for Darwin and Oenpelli has been analysed together with a shorter and discontinuous record for Adelaide River. The latter station is included to give some idea of variation in an inland direction.

The model assumes that inputs to a soil moisture store are instantaneous and consist of actual weekly rainfall less a calculated mean weekly evapotranspiration. The maximum soil moisture storage available to plants, that is the field capacity of the soil, is assumed to be 100 mm. Excess rainfall beyond that required to meet evapotranspiration demands and to replenish soil moisture up to field capacity is taken to indicate the amount of water surplus for the week. Evapotranspiration is calculated to be equal to 0.8 of evaporation from a US Class A pan evaporimeter when soil moisture plus rainfall in the previous week is above 50% of maximum storage capacity, and 0.4 when it is less than 50%. Where soil moisture is less than calculated evapotranspiration demand the latter is reduced to the amount actually remaining in soil store. The model thus produces estimates of week by week changes in soil moisture availability and of water surplus. The former can be used to provide estimates of periods of plant growth while the latter can be transformed to give indications of catchment yields. It should be noted that where soils have field capacities significantly above or below the 100 mm assumed, the drought and runoff estimates derived may be higher or lower.

Cracking clays are the most common soil in the coastal plains environment. Even though their wetting and drying characteristics are different from those assumed in the model the water balance results presented here can be taken as reasonably representative of those soils. The reason for this is that in an area with a distinct wet and dry season, cracking clays would remain in their cracked, dry state during the dry period and would have very low to nil available water in store. In the wet season, following a short period in which a change of state takes place, the soil would attain high levels of available water and generally act as a more normal clay soil.

The most significant features of the soil moisture regimes of the coastal plains are their highly seasonal distribution and their restricted spatial variability. This is demonstrated in Figure 15 which shows the mean weekly levels of soil moisture storage for three stations. There is a decrease in soil water availability inland of Adelaide River, but the difference is not great. There is also a significant degree of temporal reliability in soil moisture regimes in both timing and moisture level. This is illustrated for Oenpelli in Figure 16 which shows weekly variability expressed as a frequency analysis. The pattern shown is similar at Darwin and Adelaide River.

Temporal variability can also be indicated by calculating the time of commencement and cessation of each wetting cycle (or plant growth season) for each station and by noting the earliest and latest dates for the start of the wet and dry seasons. These are given in Table 4. Figure 17 presents the same information on a year by year basis. Table 4 shows that there is again little variation from place to place, and Figure 17 reveals that although the annual occurrence of growth periods is regular, short false starts are common. Figure 17 also indicates the somewhat greater variability in the start of growth periods than in their conclusion. False starts to the season are common, false

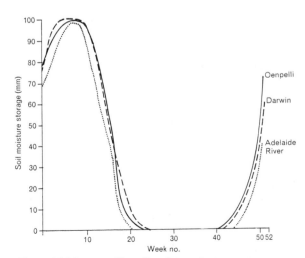

Figure 15. Mean weekly soil moisture storage.

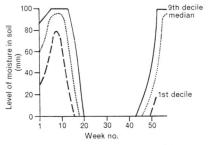

Figure 16. Weekly variability in soil moisture at Oenpelli.

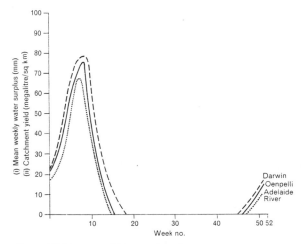

Figure 18. Mean weekly water surplus.

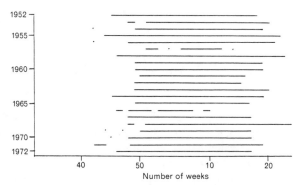

Figure 17. Length of period each year (1952-1972) for which available soil water was greater than 10 mm.

endings rare. It is noteworthy in terms of plant production that nearly every year has a short growth period marked by a good surface and immediate sub-surface wetting followed by a short dry period before the time of continuous soil moisture availability begins. It is also interesting to note that whereas false starts to the wet season are difficult to determine from rainfall data alone, they are more obvious when considered in terms of soil moisture.

WATER SURPLUS

The water balance model provides estimates of point water surplus (i.e. the amount by which weekly rainfall is surplus to the requirements of evapotranspira-

Table 4. Period of availability of soil moisture at various stations (start and cessation of pasture growth period).

	Darwin	Adelaide River	Oenpelli
Start (median data)	Nov. 2	Nov. 9	Nov. 14
End	May 16	May 4	May 12
Average length (weeks)	28	25	25

tion and replenishment of soil moisture store up to field capacity). Point water surplus can be taken as comprising the water available for surface runoff and deep percolation.

Figure 18 presents the mean weekly point water surplus for three stations. As could be expected, the amount reaches a maximum during the wet season in February, and in the dry season falls to nil. The differences between stations indicate a small decrease inland directly related to the decrease in rainfall in that direction.

In contrast temporal variability in water surplus for any given period over a number of years can be quite significant even though, by comparison with other regions, the range is relatively restricted. Temporal variability is shown in Figure 19 by means of quartile and decile distributions. The range of values is highest at the transition periods from dry to wet season and vice versa. The range lessens during the middle of the wet season and is minimal during the dry season.

Week by week estimates of point water surplus can be converted using catchment areas to provide estimates of monthly, seasonal and annual catchment yields. On Figure 18 a direct conversion scale relates point water surplus expressed in millimetres to yield in megalitres per square kilometre of catchment.

Median annual water surplus has been calculated to be 720 mm at Darwin, 670 mm at Adelaide River and 700 mm at Oenpelli. An indication of the relatively restricted range of annual variation in water surplus is shown at the latter station, where in 50% of years this surplus varies from only 500 to 850 mm, and the extreme values estimated over the sixty year period analysed, ranged from 220 mm in 1952 to 1370 mm in 1969. The actual yearly values are plotted in Figure 20.

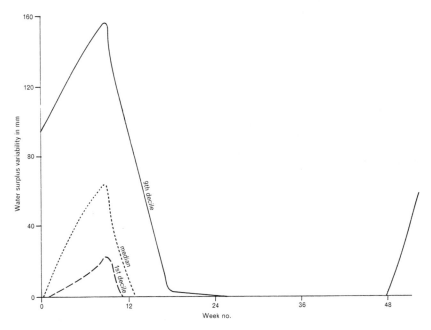

Figure 19. Weekly variability in water surplus at Oenpelli.

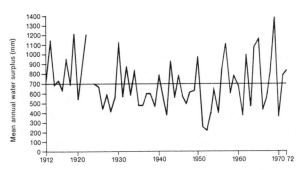

Figure 20. Annual variability in water surplus at Oenpelli.

WATER BALANCE AND BIOLOGICAL ACTIVITY

One of the most important features of the water balance in terms of biological productivity is the reliability in timing and in the amount of each of the separate components (soil moisture, water surplus) and their low degree of variability from year to year. This reliability is more apparent in soil moisture and water surplus than it is in the rainfall record. Although the growth period of vegetation is restricted on average to 25 to 28 weeks per year, it never fails. This is in contrast to other monsoon climates and in areas to the south of the coastal plains region. Year to year productivity is assured and this is a key factor in plant and animal activity and distribution.

The availability of open water in the environment throughout the year, despite the high rates of evaporation, is guaranteed by the concentration of reliable but not great water surpluses into the billabongs and depressions that are common in this riverine plain environment. It is these that provide the bulk of animal life with dry season access to water which in effect become habitat refugia.

REFERENCES

Anon. 1961. *Climatology Survey. Region 1: Darwin, Katherine, Northern Territory.* Aust. Met. Bur., Melbourne.

Bjerknes, J. 1969. Atmospheric teleconnections from the equatorial Pacific. *Monthly Weather Review* 97: 163-172.

Christian, C.S & Stewart, G.A. 1953. General report on survey of Katherine-Darwin region (1946). *CSIRO Australia Land Research Series* 1.

Fitzpatrick, E.A. 1965. Climate of the Tipperary area. *CSIRO Australia Land Research Series* 13: 39-52.

Keig, G. & McAlpine, J.R. 1969. WATBAL. A computer system for the estimation and analysis of soil moisture regimes from simple climatic data. *CSIRO Australia Div. Land Research Tech. Memo* 69/9.

Lee, P.M. & Neal, A.B. 1981. The Climate of northern Australia. First Menzies Regional Workshop on Living in the North, Darwin, 4-6 June, 1981.

McAlpine, J.R. 1969. Climate of the Adelaide-Alligator area. *CSIRO Australia Land Research Series* 25: 49-55.

McAlpine, J.R. 1976. Climate and water balance of the Alligator Rivers area. *CSIRO Australia Land Research Series* 38: 35-49.

Slatyer, R.O. 1960. Agricultural climatology of the Katherine area, N.T. *CSIRO Australia Land Research Regional Survey Tech. Pap.* 12.

Slatyer, R.O. 1970. Climate of the Ord-Victoria area. *CSIRO Australia Land Research Series* 28: 62-74.

Southern, R.L. 1966. A review of weather disturbances controlling the distribution of rainfall in the Darwin-Katherine region, N.T. *Aust. Met. Bur. Working Paper* 65/3203.

Tanaka, M. 1980. Role of the circulation at the 150 mb level in the winter and summer monsoon in the Asian and Australian regions. *Inst. of Geoscience, Tsukuba, Japan, Occ. Paper* 2.

CHAPTER 3

Hydrology of the northern wetlands

D. KINGSTON

INTRODUCTION

As recently as thirty years ago very little was known about the hydrological characteristics of the rivers which flow to the sea between Darwin to the west and the Arnhem Land plateau to the east. In particular, there was a dearth of accurate information on the flooding behaviour of the coastal plains, with their rich biota of swamp-dwelling mammals, reptiles, birds and fish. It was well known that the low, flat topography of the coastal plains and the seasonally wet monsoonal climate produced widespread and prolonged flooding; but, as Purich noted in 1965, for just how long, and to what depth the plains flooded, and with what frequency, were questions that remained unanswered (Purich 1965).

Starting in 1956, automatic water level recorders were installed at twelve stations on the flood plains. The Water Resources Branch of the Northern Territory Administration also monitored stream discharge on some of the main streams, and tide gauges were set up on certain rivers. Wet season access was never easy, and ultimately depended upon helicopters. As a result of this work we now possess reliable quantitative information on the present-day flood behaviour of the northern wetlands, the essence of which is summarised below. For further details, see Watson (1966) and Anon. (1980, 1981).

DISTRIBUTION AND DRAINAGE STATUS OF THE NORTHERN WETLANDS

Figure 21 shows the distribution and drainage characteristics of the coastal plains. The areas designated as perennially swampy are the areas of freshwater paperbark swamp mapped by Story, Hooper and Williams during the land system survey of this region by the CSIRO Division of Land Research and Regional Survey in 1965 (Story et al. 1969). Qualitative drainage status is shown by the circled letters P, S and T. Those sectors of the coastal plains where the floodwaters are derived mainly from direct inputs from rainfall are termed precipitation-dominated (P). Where tidal influence is the major cause of flooding, as in the lower right bank sector of the Adelaide River basin, the plain is described as tide-dominated (T). In the upstream reaches of the coastal rivers excess fllow from the channels is the main source of floodwater. These sectors are indicated as streamflow-dominated (S). The choice of attribute (P, T or S) is an attempt to summarise twenty years of observations by the author on the flood behaviour of these plains.

RAINFALL, RUNOFF AND SEASONAL FLOODING IN THE ADELAIDE RIVER BASIN

In the precipitation-dominated sectors of the coastal plains, flooding is a function of topography, antecedent soil moisture content, and rainfall intensity and duration. Figure 22 presents monthly streamflow data for the Adelaide River area and rainfall data for Darwin to represent the rainfall and streamflow impacts on the coastal plain of the Adelaide. (As monthly totals it would be correct to show these in histogram form but the data have been joined by lines in this figure to clarify the range of variability.)

Runoff is a residual quantity; that is to say, except when rainfall is intense, runoff is the amount of water left once evaporation, replenishment of soil moisture, use by vegetation, and infiltration are satisfied. Thus when rainfall is low, runoff is disproportionately much lower, as the bottom line of the 30-70% probability zones indicates. It is also clear from Figure 22 that runoff responds substantially two to three months after the rains begin, after the initial soil moisture needs and evaporation are adequately satisfied.

If 150 mm of rain are needed to moisten and seal the cracking clay soils of the coastal plains, then it is clear from Figure 22 that areas subject to inundation by direct rainfall (such as many of the ill-drained coastal plains), may start to pond water well ahead of peak flow in the rivers or in the creeks draining onto the fringes of the plain. The general pattern for these maritime plains is for the coastal reaches to be

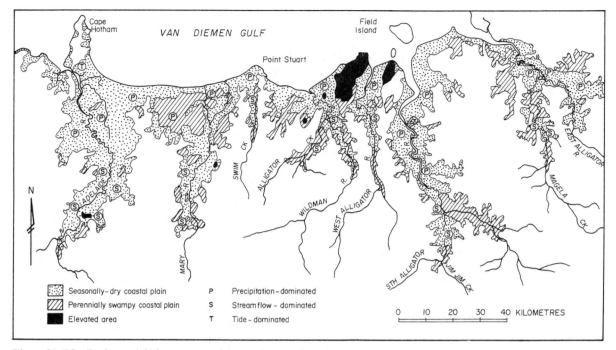

Figure 21. Distribution and drainage status of the northern wetlands (after Story et al. 1969 and author's observations).

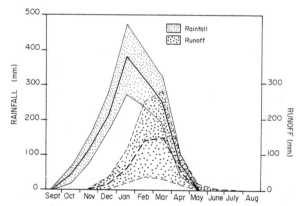

Figure 22. Runoff amount and incidence in the Adelaide River basin in relation to rainfall (rainfall for Darwin; runoff regionalised for six stations in the Adelaide River basin; zones show range from 30% to 70% probability of occurrence).

windows or waterholes sustained by groundwater inflow.

The analysis of runoff may be taken a stage further using the subdivision of the wetlands into seasonally dry and perennially swampy (Fig. 21). As a broad generalisation, the latter areas are below the general level of the plains whereas the former conform to the general surface level. Figure 23 shows the depth and duration of flooding at three areas in the Adelaide River basin. In contrast to the wetlands further east, the floods are shallow and relatively short-lived.

COASTAL PROCESSES AND SEASONAL FLOODING ON THE MARY AND ALLIGATOR RIVERS PLAINS

In addition to the normal processes of overbank flooding whenever bankfull discharge is exceeded during times of prolonged and heavy rainfall, past and present coastal processes also exert a profound influence upon flooding in the coastal wetlands.

Very characteristic features of the Mary River valley in its lower coastal reaches are the chenier ridges which mark successive stages in the northward advance of the Holocene shoreline (Clarke et al. 1979). Coastal advance began shortly after the sea had risen to its present level about 6,000-7,000 years ago.

inundated by direct rainfall, and for the headward reaches to be inundated by local runoff and periodic overflow from the principal river.

The upper reaches include areas of alluvial sands and silts, and in these areas scoured depressions created and seasonally replenished by streamflow may occur. This is the area to look for perennial fresh waterholes and in some cases these may be watertable

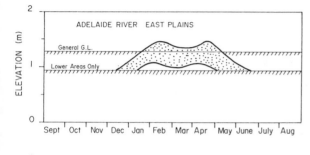

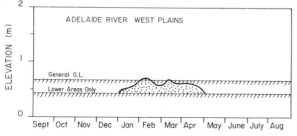

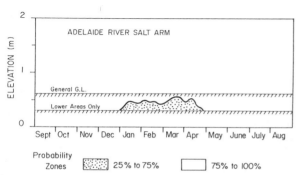

Figure 23. Flooding probability in the Adelaide River basin (after Purich 1965).

Mud brought eastwards by longshore drift is trapped by mangroves in the intertidal zone. At rare intervals, tropical cyclones cause severe coastal erosion. The resulting storm beaches of shelly grit, coral and beach-rock fragments, and some quartz sand are in due course colonised by Pandanus palms, giving rise to the distinctive wooded ridges characteristic of the lower Mary chenier plain. Mud carried by the Mary flood-waters is deposited on the coastal plains inland of the littoral zone. Over the years a layer of dark organic clay accumulates above the intertidal muds and estuarine clays. Differential scour and sedimentation during floods, and gradual compaction of the coastal plain deposits lead to a surface with considerable micro-relief, although from a distance it seems to be quite level.

A combination of coastal and fluviatile processes created the coastal plains. Ground levels in the lower reaches of the Adelaide River plains are frequently 1-2 m above Australian Height Datum (AHD). Mean High Water Spring Tide at Darwin is 2.97 m AHD, or nearly a metre higher than on the plains. Likewise, high tide in the East Alligator is up to one metre higher than on large tracts of the Magela plain.

Prevailing north-westerly winds generate longshore currents which flow from west to east. Littoral drift of sediment across river mouths will modify the impact of tides. The channel resistance created in the mouth will attenuate the height to which the tide rises on the landward side in response to the harmonic motion in the open sea. Further constriction by the growth of coastal barriers may inhibit tidal penetration almost completely, the tidal meanders becoming relics. This is now seen on the lower Mary River. The lower Adelaide River was like this once, as can be seen just west of Woolner. However, headward retreat of a channel in what is now the Narrows on the Adelaide led to river capture and rejuvenated the tidal performance in the lower Adelaide, re-establishing the tidal meander system as far back as Beatrice Hill, with dry season saline intrusion even further.

The South and East Alligator Rivers have substantial catchments of high runoff country, the escarpment country, and enough net seaward flow to prevent closure of their channel mouths.

The barrier across the Magela plains embayment was formed, not so much by littoral drift directly, as by the spill of the East Alligator. With a high tide of 3.8 m, this is the level to which the ground has been raised at the Magela outlet by sediment of estuarine origin. Wherever the constraint on tidal movement is effective, freshwater swamps may be expected to come into existence as in the Mary and Magela, and thus provide a habitat for a wide range of semi-aquatic plants and animals. Figures 24 and 25 show the probable depth and duration of flooding on the West, South and East Alligator River plains, as well as two localities on the Wildman River plains. It is immediately apparent that except on the East Alligator plains, where the flood regime is similar to that of the Adelaide River in being quite short, flooding of the coastal plains is prolonged and deep.

Levees along the South Alligator have occluded the outlets of Jim Jim Creek and Nourlangie Creek, creating the extensive paperbark swamps shown on Figure 21. The mouth of the Wildman is silting up progressively. As tidal capacity is lost to siltation the semi-diurnal tidal flow will decrease and the tidal channel will, in turn, fail to maintain itself.

Tidal meanders will become relics as on the Mary River plains. The alluvial deposits of the Wildman in this aggraded zone have built a barrier across Alligator Creek, creating extensive swamps (Fig. 21). The West Alligator River also appears to have silted up in the corresponding zone, blocking itself and creating the

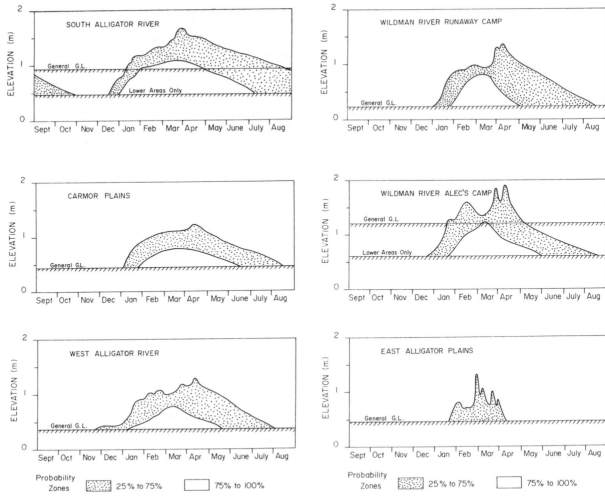

Figure 24. Flooding probability in the West and South Alligator basins (after Purich 1965).

Figure 25. Flooding probability in the Wildman and East Alligator basins (after Purich 1965).

narrow swampy zone shown on Figure 21.

Blockage of the mouth of the Mary is virtually complete. The small water-courses that cross the coastal barrier are merely the overflow of the whole coastal lagoon system and probably only have significant aggregate flow in above average wet seasons. In fact the most likely outflow of the system does not even enjoy the name of the Mary but is called Sampan Creek. The coastal barrier has created the large swamp (marked P on Figure 21) which has its inundation guaranteed by precipitation. The coastal barrier is of coarser material than the plain behind it and probably allows ground water flow to the sea.

The East Alligator itself has adequate runoff and tidal capacity for channel maintenance. As noted earlier, its flood overflow deposits sediment across the mouth of the Magela embayment to a sufficient depth to occlude the system and create permanent swamps.

The overflow of the Magela basin has been studied in some detail in what happens to have been a series of wet years. The system overflows the alluvial barrier in above average wet seasons. If a below average wet season occurs with a uniform intensity through the summer the outflow could conceivably fail. The lagoon when full has an area of roughly 170 km² from which the evaporative output during February will be about 30 million cubic metres or ten cubic metres per second. With the lagoon systems full a steady flow of this rate could enter and not emerge at the other end. The flow balance is of course more complicated than indicated above, since there are also surface slopes, storage changes and infiltration components to be considered. However, the figure serves to illustrate the phenomenon of river occlusion by littoral or alluvial sediments which is so characteristic of the northern wetlands.

REFERENCES

Anon. 1980. *Uranium Province Hydrology.* Department of Transport and Works, Northern Territory, 10 vols.

Anon. 1981. *Water level and discharge data records of the Hydrographic Section of the Water Division, Northern Territory (1953-1981).*

Clarke, M.F., R.J.Wasson & M.A.J.Williams 1979. Point Stuart chenier plain and Holocene sea levels in northern Australia. *Search* 10: 90-92.

Purich, P. 1965. *A study of flooding of the sub-coastal plains in northern Australia.* Northern Territory Administration, Water Resources Branch (Mimeo).

Story, R., M.A.J.Williams, A.D.L.Hooper, R.E.O'Ferrall & J.R.McAlpine 1969. Lands of the Adelaide-Alligator area, Northern Territory. *CSIRO Land Research Series* 25.

Watson, I.S. 1966. *Soil Salinity on the sub-coastal plains of the Adelaide River.* Agricultural Development Advisory Committee, Darwin (Mimeo).

Part 2:
Vegetation and fire

Introduction

Throughout Australia and many other parts of the world these two topics are inextricably bound together. This link is even stronger in the monsoonal lowlands where high temperatures and lack of rainfall during the dry season leave almost all parts of the landscape desiccated and subject to fire. Fire is both a determinant of and a factor influenced by the vegetation; and in both chapters attempts are made to unravel the complexity of this interaction, and to explore the effects of such interaction upon the animals of the various ecosystems.

In Chapter 4 Dunlop and Webb introduce several new approaches to the treatment of the flora of this region. In the analysis of the five major habitats, the authors have identified a recurring theme of vegetation comprising elements which are representative of the autochthonous (typically 'Australian') flora and taxa found throughout Southeast Asia. Some habitats, such as the tidal mudflats, coastal floodplains and monsoon forests, are dominated by taxa found also outside Australia. Others, for example the woodlands and open forests of the subcoastal well-drained lowlands, include 'Australian' elements such as eucalypts and species of *Grevillea* and *Acacia*. Such is the dominance of the eucalypts, which are typically taller than most other species, that the open forest and woodland give the impression of being much more 'Australian' than they really are. Dunlop and Webb show that there is a large number of species within this vegetation which have affiliations much more to the tropics of the rest of the world than to Australia.

Although the sandstone plateau is not really part of the region discussed in this book, we considered that the chapter should include some discussion of its flora. This flora is quite different from that of the well-drained lowlands, being highly autochthonous. Many species from such genera as *Grevillea, Calytrix* and *Jacksonia* have evolved there. Whilst a small number of species of these genera are found in well-drained lowlands, Dunlop and Webb argue that these are escapees from their original habitats in the sandstone.

The position of the small but important monsoon forest is also interesting floristically. Although the present day climate is about as humid as it has been for tens of thousands of years, it is still a harsh one for the development of rain forest. Typically small patches are confined to well-watered and relatively fire-proof localities. By comparison with Indonesia and North Queensland development has been quite truncated, not only with respect to area but also in complexity, possibly because of moisture stress and of fire which may enter these forests from surrounding vegetation. A few species are endemic and poorly dispersed, and most are found widely in other areas like Cape York, New Guinea, and Indonesia. Another interesting feature of monsoon forest floristics is the adaptation of some typically monsoon forest taxa to open forest conditions. Species of *Terminalia* and *Buchanania*, for example, have developed in this direction, although the genera are typically mesic. There has thus been some adaptation to the predominantly xeric vegetation which has generally prevailed in this region for hundreds of thousands of years.

The discussion of fire in a separate chapter treats the subject initially from a prehistoric perspective, and argues that fire is, and for several million years has been inevitable in such a hot and dry part of the world. Only in very rare examples of monsoon forests are there sufficient moisture and natural barriers to keep fire away from at least some parts of them. The question is then rather one of how fire should be most effectively used, and what policy we should adopt to minimise damage to the natural flora and fauna, cultivated and pastoral areas, and developed areas such as settlements and other built-up areas. In Chapter 5 Haynes reviews the policies thought to have been quite generally used by Aboriginal people, and suggests that these policies may be useful and appropriate where some objectives, such as nature conservation, are being pursued.

The interactions between fire and vegetation are of course still not well understood in most places in the world, and much research work will be required before these interactions are understood here. Nevertheless some useful experiments and observations have been made and from these further inquiry is developing.

One particularly useful concept is developed here, argued from a phytogeographic perspective. Dunlop and Webb conclude that almost all species in the subcoastal well-drained lowlands (open forest and woodland), albeit comprising a majority of taxa commonly found outside Australia, are well adapted to fire. They suggest that most species behave the same way, regardless of intensity, season of burning, or periodicity of the fire regime. The exceptional species in the open forest and woodlands are viewed as being part of the autochthonous flora of the sandstone plateau. The sandstone flora, regarded as discrete in origin, is certainly considered to be vulnerable to fire regimes which preclude adequate reproduction.

Another perspective, argued by Haynes in Chapter 5, is that the whole biota had been moulded by an equilibrium established by the Aboriginal people over several thousand years. Whilst the distinction between the sandstone and lowland floras is recognised, the sensitive species of the former area had become widespread: their presence had to be maintained by some form of benevolent fire regime which the Aborigines are believed to have provided.

One interesting direction to research is the analysis of life histories of the suspected sensitive species. In some cases this will be relatively simple, but with others, particularly the rarer birds and mammals, this task may be very exacting and complex. We hope that such research may be stimulated by some of the ideas outlined here.

CHAPTER 4

Flora and vegetation

C. R. DUNLOP & L. J. WEBB

INTRODUCTION

We do not know anything of the attitude of the early Aboriginal settlers to the landscape of the north-western tropical coast. Whatever it was, it did not prevent the acceptance and development of the coastal and then inland zones as Aboriginal homelands. For these ancient settlers faced with a strange new ecology, 'botany began anew' (cf. Webb 1973, Webb & Smyth 1984). Some tens of thousands of years later, the same landscapes deterred and apalled European mariners who made sporadic and inadvertent landfalls in an evidently barren and skeletal terrain alleviated by a few estuarine and riverine swamps.

The land had been sculpted by the severe erosion of monsoonal and cyclonic rains, so that there were no deep soils except on flat areas and gentle slopes. The parent rocks were in general extremely ancient and siliceous, of Proterozoic age or older. The soils of the low plateau areas fringing the coast were lateritic, and leached of essential plant nutrients during millions of years. Thus the sclerophyllous vegetation that graced the torrid arid-looking landscape was, at least to the first European eyes, disturbing and anomalous indeed.

Even in modern times, there is no coherent account of this, for most people, remote vegetation: its ecological and biogeographic relationships, and its intrinsic features that we may value. Since World War II, only a few decades ago, the first systematic biological studies of the region followed the colonialist tradition of expeditions to 'undeveloped' areas by visiting scientists (e.g. land use surveys for which references are given in Story 1976). With few exceptions, (e.g. Specht 1958 a, b, c) the reasons for increasing scientific knowledge of the area were based on a wish to develop agricultural and pastoral industry. Despite some recent expensive enterprises, however, the north-western Australian environment has not yet been tamed by the modern technology of renewable resource development. The spectacular success of mining and the extraction of non-renewable resources do not concern us here, except to note that its vagrant people provide additional justification for an investment in understanding the natural environment of a slowly growing population.

Given our so far imperfect knowledge of the biological and ecological relationships of the plants and plant communities of the Northern Territory coastal lowlands, what are some of the essential and vital questions to be answered? How different are the ecosystems of the Northern Territory, and how far and in what form do they extend westwards and eastwards in Australia? Are there many species peculiar to the region? How are they related to taxa in other countries northwards, and westwards across the Indian Ocean? How old are the 'jungle' (monsoon forest) pockets – both in terms of occupation of a particular site, and origin of their flora? How different is the fire regime and its differential effects on vegetation and wildlife from the fires of south-eastern Australia? Is there any correlation between Aboriginal ownership and land use and vegetation units classified by western science? Are there any consistent patterns of regeneration or succession in the monsoon forests, eucalypt forests, and their intergrades; and in the wetlands? How viable are vegetation isolates, i.e. relatively small patches (e.g. monsoon forests) in unrelated vegetation and what is their rate of species extinction?

These are only a few of the questions that have recently begun to attract the interest of botanists and others. This is not the place to attempt to assemble the scattered literature on the vegetation of the tropical monsoonal coastal lowlands. Instead, some salient features will be described in a way which may point to further essential studies.

SUMMARY OF THE MAIN VEGETATION HABITATS

It is now well appreciated that there is a very subtle diversity in ecological features throughout the region. These should be considered not only in space but also according to the seasons of the year. Climatic variations associated with the strongly marked wet and dry seasons are superimposed on more obvious habitat features. To set the scene, it is convenient to view the vegetation of the coastal lowlands of the Northern Territory in five broad categories of land systems and habitat types:

1. Coastal sands and tidal mudflats (saline wetlands),
2. Coastal flood plains (freshwater wetlands),
3. Subcoastal well-drained lowlands,
4. Rocky substrates, mainly sandstone plateau outliers,
5. Various niches, generally fire-proof, in the above habitats with pockets of monsoon forest or rainforest.

The coastal system, incorporating the saline and freshwater wetlands, is the most recently formed and the simplest in terms of species richness and life form diversity. The eucalypt-dominated communities of the well-drained lowlands are more complex alliances that occur widely throughout tropical Australia, comprising diverse mixtures of indigenous and extra-Australian taxa which contain a broad spectrum of life forms.

The sandstone, formerly much more widespread, supports a flora which has features in common with other sandstone floras of Australia. This indicates that at least some component elements are relictual. The monsoon forest pockets or 'jungles' also have many floristic similarities to similar vegetation types farther eastwards in the Australian tropics and in the tropical Indo-Malesian region farther north.

BRIEF DESCRIPTION OF THE VEGETATION TYPES

Coastal sands

Strand vegetation

The strand is here considered to comprise the beaches and coastal dunes. As noted by Good (1974), the tropical coasts of the Old World share a suite of specifically strand species, included among which are *Cordia subcordata, Guettarda speciosa, Casuarina equisetifolia* and *Scaevola sericea.* Good lists a further group of pan-tropical species, all of which occur on northern Australian shores. The more prominent of these are *Hibiscus tiliaceus, Ipomoea pes-caprae* and *Thespesia populnea.* An island habitat which is largely absent from the north coast is the coral cay with its limestone flora. A common species of these cays, *Pisonia grandis,* has however been recently discovered on two small islands off Cobourg Peninsula.

Burbidge (1960) and Beadle (1981) have remarked on the strong affinities of coastal species with the halophytic assemblages of inland Australia. The chenopods *Halosarcia indica* subsp. *leiostachya* and *Salsola kali* occur in coastal and inland saline habitats

as does the grass *Sporobolus virginicus.* Examples of closely related taxa with a similar distribution are the dioecious grass genera *Zygochloa* and *Spinifex* – both grow on sand dunes, the former on inland dunes, the latter on coastal. In Beadle's view the halophytic element in the arid zone flora developed as a consequence of invasion of species from the strand and tidal mudflats.

Tidal mudflats

Mangrove forest and scrub

Northern Australia is rich in mangroves and is considered by Specht (1981a) to be their centre of origin. Although the species number for the north coast is relatively high, the areas of mangrove swamp are not as extensive as in New Guinea (Paijmans 1976) or North Queensland (Love 1981). The mangrove forests along the north Australian coast are strongly demarcated, usually abutting sand dunes, open eucalypt forest and even monsoon vine-forest patches. They are rarely associated with extensive freshwater swamps, as described for the New Guinea lowlands by Paijmans.

A recent survey (Wells 1982, 1983) records 25 species of mangrove from the Northern Territory. To these may be added *Rhizophora lamarckii* (McCusker 1984), leaving only three Queensland species which as yet are unrecorded for the region, namely *Rhizophora mucronata, Heritiera littoralis* and *Cynometra ramiflora.* The account of Wells (1982) may be referred to for the list of species present and for details of distribution, habitat and species associations for all the major river systems of the region. The mangrove palm, *Nypa fruticans,* however, deserves special mention. It is known from three sites, one on Cobourg Peninsula and two on Melville Island. At all sites it occupies what appears to be a specialised habitat, that of a perennial, freshwater stream flowing strongly through mangrove flats. Associated species are mangroves which tolerate freshwater such as *Xylocarpus granatum* and *Rhizophora apiculata* as well as *Diospyros ferrea* var. *reticulata,* a species not usually regarded as a true mangrove.

Besides *Diospyros,* there is a number of other salt-tolerant species commonly associated with the mangroves. These are the fern *Acrostichum speciosum,* vines of the genera *Derris, Gymnanthera* and *Ischnostemma,* two species of *Amyema* (mistletoe), the lichen *Roccella belangeriana* and the succulent subshrubs *Batis argillicola* and *Suaeda australis.*

Coastal riverine flood plains

The plains present an aspect of flat, treeless uniformity,

here and there broken by river channels, tree-ringed billabongs and paperbark (*Melaleuca* spp.) swamps. Height with respect to sea level varies within a narrow range: in some locations land in these systems is lower than high tide level and is protected by levees; in others it is slightly higher. General flooding of the plains occurs during the wet season, reconnecting the strings of waterholes with the major river channels. There is evidence of considerable influence from feral water buffaloes (Chapter 11).

Recent evidence from pollen cores and radiocarbon dating suggests an age of 1,000-6,000 years for the riverine flood plains (Russell-Smith, pers. comm., Chappell & Thom 1977). In geological terms this is a very young landscape, in strong contrast to the sandstone and adjacent lateritic formations of the sub-coastal lowlands.

Accordingly, the vegetation of the plains is inferred to be relatively recent, having originated by dispersal from geographically distant areas with similar habitats and associated floras, as well as by an expansion of a flora that had managed to maintain itself during the marine transgressions in small coastal niches. Many of the species occur throughout the Old World tropics suggesting that the few endemics present, for example *Bambusa arnhemica*, may well be shown in future taxonomic revisions to be conspecific with Asian taxa. As noted separately, components of the strand, mangroves and floating mats are similarly widespread tropical species.

Conditions for plant growth on the plains are extreme. During the wet season and for several months afterwards the soils are waterlogged. Towards the end of the dry season when the free water has evaporated, the cracking clays dry out and shrink. Any moisture that may be present in the profile is bound in the clay and is largely unavailable. Tinley (1982) argues that it is this cycle of waterlogging and excessive drying which excludes woody plants from this soil type. Woody plants are certainly absent from the plains proper, occurring only along permanent billabongs and in shallow swamps where water is available throughout the year.

Species numbers in the plains communities are relatively low (Taylor & Dunlop 1985) with members of the family Cyperaceae (sedges) and Poaceae (grasses) being dominant. Typical of the former are *Eleocharis* spp., *Fimbristylis* spp. and *Cyperus* spp., and of the latter, *Pseudoraphis spinescens*, *Hymenachne acutiglume* and *Oryza meridionalis*. Aquatic groups are well represented by the water lilies (*Nymphaea* spp., *Nymphoides* spp. and *Nelumbo nucifera*), submerged species (e.g. *Blyxa* spp. and *Maidenia rubra*) and floating species such as *Utricularia aurea* and *Azolla pinnata*.

Seasonal sedge and grass swards
The success of certain species is apparently linked to their ability to form large colonies, often to the exclusion of all competition. Species forming monospecific swards may be annual (*Oryza meridionalis*), tufted perennials (*Fimbristylis tristachya*), perennial grasses (e.g. *Hymenachne acutiglume* and the exotic *Brachiaria mutica*) and the various species of *Eleocharis* which are often rhizomatous and may persist by perennating tubers (e.g. *E. dulcis*).

Observations at Kapalga on the South Alligator River indicate that the mosaic of monospecific populations does not have a constant pattern in space, and may change from wet season to wet season (Taylor, pers. comm.). Annual species which regenerate each year by seed are the most susceptible to variation, appearing *en masse* in one year and being completely absent in another. Populations of the Lotus Lily, *Nelumbo nucifera* (Plate 4) which is typically found in permanent swamps, appear to have a cycle of establishment to moribundity to apparent dormancy over several seasons. However this observation may be due in part to the effects of intense grazing by feral buffaloes (see Chapter 11). Examples of cyclic annuals are the suffrutescent legumes *Sesbania cannabina* and *Cassia obtusifolia*, the latter an introduced weed.

Permanent swamps and floating mats
Permanent swamps on the plains are usually spring-fed and are therefore usually located on the landward edge. Paper bark trees of the genus *Melaleuca*, especially *M. leucadendra*, *M. cajuputi* and *M. viridiflora*, frequently form dense stands on these swamps. In less aerated soils sedges such as *Scleria poaeformis* may dominate: species of the same habitat include taro, *Colocasia esculenta*, *Hymenochaeta grossa* and various species of *Cyperus*. Rainforest thickets are also prevalent on the slightly higher, better drained soils fringing the swamps and rivers (Plate 5). Typical species of the wet rainforests are the endemic palms, *Livistona benthamii* and *Carpentaria acuminata* (Photographs 1 and 2).

A vegetation type which is termed here 'floating mats' is an ecologically interesting and fragile system (Photograph 3 and Plate 6). Mat development is dependent on reasonably large, permanent bodies of still or slow moving fresh water such as are found in the sluggish streams and seasonally land-locked billabongs of the coastal flood plains. Under natural conditions the mats may be set adrift by flood waters but they are particularly susceptible to damage by the introduced water buffalo which graze and wallow in the billabongs.

Recent work on floating mats in northern Australia

is reported by Kershaw (1978) and Hill & Webb (1982). The account given here is largely based on the unpublished observations of E.R.Petherick who has studied the mats of several river systems south-west of Darwin, over many years.

The mat communities vary in composition and complexity with over forty species from twenty families being recorded by Petherick from mats in four river systems. Three elements may be distinguished in the process of mat formation: first, the floating aquatics, many of which live in large colonies and form rafts (e.g. *Pistia stratiotes, Azolla pinnata*); secondly, the rooted aquatic vines which, in lying across the surface of the water impede the progress of the rafts and hold them against the banks (e.g. *Ludwigia adscendens, Ipomoea aquatica*); thirdly, a succession of species including vines, shrubs, trees and ferns which colonise the mat as the substrate of detritus from the decomposing floating aquatics accumulates (e.g. *Phragmites karka, Cyclosorus interruptus, Nauclea orientalis*).

With a few exceptions, the species of the mats are known from similar habitats extending from New Guinea through to southern Asia, Africa and India (Paijmans 1976, Walter 1971, Kern 1974). Most of the species are common inhabitants of swampy habitats across northern Australia but at least one species, *Echinochloa praestans* is known only from the mats. It grows in similar situations in New Guinea (Michael 1980).

Subcoastal well-drained lowlands

Classification of the vegetation

The pioneering work of CSIRO land system surveys (e.g. Christian & Stewart 1953, Story 1969, 1976) remains today a basic source of data on vegetation from many areas in northern Australia. Since the discovery of uranium in the 1960s, the Alligator Rivers region in particular has received a great deal of interest from plant ecologists, building on the original work of Specht (1958a). References to the classificatory work published on this region may be found in Taylor & Dunlop (1985). Vegetation which is floristically and structurally similar in northern Western Australia and northern Queensland is described in the accounts of Hnatiuk & Kenneally (1981) and Specht et al. (1977) respectively.

Recent volumes on the major tropical savanna regions of Africa, South America and Australia (Bourlière & Hadley 1983, Huntley & Walker 1982) have included a reappraisal and reclassification of the eucalypt-dominated communities of monsoonal Australia by Gillison (1983) and Walker & Gillison (1982). Savanna, in the sense used by these authors for monsoonal Australia, follows broadly the definition of

Bourlière & Hadley (1983): '(1) where the grass stratum is continuous and important, occasionally interrupted by trees and shrubs; (2) where bush fires occur from time to time; and (3) where the main growth patterns are closely associated with alternating wet and dry seasons'. Certainly in north Australia the denser types of eucalypt forest near the coast are more than 'occasionally interrupted by trees. . .' but as Walker & Gillison (1982) have pointed out, the Australian savannas are broadly similar to those of the other southern continents and for that reason alone, the use of a 'savanna nomenclature' serves to place the monsoonal eucalypt forests and woodlands into a broader context. Furthermore, inherent in the term savanna is the concept of a continuous graminoid layer with its annual cycle of growth and drying off – probably the most distinctive and unifying feature of all the eucalypt (and grassland) communities in the region.

Indeed it is the intention in this chapter to emphasise the integrity of these eucalypt communities as a unit distinct from the coastal and sandstone escarpment ecosystems and the monsoon forests. Descriptions of component categories should be sought in Blake (1953b) and Story (1969, 1976). Detailed data on life forms in the eucalypt community of the Alligator Rivers area are provided by Taylor & Dunlop (1985).

Eucalypt communities

The most distinctive and wide-ranging eucalypt community in northern Australia, extending from the Kimberley Region to the Cape York Peninsula, is dominated by *Eucalyptus tetrodonta* and *E. miniata*, either singly or in combination. As defined by Story (1969, 1976) in the Northern Territory, canopy height is over 13 m, with crowns mostly touching. In the wetter northern part of Cape York, it is much taller, over 30 m.

Smaller trees in the understorey are patchily distributed, e.g. *Erythrophleum chlorostachys, Terminalia grandiflora, Acacia* spp., *Planchonia careya, Petalostigma pubescens*. Although the community appears to be uniform in structure and species composition, there are conspicuous variations in the physiognomy of the understorey and composition of the herb layer at different seasons.

Notable amongst the non-eucalypt understorey is a suite of obligately deciduous and evergreen species which have taxonomic affinities with taxa of the monsoon forest. They are here referred to as 'interspersed elements' and are discussed in detail in the section on monsoon forest patches (below), which comprise many closely related taxa.

On shallower and rocky soils, stunted eucalypt communities with a dense grass layer, canopy less than

7 m in height and with crowns separate, are dominated by *Eucalyptus foelscheana* and *E. tectifica*. These are the 'eucalypt grassy woodlands' of Story (1969, 1976) which also include woodlands on seasonally inundated soils in which *Eucalyptus latifolia* and *E. polycarpa* form the overstorey. Irrespective of which species of eucalypt are dominant, in all the savanna communities there is ample penetration of light to the lower layers to support intermediate tree and shrub layers, as well as a dense growth of annual and perennial grasses. Competing with the grassy layer, and often obscured by it, is a large variety of annual and perennial herbs, vines and subshrubs.

The sharp seasonality of the climate with its cycles of winter-dry and summer-wet seasons is surely the overwhelming factor in determining adaptations in the eucalypt savanna: species must survive the long dry season and then respond quickly and compete effectively during the relatively short but intense wet season. Adaptations of the various life forms reflect two primary strategies for coping with the dry season: obtaining moisture from the lower layers of the soil (i.e. remaining active) or going into dormancy. Life forms include annual herbs (e.g. *Mitrasacme connata, Stackhousia intermedia*) which survive the dry season in the form of seeds; herbs and vines with annual aerial parts and perennial, often tuberous, roots (e.g. *Ipomoea graminea, Microstemma tuberosa*); perennial evergreen shrubs (e.g. *Petalostigma quadriloculare*) and trees (e.g. *Eucalyptus porrecta*); deciduous trees (e.g. *Terminalia ferdinandiana*), shrubs (e.g. *Croton arnhemicus*), and vines (e.g. *Tinospora smilacina*); parasitic trees (e.g. *Exocarpos latifolius*); herbs (e.g. *Striga curviflora*) and mistletoes (e.g. *Amyema* spp., *Lysiana* spp.). Conditions for epiphytic species could be termed marginal but two hardy orchids, *Dendrobium affine* and *Cymbidium canaliculatum* are reasonably common. In the herb layer the life forms most favoured are the true annuals and perennial-rooted species with annual tops. The latter category includes many of the vines of the families Fabaceae, Convolvulaceae and Asclepiadaceae which are common on the forest floor. Palms and cycads, one species of which is deciduous, are also widespread throughout the region and in areas may form the dominant vegetation. On Cobourg Peninsula and Melville Island localised populations of *Gronophyllum ramsayi* occur as co-dominants with *E. tetrodonta* while *Livistona humilis* (Plate 7), which is normally seen as an understorey to *E. tetrodonta-E. miniata*, forms dense monospecific stands on sand plains south-west of Darwin.

All evergreen species or those which are leafless for only a brief period are dependent on a deep root system. One such species is *Brachychiton diversifolius*, a relatively tall tree which spends much of its early life as a seedling or small shrub, developing all the while a deep and massive tap root. A remarkable variation in leaf shape, about six distinct types in all, is exhibited by *B. diversifolius* as it passes from the seedling to adult stage. The ability of the eucalypts to seek moisture at great depths in the soil is a major attribute ensuring their dominance in the tree layer. Several eucalypts are able to lose their leaves and remain dormant for an extended period and it is these species which can compete most effectively on the shallow soils; on the deeper soils, however, the evergreen Woollybutt (*E. miniata*) and Stringybark (*E. tetrodonta*) are dominant.

Storms late in the dry season, from about September onward, herald the coming wet season. The effects of these storms on the tree layer are minimal: the moisture, which penetrates only a few centimetres is soon lost during the hot cloudless days which follow. The perennial grasses are one group of plants which can readily take advantage of these patchy early storms, responding with green shoots within a few days. Growth is generally slow until the monsoons arrive although a rather exceptional species, *Allopteropsis semialata*, is usually in full flower before the end of November. This is not to say that there is little activity in the rest of the flora during the late dry season. As if in anticipation of the 'wet', the deciduous trees flower and put on new foliage, flowers appear at ground level from perennial roots and there is a general flush of growth throughout the forest. Although temperatures are high year-round, this slight temperature rise (Fig. 26) during the period from early August to October is significant enough to produce a northern spring.

Dry season fires early in the year serve to imitate this temperature rise and will trigger a premature response from perennial species such as the deciduous *Cycas armstrongii* which produces new leaves immediately after a fire. The subshrub *Pachynema junceum*, if left unburnt, will flower in December but will sprout new stems and flowers at any time after being burnt.

Of the annual grasses, the various species of *Sorghum* (of which there are about 12) are the most conspicuous. They are generally tall with individual plants reaching 4 m although heights in the region of

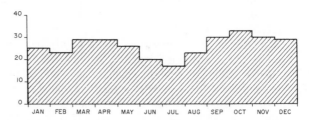

Figure 26. Number of days over 30°C at Darwin (from Plumb 1977).

2.5 m are more usual. As with all members of the tribe Andropogoneae, a group well represented in the monsoon tropics, the seeds of *Sorghum* are protected from fire and predation by the tough persistent glumes. At the top of this disseminule is attached a corkscrew-like awn which dries out completely and is sensitively hygroscopic. On absorbing moisture, the awn twists vigorously, driving the sharp pedicle into the upper layers of the soil in preparation for germination. *Sorghum's* facility for self-sowing no doubt gives it some competitive advantage: success can be judged by its occurrence on many soil types from temporarily inundated laterite to beach dunes and by its wide distribution from the north coast to south of 15°S latitude.

Sandstone scrub and heath

Descriptions of the vegetation of the Arnhem Land plateau are provided by Story (1976) who recognized sandstone scrub, sandstone woodland and *Allosyncarpia* forest communities: further accounts by Dunlop & Begg (1981), Burgman & Thompson (1982), Taylor & Dunlop (1985) and Russell-Smith & Dunlop (1985) are narrower in scope but more detailed, particularly for the monsoon forest communities. Specht (1981c) has classified all the sandstone heath across the continent including those under discussion here as 'Sclerophyll communities'. In this section attention will be focussed on the flora rather than the plant communities, information on which is available in the references cited above.

The sandstone flora belongs to what is known as the autochthonous or 'Australian' element (Burbidge 1960). This element consists of various groups which originated in the eastern part of the ancient conglomerate landmass, Gondwanaland, now represented by the Australian continent and adjacent islands. These taxa are mainly endemic to Australia or the Australasian region. The best examples of these groups are the subfamily Leptospermoideae of Myrtaceae (e.g. *Eucalyptus, Leptospermum, Lophostemon*) and various groups within the Proteaceae (see Johnson & Briggs 1975: e.g. *Grevillea, Banksia*) and Rutaceae (tribe Boronieae: e.g. *Boronia*). Features of the Australian element include the almost complete lack of obligately deciduous species and development of scleromorphy in the form of sclerophyllous foliage (e.g. *Grevillea* spp.), phyllodes (e.g. *Acacia* spp.), cladodes (e.g. *Pachynema* spp.) and highly reduced foliage (e.g. *Calytrix exstipulata*). While these characteristics would appear to be adaptations to arid climatic conditions, it is generally agreed they have developed in response to nutrient-deficient soils (Barlow 1981 with references, van Steenis 1979). An additional feature of the flora is its relationship to fire: in contrast with the flora of the lowland eucalypt commu-

nities, as will be further discussed below, it appears to promote and is adapted to hot, relatively infrequent fires. Flammable characteristics, besides those already mentioned for *Eucalyptus*, include viscidness in species of *Acacia, Grevillea* and *Triodia*: Gill (1981) should be referred to for the manifold adaptations which involve seed release, germination, establishment, recovery and flowering.

Due to the rugged and inaccessible nature of the sandstone plateau, the sandstone flora remains one of the last frontiers for botanical exploration in Australia. The highest, most dissected and the best watered part of the plateau is in the north-west portion of Arnhem Land, the western edge of which lies in Kakadu National Park. Not surprisingly, this area has the greatest diversity of habitats and the richest flora in the region. Helicopter surveys in this area in the last decade, starting with the CSIRO surveys of 1972-1973 (Story 1976), have yielded many undescribed genera and species as well as the discovery of unexpected distributional records. These have provided links between the Arnhem Land plateau, other sandstone regions of north Australia and areas in the south of the continent. The genus *Borya* (family Xanthorrhoeaceae) is just such a record. It is probably also the most disjunctly distributed genus in Australia. Besides the single species in Arnhem Land, other species occur as isolates in the Kimberleys, south-west Western Australia, the Grampians in Victoria and north-east Queensland (Specht 1981c). There is also a number of species of *Eucalyptus* (*E. phoenicea, E. brachyandra, E. herbertiana*) which usually occur on sandstone or on soils recently derived from sandstone and which are distributed, disjunctly, across the Northern Territory and the Kimberleys. The connection between the sandstone floras of the Kimberleys and the Northern Territory seems to be stronger than the Northern Territory-North Queensland one. This is in parallel with the general eucalypt distribution discussed elsewhere in this chapter. There is, however, a number of species, such as in *Choriceras, Allopterigeron* and *Monotaxis*, which are shared between Northern Territory and Queensland but which are not known from the Kimberleys.

Although the point has been made that there is a suite of species common to the sandstone plateaux across northern Australia, it should be noted that each region has a large body of endemics. Most of these endemics, some of them monospecific genera, are generally restricted in range within a region and in many cases are confined to very specific habitats.

Of interest in the latter category is *Micraira*, an exceptional genus in the Poaceae which until its discovery on the Arnhem Land plateau was considered monospecific: it was known previously only from south-east Queensland. The genus now has thirteen named species, many of which are localised in one or two places in Arnhem Land and the Kimberley region of Western Australia (Lazarides 1979, 1984). Besides its disjunct distribution, *Micraira* is exceptional in being the only grass genus with spirally arranged leaves (Clifford 1964). Perhaps the most notable feature of the genus is its adaptation to a very narrow ecological niche on the sandstone. As will be mentioned later, there are many varied habitats through the sandstone but the most inhospitable are the bare stone pavements, the microrelief of which is sufficient to trap water for short periods but which is too shallow to accumulate soil. *Micraira* appears to be alone in utilising these habitats to any extent and does this through its facility for 'resurrecting' (Gaff & Latz 1978). Species capable of resurrection have been reported from a number of families from Australia, mainly from arid environments. The leaves of resurrection plants have the ability to dry out and lose their chlorophyll yet remain attached to the plant where they will green up and become functional again within twenty four hours of soaking rain. In a region where the wet season may be punctuated by varying periods of drought, the ability to switch on or off according to moisture availability confers extraordinary advantage in these most tenuous of niches.

Further endemics which are restricted in range are *Pandanus basedowii* and *Allosyncarpia ternata*, which although locally common, are found only in the northern part of the Arnhem Land plateau, neither extending to Katherine Gorge, 200 km south on the same sandstone block. A number of other species, among them *Neobyrnesia suberosa*, are restricted to rock faces in sheltered ravines and are known only from a handful of sites in the Alligator Rivers area. Similarly, species of *Stylidium*, *Hibiscus* section Furcaria, *Pityrodia*, *Micraira* and a number of other genera, have proliferated in discrete habitats, with several in each genus known only from one collection site. Carlquist (1979), commenting on evolution in *Stylidium*, observed that the Arnhem Land plateau provides a mosaic of suitable *Stylidium* habitats, not separated necessarily by large distances but effectively isolated by areas of bare rock and ravines – a situation conducive to speciation in this genus.

Partitioning between the floras of the sandstone and eucalypt savannas, on the basis of floristics and life forms present, is reasonably well defined (Taylor & Dunlop 1985). Furthermore, the sandstone flora has more taxa in common, at the generic and tribal levels, with some temperate and sub-tropical areas of Australia, than it does with the adjoining eucalypt savanna (Specht 1981c). The following analysis of the tribes and genera of family Papilionaceae represented in the two floras best illustrates this point.

Approximately 20 tribes (Hutchinson 1964), none of which are endemic, occur in the region. These fall into three broad categories based on geographical occurrence:

1. Africa, Australia;

2. Old World tropics. This may include (1) above in association with parts of New Guinea and Asia;

3. Old and New World Tropics.

Tribes of the first category include the Podalyrieae (*Daviesia*, *Jacksonia*, *Leptosema*, *Burtonia*), Lotononideae (*Plagiocarpus*) and the Bossiaeae (*Bossiaea*, *Hovea*, *Templetonia*). All are confined to the sandstone or to sandy soils derived from the sandstone. The genera mentioned are all Australian endemics and all are represented elsewhere on the continent, especially in south-west Western Australia where in *Jacksonia*, *Templetonia*, *Daviesia* and *Bossiaea* alone there are over 100 species (Green 1981). Members of these tribes are also predominantly xeromorphic: *Bossiaea*, *Jacksonia* and *Leptosema* are leafless shrubs with expanded stems while *Burtonia*, *Daviesia* and *Templetonia* have reduced, narrow leaves. Other genera of the sandstone include *Atylosia*, *Tephrosia*, *Crotalaria* and *Indigofera*, all of which are wide ranging, occurring on a variety of soil types throughout the region. *Atylosia* is an Old World genus and the last three have species in the tropical regions of both hemispheres.

The legumes of the eucalypt savanna are much more numerous than their sandstone counterparts (26 genera against 11) and exhibit a greater variety of life forms. Annual vines and subshrubs with perennial roots are common (e.g. *Vigna*, *Pycnospora*, *Eriosema*, *Clitoria*), as are annual herbs (*Desmodium*, *Alysicarpus*) and evergreen shrubs (*Dicerma*, *Galactia*). One genus, *Erythrina*, is a deciduous tree. The most common and widespread species belong to the annual aerial parts - perennial root category which is seen as a life form perfectly adapted to the seasonal climate and the prevailing fire regime. *Vigna lanceolata* goes one step further and produces two types of pods, an aerial version with several seeds and a cleistogamous, subterranean one with a single seed. Such an adaptation guarantees seed production against damage to aerial parts by insects and fire.

Only one legume genus in the savanna, *Austrodolichos*, is endemic. The majority of genera are confined to the Old World (e.g. *Dunbaria*, *Flemingia*, *Glycine*, *Smithia*) with several having representatives in tropical America (e.g. *Alysicarpus*, *Desmodium*, *Clitoria*).

From the above analysis we may conclude:

(a) that the papilionaceous flora of the sandstone have strong connections, at the generic level, with temperate and especially south-western Australia,

(b) that the sandstone legumes have connections at the tribal level with Africa. It is reasonable to suppose (as Johnson & Briggs 1975, for Proteaceae), that the

Afro-Australian connection was established before the breakup of Gondwanaland and that the floras have since diversified independently,

(c) that life forms of the sandstone legumes are predominantly xeromorphic, evergreen shrubs, without especial adaptations to a strongly seasonal climate. Conversely, many taxa of the subcoastal lowlands go into dormancy during the dry season, suggesting they have evolved under a wet-dry climatic regime,

(d) that the leguminous genera of the sandstone flora are endemic while those of the eucalypt savanna are widespread through the Old World tropics. These latter genera are undoubtedly recent immigrants from the Malesian flora (van Steenis 1979).

The conclusions reached here with respect to the integrity of the sandstone communities and their continued maintenance through time have been well expressed by previous authors, particularly Johnson & Briggs (1975). These authors have also noted that these 'low-nutrient scleromorphs' have moved into the arid regions. This is particularly true of the spinifex grasses, *Triodia* and *Plectrachne* which dominate much of the tropical sandstone plateaux as well as the red sand plains and sand dunes of central Australia.

Sandstone elements are also present in the eucalypt savanna where they are generally confined to sandy and skeletal soils (e.g. *Verticordia* spp., *Jacksonia dilatata, Callitris intratropica*) though there is a number of species in *Hibbertia, Pachynema, Calytrix* and *Grevillea* which show greater adaptability, occurring over a range of soil types throughout the savanna. Where the sandy soils are extensive the autochthonous flora increases in richness and diversity, showing its best development in the region on the sand sheets in the Murgenella area of Arnhem Land. Here the ground layer is dense and shrubby, containing species of *Leptocarpus, Grevillea, Leucopogon, Hovea* and other typically autochthonous elements. The overstorey is dominated by *Eucalyptus tetrodonta*, the palm *Gronophyllum ramsayi* and stands of Cypress pine (*Callitris intratropica*). Here the sandstone species form a distinct assemblage, quite unlike the savanna on laterite in respect of composition, structure and fire behaviour.

Monsoon forest patches and interspersed elements

The contemporary realization that, in many vital respects, Australia and the South-east Asian region share ancient biological origins signals the need for comparative ecological studies in various fields such as geography, economics, culture, and the natural ecosystems to be conserved or converted for economic production. Beard (1967) had noted the lack of information for comparison of tropical Australia with other continents, and suggested some correspondences at the level of formation-type with tropical Africa and America. Despite linguistic, economic and cultural differences, it may be possible to develop a 'common ecological language' (Smitinand et al. 1982) about the tropical forests and related resources, so that conservation and development can be ensured in the long term.

For such a comparative study, the 'jungle pockets' (monsoon forests) that have survived since ancient times in tropical Australia (Webb & Tracey 1981a) provide a unique and indispensable link. Apart from the coastal wetlands (Groups 1 and 2, this chapter), the monsoon forests and the rainforests are the only terrestrial vegetation shared with South-east Asia, which has very few counterparts or outliers of the Australian sclerophyll woodlands and savannas characterized by endemic species of *Eucalyptus, Melaleuca, Casuarina, Acacia, Grevillea,* etc.

The CSIRO land resources survey mentioned earlier, which began in the 1940s, had scientific goals narrowly dictated by economic development. The restricted patches of monsoon forest and rainforest were noted as 'ecologically interesting but economically unimportant because of their limited extent' (Speck 1965). In this section we shall examine some of these features of ecological interest, and indeed of singular scientific importance in biology and biogeography. Four botanists made the first contributions: Blake (1953b), Perry & Christian (1954) and Specht (1958a, b). Specht (1958c) also traced the fragmentary history of botanical exploration of Arnhem Land, i.e. the north-eastern part of the Northern Territory, but this was essentially taxonomic and does not directly concern us here.

Gardner (1942) had earlier emphasized the distinctive character of the many vestiges of the so-called Indo-Malesian element in tropical Western Australia: the deciduous foliage of many taxa found also in the tropics to the north, and even some deciduous species of *Eucalyptus*, gave the vegetation a unique appearance. However, mixed deciduous species forming separate communities among the all-pervasive sclerophylls occurred in very scattered and relatively small areas. Blake (1953b) referred to these small patches, locally known as 'jungles' in the Northern Territory, as monsoon forests. In the Barkly region straddling the Queensland border, the patches on limestone were often only about 20 m in diameter (Perry & Christian 1954). Thus, except for some small stands of *Eucalyptus* and certain other taxa, no other community showed such complete disjunction in its distribution as the monsoon forests (Specht 1958b).

A further salient feature was the very special nature of their habitats, of which Blake (1953b) identified five main types: at headwaters of spring-fed creeks, on levees, in the open forest, on rocky outcrops, and as belts behind littoral vegetation, i.e. in the lee of coastal foredunes. Specht (1958a) noted that the stands of monsoon forest grew over a wide range of soils provided moisture levels were high, and listed seven main habitat types: (1) promontories or islands of lateritic red earths; (2) lateritic podsolics; (3) skeletal quartzites; (4) lee of coastal sand dunes and sheltered hollows within the dunes; (5) along permanent freshwater streams through lateritic soils; (6) around sandy freshwater pools remaining in gorges of quartzitic hills; and (7) immediate bases of the sandstone hills draining rain from sheer bare rocks above. Specht considered that moist habitats near swamps and streams supported more mesophilous elements that approached rainforest in structure.

Another unusual ecological feature, already noted by Gardner (1942) in what he called the monsoon woodland, was the sporadic presence of monsoon forest species, generally different from those of the same genus in the closed monsoon forests, as an understorey in eucalypt woodland (q.v.). This interspersion of elements of Gondwanan and derivative (sclerophyll) flora was recently suggested as of possible significance in understanding the evolution and adaptation of the different Australian floras with a common origin (Webb & Tracey 1981a, Kenneally 1983). Blake (1953b) classified the mixtures as 'deciduous low mixed open forest'. Different combinations of taller eucalypts on certain soils such as deep sands and red or brown earths were described with a scattered and mostly deciduous broad-leaved understorey including species of *Xanthostemon, Cochlospermum, Buchanania, Croton, Erythrophleum, Petalostigma, Gardenia, Syzygium, Terminalia, Erythrina, Ficus, Gyrocarpus, Planchonia, Brachychiton, Alphitonia, Owenia.*

Perry & Christian (1954) recognized similar mixtures, e.g. 'shrub woodland' that included the *Ventilago viminalis* alliance with *Capparis, Celastrus, Bauhinia, Erythrina, Acacia, Carissa; Lysiphyllum-Gyrocarpus* with *Ventilago, Atalaya, Acacia; Eucalyptus papuana-E. tectifica* with *E. confertiflora, E. grandiflora* and *Alphitonia, Planchonia, Owenia, Celastrus, Canarium, Ficus, Gyrocarpus, Carissa* etc.; *Terminalia, Lysiphyllum, Cochlospermum*; plus *Eucalyptus camaldulensis, Melaleuca saligna, M. mimosoides* with *Nauclea, Terminalia, Ficus, Piliostigma, Brachychiton, Cathormion, Barringtonia, Lophostemon, Livistona*, etc., as tall fringing forest along perennial streams.

During later surveys in the adjacent West Kimberley and North Kimberley areas, Speck (1960) and Speck & Lazarides (1964) remarked on the biogeographic importance of wide-ranging species of *Adansonia, Cochlospermum, Lysiphyllum* and *Terminalia*

among the eucalypt-dominated vegetation. The monsoon woodlands were composed of mostly deciduous low trees (e.g. *Brachychiton, Terminalia, Vitex, Planchonella, Ficus, Strychnos, Ehretia, Gardenia, Ixora, Atalaya, Buchanania*) on sandstone scarps and gorges; an *Adansonia gregorii* alliance with *Cochlospermum, Ficus, Terminalia, Wrightia, Celtis* on rugged limestone country; and *Adansonia, Terminalia, Brachychiton, Lophostemon* along streams and in channels and drainage floors. Mixed eucalypt and deciduous species representing interspersed floras were also described in a variety of special habitats including deep sands, quartzite ridges, levee crests of flood plains, and volcanic red earths.

In the Tipperary area, in addition to deciduous species mixtures, Speck (1965) distinguished 'rain forest' as several-layered, evergreen, with mesomorphic trees up to 30 m in height and restricted to certain depositional areas less than about 1 km^2.

A more recent phase of plant taxonomic and ecological surveys in the Northern Territory can be dated from the late 1960s. It involved the final CSIRO Land Research Series of Story (1969, 1976); and new studies in adjacent north-western Australia by the Western Australian Herbarium, Museum, and Department of Fisheries and Wildlife; and by Beard (1976, 1977) of Perth. These are conveniently listed by McKenzie & Kenneally (1983). Most recent contributions by visiting and resident groups of the Darwin-Kakadu areas are Fox et al. (1977), Webb & Tracey (1979), Kikkawa et al. (1981), Webb et al. (1984, 1986), Kershaw (1985), Taylor & Dunlop (1985) and Russell-Smith & Dunlop (1985).

In the Alligator and Adelaide Rivers area, the work by Story (1969, 1976) refined the classification of the local vegetation and its environmental relationships, and raised a number of interesting ecological questions, some of which are discussed below. He distinguished both rainforest (unbroken canopy, evergreen, to 25 m high) and 'semi-deciduous forest' or monsoon forest (broken canopy, up to 12 m in height). The substantial background about the climate, geomorphology, and soils in which the rare vine forest pockets occur has been extremely useful in identifying these special niches in Table 5 below.

Stocker & Hyland (1981) briefly described the deciduous monsoon forests or 'semi-evergreen vine forests' (sic) of north-eastern Australia. Several floristic assemblages recorded by previous authors including Stocker (1968) were noted for the north-western region. The monsoon forest flora was interpreted as having been 'selected from the wetter eastern types and most of the eastern families have persisted in the drier localities where they are represented by fewer genera and species' (Stocker & Hyland 1981).

Table 5. Environmental and structural classification of habitat-types of monsoon forests in the Northern Territory coastal and sub-coastal areas (Webb & Tracey 1979).

Habitat-type	Structural type*	Topography and soils
Sandstone ravines: With seasonal or perennial spring-fed streams, e.g. West Arnhem and Escarpment, Katherine Gorge	Simple notophyll semi-evergreen and microphyll vine forest, plus sclerophylls (typically *Allosyncarpia ternata* in West Arnhem Land)	Steep-sided narrow boulder-strewn ravines and gullies in Kombolgie sandstone. Skeletal soils of Buldiva LS** (1, 38), and Bend & Baker LS (38). Low nutrient status. Occasional fires intrude down the bouldery slopes
Soakage pockets: At head of perennial spring-fed streams, e.g. Radon Ck (West Arnhem Land), McCallums Ck (northwest Mt Tolmer), Taracumbie Ck (Melville Is.)	Notophyll evergreen vine forest with palms, tending to complex notophyll vine forest on everwet organic soils, and fringed along drier fire-prone edges by semi-deciduous notophyll and deciduous microphyll vine thicket, plus sclerophylls	Perennial seepage pockets associated with aquifers in sandstone, forming strips in shallow gullies, on toe-slopes and alluvial fans, and influencing narrow alluvial riparian fringes downslope. Noted in Koolpinyah LS (1). Low to medium nutrient status. Fires intrude along edges
Lateritic red earths: With deep brown sandy/loamy topsoils e.g. East Pt. near Darwin, north-eastern Cape Van Diemen (Melville Is.)	Semi-deciduous notophyll vine forest, buffered seaward by stunted and salt wind-pruned scrub of tolerant species of *Syzygium, Pouteria, Diospyros* etc. Rare fragments among extensive open eucalypt forest	Coastal edge of lateritic plateau of Charles LS (1) i.e. edge of plain of Littoral LS (1), and on coastal cliffs of Van Diemen sandstone illustrated by Hughes (1978, Pl. 3). Also recorded on gentle slopes of Bynoe LS, steep sheltered slopes of Mullaman LS, and on uplands associated with 'amphibolite red and various lateritic soils' of Batchelor LS (1). Medium nutrient status. Fires intrude along edges
Lateritic colluvial loams above floodplains: e.g. Kapalga, South Alligator River	Semi-deciduous notophyll vine forest	Upper and middle wash slopes with deep loams/loamy sands between lateritic plateaus and floodplains of Kosher LS (25, 38). Low to medium nutrient status. Fire effects marginal
Coastal dunes fringing mangroves: e.g. coastal plains east of Darwin, south of Cape Van Diemen (Melville Is.)	Semi-deciduous mesophyll and notophyll vine forest, typically fringed by *Melaleuca leucadendra* at edge of saline flats	Deep organic calcareous sands and loamy sands around low lateritic headlands fringing saline flats and typically overlying truncated lateritic profiles, sometimes adjacent to paperbark flats. Unit in Littoral LS (25, 38). Medium to high nutrient status. Little or no evidence of fire
Old Strand lines and fluviatile deposits: e.g. coastal plains north of Daly R.	Semi-deciduous mesophyll and notophyll vine forest on drier well-drained sands, tending to complex mesophyll vine forest with palms on clayey but moderately well-drained sites	Deep coarse gritty calcareous sands on former strandlines and old beach ridges several km inland, associated with fluviatile deposits of variable drainage status. Small headlands in tributary floodplains noted in Green Ant LS (13). Nutrient status mostly high. Fire intrusion very limited
Rocky outcrops: e.g. Mt Bundey area (granite), Murrenja Hill (sandstone mixed with gabbro), 16-mile Cave Reserve south of Katherine (limestone)	Deciduous microphyll vine thicket and semi-deciduous notophyll vine forest	Generally fire proof bouldery and rocky outcrops as in Litchfield LS (1), Budbudjong LS (13). Medium nutrient status

*The different floristic elements within these types mostly belong to the tropical (megatherm) highly seasonal subhumid and humid ecofloristic provinces (Webb et al. 1984).
** Numbers in brackets refer to CSIRO Land Research Series Reports 1, 13, 25 or 38 in which units of relevant Land Systems (LS) are noted as supporting monsoon forest patches.

Habitat types of the monsoon forests

The unpublished report by Webb & Tracey (1979) on the monsoon forests of the north-west area of the Northern Territory serves as a preliminary summary of ecological relationships and a convenient basis for discussing some problems that deserve future research. Table 5 lists the main habitat types identified during a month's survey of selected sites on the West Arnhem Land Escarpment, the coastal plain of the Alligator Rivers area, Melville Island, and in the Daly River and Katherine areas. To assist interpretation of the structural types of rainforest (including monsoon rainforest) in Table 5, a field key is given in Appendix 1. Definitions of the structural features used in the key are given in Webb (1968, 1978) and in Webb et al. (1976).

Topographic features, while influencing soil depth, mineral nutrient accumulation, and permanence of soil moisture, are also relevant to the protection of monsoon forest species from fire. These species vary from fire-tolerant (*Buchanania, Canarium, Gardenia*, etc., often with corky barks in monsoon woodland; *Allosyncarpia* in sandstone gullies) to fire-sensitive (numerous species restricted to interior of vine forest patches). Thus narrow ravines, bouldery outcrops and stony surfaces, and permanently swampy patches of minimum 'critical mass' are typically fire-proof (see Table 5).

The general floristic relationship of the monsoon forest types are shown in Table 6. The ecofloristic provinces B_1 (humid tending to subhumid) and B_3 (subhumid and drier) are both well represented in the Northern Territory. Higher rainfall areas near the coast on sandy soils with permanent groundwater and nutrient accumulation, and protected from fire, tend to support fragmentary associations of the more humid B_2 province that characterises North Queensland rainforests (Webb et al. 1986).

Although more intensive botanical survey and collecting in the many disjunct patches of 'jungle' will be necessary to provide sound data, some observations are possible at this stage.

The rocky outcrops with deciduous vine thicket and semi-deciduous notophyll vine forest have a relatively large number of species exclusive to them, although these are also found in similar habitat-types outside the region, e.g. species of *Ficus, Diospyros, Gyrocarpus, Lysiphyllum*. These vegetation types are attenuated versions of the floristic province B_1 centred in northeast Australia, as well as part of the core area of the drier province B_3 (see Table 6).

The soakage pockets also possess exclusive species, for instance, of *Horsfieldia, Euodia, Elaeocarpus, Ilex, Syzygium* and *Xanthostemon*. Some of the species are endemic to the Northern Territory, but most of them are found in similar sites in eastern Cape York Peninsula such as the Olive River and Jardine River areas on coastal sand dunes. On Melville Island, at Taracumbie Creek and elsewhere, the narrow riverine soakage pockets support probably the most highly developed monsoon forests of the region, with a greater variety of species than those of the West Arnhem Land Escarpment, including filmy ferns, and woody species such as *Strychnos colubrina* found otherwise only in the humid tropical lowlands of North Queensland.

There are many species common to the lateritic red earths, lateritic colluvial loams, coastal dunes, and old strand lines and fluviatile deposits, but the frequency of species varies considerably in these different communities. Ecological 'wides', also found in northeastern Australia, includes *Celtis philippensis, Cupaniopsis anacardioides, Diospyros maritima, Dracaena angustifolia, Exocarpos latifolius, Memecylon pauciflorum*.

The most extensive and highly developed monsoon forest area in the Northern Territory is on the Arnhem Peninsula and on Melville Island (Stocker & Hyland 1981). The latter is evidently related to the relict patches now rare or destroyed since European settlement on coastal cliffs of the mainland (e.g. East Point, Darwin). Radiocarbon dating of jungle fowl mounds, originally inside monsoon forests but now isolated in eucalypt woodland near Snake Bay, Melville Island, indicates that monsoon forests of this habitat-type were more extensive in the climatically recent past (Stocker 1971).

The 'wet end' of the habitat-type of old strand lines and fluviatile deposits harbours some unusual species such as *Syzygium nervosum*. This tree has not been recorded from north-eastern Australia, but occurs through southeast Asia to India.

There is also a peculiar monsoon forest type adjacent to low coastal dunes, dominated by a species of *Diospyros* and fringing the mangroves, which has scattered counterparts in Cape York Peninsula (e.g. Lockhart River).

Perhaps eight taxa were recorded for the first time in the Northern Territory during the survey by Webb & Tracey (1979). However the monsoon forests of the region would seem to be comparatively poor in new taxonomic prospects, suggesting that the size, variety and conditions of isolation of the monsoon forest pockets have not permitted speciation or even the development of vicariant species in recent times. The successions of the contemporary monsoon forests have been truncated or 'cornered' in the sense that Dansereau (1957) used the term during cliseral shifts, so that the more specialized and mesic taxa of advanced seral stages have generally been lost. Exceptions and survivors recorded during this brief survey include *Ternstroemia cherryi*, a rare tree found in relatively deep gorges of the Western Arnhem Land Escarpment, which is otherwise common in the tropical humid rainforests of North Queensland. These gorges, which are not easily accessible during a reconnaissance survey, would repay more thorough search botanically, and an evaluation of species vagility as in Webb et al. (1984) to provide keys to past climatic history and biogeographic disjunctions.

A follow-up survey of the animal ecology of the sites of monsoon forests selected by Webb & Tracey (1979) was made by Kikkawa & Monteith (1980), and biological aspects of the gradients and boundaries of selected monsoon forest patches were discussed by Kikkawa et al. (1981).

Story (1976) raised a variety of questions about the Northern Territory rainforest and monsoon forest

Table 6. Climatic and edaphic features of core areas of the rainforest ecofloristic provinces (species classification) and their equivalent structural types (from Webb et al. 1984).

Eco-floristic province	Structural type	Climatic type	Edaphic type	'Average' climatic station	Alt. (m)	Mean annual rainfall (mm)	Mean annual raindays	Mean rainfall driest six consecutive months	Mean annual temp. (°C)	Mean minimum temp. coldest month (°C)	Inferred soil mineral nutrient status
B_1	Semi-deciduous mesophyll and notophyll vine forest (SDMVF, SDNVF)	Tropical (megatherm) seasonal humid/subhumid (monsoonal)	Eutrophic/ mesotrophic/ oligotrophic	Darwin (NT) 12°25'S 130°52'E	29	1594	109	110	27.5	18.9	High/medium/low
				Oenpelli (NT) 12°20'S 133°03'E	7	1360	92	51	27.8	17.9	
				Iron Range (Q) 12°47'S 143°18'E	19	2049	202	215	25.4	18.4	
				Moreton (Q) 12°27'S 142°38'E	39	1362	102	66	26.2	17.0	
B_2	Complex mesophyll vine forest (CMVF) at lower altitudes in coastal zone	Tropical (megatherm) aseasonal/ seasonal humid (± cloud)	Eutrophic/ mesotrophic	Innisfail (Q) 17°32'S 145°58'E	40	3644	155	760	23.5	15.1	High/medium
				Cardwell (Q) 18°16'S 146°02'E	5	2127	122	289	24.1	13.3	
				Kairi (Q) 17°12'S 145°34'E	715	1260	113	189	20.2	10.8	
B_3	Deciduous micro-phyll vine thicket (DVT)	Tropical (megatherm) highly seasonal sub-humid (monsoonal)	Eutrophic/ mesotrophic/ oligotrophic	Coen (Q) 13°57'S 143°12'E	193	1146	86	44	25.4	16.7	High/medium/low
				Kowanyama (Q) 15°28'S 141°25'E	13	1222	71	36	26.9	14.8	
				Katherine (NT) 14°28'S 132°16'E	107	952	62	46	27.2	12.9	

Photograph 1. Patch of rainforest at the mouth of the South Alligator River, with *Eucalyptus* open forest in the background. Note the emergent palms *Carpentaria acuminata* and abundant vines (Photograph: D.G.Tulloch).

Photograph 2. Isolated palms *Livistona benthamii*, after surrounding monsoon forest (rain forest) had long disappeared as a result of the activity of buffaloes near Woolwonga, April 1984 (Photograph: R.W.Braithwaite).

Photograph 3. Floating mat of aquatic vegetation, untouched by buffaloes. Such a raft is sufficient to hold the weight of an adult person, and is an important nesting habitat for saltwater crocodiles (Photograph: G.J.W.Webb).

communities 'as a basis for ecological studies', and detailed 'some puzzling features of the ecology' dealing with soil and water relations of semi-deciduous forest and several grassland types. Except for some climatological studies, the eco-physiology of plants in relation to water is still a virgin field in tropical Australia. Even at the broad biogeographical level, there is a complex interaction of ecological factors to produce intricate mosaics of vegetation, especially when regularly burnt (Whitmore 1975). Under the latter conditions, boundaries are generally sharp. Rain forest (usually spelt rainforest by Australian authors) has been adopted as an omnibus term for closed forests with broad mesomorphic leaves in the better known regions of eastern Australia, even when they contain a conspicuous proportion of completely leafless (deciduous) or partly leafless (semi-evergreen) species during the dry season (hence also the term 'dry rainforest' adopted in New South Wales for drier subcoastal types). However, the term 'monsoon forest' was adopted early in tropical north-western Australia, as was monsoon woodland to describe the communities of eucalypts with a scattered understorey of mainly broad-leaved deciduous species. Rain forest was generally reserved for patches of tall evergreen mesomorphic trees in relatively wet sites, which are uncommon in this region.

Since eucalypt-dominated vegetation characterized by genera of Myrtaceae, Proteaceae and Leguminosae is scarcely developed outside Australia in the tropics, the monsoon forests and rainforests of northern Australia occupy a unique place for comparative ecological studies.

The dynamic relationships of patches of monsoon/rain forests and the extensive eucalypts and other types of 'sclerophylls' have so far been little studied. Story (1976) made no observations except in some regularly burnt areas of monotypic *Allosyncarpia* forest on shallow sandy oligotrophic soils, which would not be included in either rain forest or monsoon forest. In northern Australia, each vine forest habitat can be broadly interpreted as either residual with a core of 'original' species of long standing, or a more recently colonized favourable niche. The most favourable conditions for both include certain combinations of topographic and edaphic factors that ensure, above all, adequate soil moisture and the exclusion of fires. Mineral nutrient accumulation, characterized especially by phosphorus and calcium, is also probably relevant for more diverse communities.

Monsoon forest patches considered to be recent are located on recently evolved landforms such as beach dunes and on springs associated with the coastal flood plains (Russell-Smith & Dunlop 1985). Such habitats were non-existent during the last ice age when the present coastal zone would have been well inland,

experiencing a relatively arid, continental climate. As mentioned earlier, monsoon forest species would presumably have persisted during this period in the damp gorges of the sandstone escarpment, the more vagile species migrating to available sites with the onset of a more favourable climatic regime.

The segregation of monsoon forest and adjacent sclerophyll vegetation is not, however, everywhere complete, resulting in the mixing of apparently botanically unrelated floras; the Gondwanan flora being most ancient, and the sclerophyll flora more recent and derivative (Webb & Tracey 1981a, Webb et al. 1984). The conditions for evolution of the sclerophyll flora remain controversial, especially in the absence of Tertiary and earlier fossils in the tropical Australian record. Matched habitats of deciduous vine forests in northern Australia and of monsoon forests in the Indo-Malesian region such as Thailand share many related genera of trees (Smitinand et al. 1982). Yet there is no counterpart of the mixed eucalypt-dominated monsoon woodlands that have different interspersed floras as in tropical Australia. As noted earlier in this chapter, the savanna-type communities, besides having the interspersed rainforest elements, also have a ground flora of apparently recent Indo-Malesian immigrants including members of families Leguminosae, Convolvulaceae, Gramineae and others. Of the grasses, tribe Andropogoneae is especially prominent in monsoonal Australia (Simon 1981) sharing many species with the Malesian region and beyond. This ground flora must surely represent an additional element which, with the deciduous broad-leaved arborescent flora, is a product of a wet summer/dry winter climate.

Thus where boundaries of monsoon forest patches are sharp and moulded by fire, the adaptations of their species and of the sclerophylls do not overlap as in the mixed monsoon woodlands. Perhaps, as with other examples of interspersion of rainforest and sclerophyll forest elements (Webb & Tracey 1981b), the mixtures represent relics of rainforest taxa as in the monsoonal fire-prone open vegetation of southeast Asia, except that in northern Australia the eucalypts and similar sclerophylls now dominate such situations. It is as though a kind of 'co-action biogeography', following the evolution of different floras and their separate community-types, preserves a tension zone in time for some of the different floral elements.

THE ROLE OF FIRE AS A DETERMINANT OF VEGETATION

Fire is associated with all the major tropical savannas of the world (Huntley & Walker 1982). These authors regard fire as one of the determinants of savanna ecosystems, second only to water availability.

Sarmiento & Monasterio (1975) maintain that all elements of the South American savannas must be adapted to a regime of frequent fire but that the essential structure of the ecosystem is not directly related to fire frequency. They have discounted theories of human-induced savanna ecosystems for the South American continent, citing palynological evidence for the existence of savanna species as far back as the Eocene. There is also general support in the literature for the occurrence of natural fires (mainly lightning) before the arrival of humans in Australia (Beard 1977, Galloway & Kemp 1981, Kemp 1981). Studies of past climatic regimes and evidence from the fossil record suggest that eucalypt-dominated communities were widespread on the continent since the early Pleistocene or before (Kemp 1981).

As northern Australia has been subjected to an arid climate and natural fires for about 15 million years (Kemp 1981), it is almost certain that eucalypt savanna was dominant well before people arrived on the scene. If Aboriginal fires have had an effect on the vegetation of the monsoon tropics it has been an ameliorating one (see Chapter 5).

Eucalypts dominate the woody flora on the Australian continent and they are second only in species number to *Acacia* in the tree and shrub layer. Both *Eucalyptus* and *Acacia* undoubtedly owe their prominence to their ability to cope with fire, although the former employ a much greater range of strategies. The strategy employed by a eucalypt community is very much dependent on the fire regime to which that community is subjected. In the broad range of plant communities throughout Australia it is usually unclear whether an adaptive trait is a response to the fire environment or to other stressful factors such as drought and infertile soils. The southern ash forests (*Eucalyptus regnans*) are an exception: fire is a critical element in the life cycle of *E. regnans* and the development of serotiny is directly related to it. It is proposed here that at the other extreme, the traits of the eucalypts of the tropical savanna have probably developed in response to other environmental factors with fire playing a minor role.

It has been generally accepted that eucalypt dominated communities throughout Australia are fire-dependent, i.e. they follow, in various ways, the pattern for the ash forests. They depend on recurrent fire events to ensure regeneration by eliminating competition and preparing a suitable seed bed which is free of litter and is unshaded. The hypothesis of Mutch (1970) is seen to hold true for these communities: 'Fire-dependent plant communities burn more readily than non fire-dependent communities because natural selection has favoured development of characteristics that make them more flammable'.

The eucalypts of the wet sclerophyll communities of southern and eastern Australia do possess these characteristics and are surely fire-dependent. Outstanding traits of the wet sclerophyll species which make them highly flammable and fire prone are high content of volatile oils in the leaves (and bark in some species, Carr & Carr 1969, 1970) and suspended strips of shed bark which effectively extends the litter layer (and the fire zone) from the ground to the crown (Gill 1975). The interval between fires is measured in decades, allowing litter to accumulate to the order of 150-500 tonnes per hectare (Luke and McArthur 1978). A fire event will thus involve the total forest from the ground to the crown and will be of the highest intensity.

Serotiny, the facility whereby the mature indehisced fruit is held in the tree until opened by a suitable stimulus such as fire, is well developed in many eucalypts and is seen to be critical to species such as *Eucalyptus regnans* which is killed by intense fires. The majority of species will persist and develop a new crown from epicormic buds but *Eucalyptus regnans* is dependent on regenerating by seed.

The tropical eucalypts, by contrast, do not possess features which make them highly flammable. They are mainly members of the subgenera Corymbia, Blakella and Eudesmia (Pryor & Johnson 1971), having a lower and less volatile oil content than their more advanced southern counterparts (i.e. ashes, peppermints, stringybarks). None of the northern species decorticates in long ribbons nor has the long, vertical, flammable strands seen in some of the true stringybarks of the sub-tropical and temperate forests.

In the Northern Territory, fire frequency is high (1-3 years) and fuel loads are in the region of 1-9 tonnes per hectare (Hoare et al. 1980). Fires therefore are of low intensity and rarely, if ever, are carried in the crowns. The fires of the tropical savanna are, in effect, grass fires (Story 1976, Lacey et al. 1982). Irrespective of whether these fires occur in the early dry or the late dry season (hottest period) where crown scorch may be total, the trees very rarely suffer enough crown damage to be killed or to produce epicormic response on the stem, although in the subgenus Blakella crowns damaged by fire may revert to an intermediate or juvenile type foliage (Blake 1953a). The 'greening up' of the forest which commonly follows a fire event in extra-tropical eucalypt forests is absent in the northern savanna. Presumably the fires are not hot enough to damage the crown. *Eucalyptus miniata* has reportedly been killed by exceptionally hot fires (Stocker 1968, C.Haynes, unpubl. inform.), evidence perhaps that it has evolved with a regime of low intensity fires.

There appears to be no parallel in the tropical savanna for seed release and seed bed preparation as seen in the wet sclerophyll forest of subtropical and temperate Australia. None of the eucalypts hold their capsules beyond maturity nor are there other examples of serotiny in the tree layer. The single species of *Banksia* in the region, *B. dentata* is one of the few in the genus which releases seed without fire (George 1981).

As noted below, seed of subgenus Blakella tends to fall during the rainy months but there is no pattern for the other subgenera. In any event, unlike the mountain ash seedlings of temperate Australia which have only other seedlings to compete against, the northern eucalypt seedling will be competing, when the rain comes, with the full complement of established ground flora. This usually includes a large proportion of perennial grasses, fast growing annuals, and species regenerating from underground storage organs, none of which has been adversely affected by the previous dry season fire. Furthermore, within the first year of establishment the seedling will face a long period of drought and a possible grass fire.

This shortfall in the provision of a competition-free seed bed is most evident in the tropical savanna but is nevertheless a feature of the dry sclerophyll forests generally. Purdie & Slatyer (1976) have shown that there is a prolific regeneration from non-eucalypt rootstocks in dry sclerophyll forest near Canberra in southeastern Australia in the first year following fire. There is a concomitant high mortality in eucalypt seedlings. Specht (1981b) has also noted that in sclerophyllous heath vegetation in South Australia, post-fire seedlings compete unfavourably with growth from epicormic buds and lignotubers. It is not surprising that this state of insecurity in regeneration by seed in *Eucalyptus* is compensated for in varying degrees and by an array of vegetative mechanisms. These mechanisms are critical for the establishment (regeneration) and persistence of the genus under stressful environmental conditions such as periodic drought, frequent fires, infertile soils and soils of extreme alkalinity and acidity, and competition from an established ground flora.

The ability to recover from dormant buds in the stems and lignotubers is widespread through the genus and lignotubers in particular are important in the establishment phase in the tropical eucalypt communities. Additionally, many of the woody fruited bloodwoods (subgenus Corymbia, Pryor & Johnson 1971) produce rhizomes (Lacey 1974) and the widespread Darwin Stringybark (*E. tetrodonta*) can reproduce by root suckers (Carr 1972). Rhizomes, root suckers and lignotubers are not confined to *Eucalyptus* but occur in many other genera of trees and shrubs in the tropical savanna (Lacey & Whelan 1976). The extreme development of these vegetative mechanisms is the formation of large clonal patches of short leafy stems which may occupy areas to 55 m² (Lacey & Whelan 1976) and are usually dense enough to exclude grasses and other herbs. They are thus self-protecting with regard to fire.

Although attributed to frequent fires (Lacey & Whelan 1976, Lacey et al. 1982), the ecological significance of these clonal patches has never been investigated. The assumption has been that rhizomatous and root suckering mechanisms are elaborate traits which developed in addition to lignotubers and have been maintained because of their ability to cope with the regime of frequent fires.

In view of the adverse conditions for seedling establishment described above, it is suggested here that rhizomes and root suckers could have developed in the tree and shrub flora of the tropical savannas irrespective of fire. There is clearly a competitive advantage for individuals which receive their early sustenance via a rhizome or root connection to an established plant. Continued attachment to the parent or to surrounding individuals on the same root or rhizomatous system would appear however to offset any advantages gained: Lacey & Whelan (1976) noted that stems of *E. tetrodonta* and *E. porrecta* in such patches did not develop into mature trees when given fire protection. This is almost certainly because in a mature undisturbed forest, all regeneration, whether seedling or clonal, will stay in a suppressed state on the forest floor until released through loss of individuals from the overstorey. When the opportunity arises, one or more clonal stems will assume dominance and be recruited to the overstorey.

The role of rootsuckers and rhizomes in the monsoon tropics is thus seen primarily as an alternative to seedling regeneration in the face of annual drought and excessive competition in the establishment phase. A similar conclusion has been reached by Johnston & Lacey (1983) referring to two temperate rainforest species which, in compensating for a lack of opportunity for seedling regeneration, produce multiple stems from a lignotuber-like base.

Data on the frequency of fire on the sandstone plateaux are not available, though it may be assumed that fires occur at intervals of 2-10 years. The sandstone country consists of a mosaic of plant communities, some of which are subject to fire and others of which occupy fire-proof niches. However even the most sheltered gullies such as those occupied by dry closed forest of *Allosyncarpia ternata* are subject to occasional fires (Story 1976). Some data on the behaviour of fire in sandstone communities are given by Begg et al. (1981). High fire intensities reported by Begg et al. are due mainly to resinous and highly flammable species of *Triodia, Plectrachne* and *Acacia*. Following such fires, fire-weeds appear (Dunlop & Begg 1981) including *Acacia* spp., *Solanum clarkiae, Clematis pickeringii, Muehlenbeckia rhyticarya* and species of *Hibiscus*, section Furcaria. Most of these species are short-lived perennials which eventually die out without continued fire, leaving a dense and continuous layer of spinifex.

Although much has been written on fires in the Australian context, it is obvious from what has been said above that very little is known on the fire-biota relationship in the monsoonal tropics. The juxtaposition of the sandstone plateau-subcoastal lowlands with their different fire regimes, post fire-responses, suites of taxa and life form spectra (see Taylor & Dunlop 1985), presents an ideal opportunity for comparative ecological studies on the effects of fire.

A NOTE ON DISTRIBUTION OF *EUCALYPTUS* SPP. IN THE AUSTRALIAN TROPICS

The classification of Pryor & Johnson (1971) and the recent work of Chippendale & Wolf (1981) provide a basis for a discussion on the relationships and distribution of the eucalypts. Table 7 has been compiled using data from these publications. The regions analysed are north of 16°S latitude in Western Australia and the Northern Territory and north of 18°S in Queensland. This zone approximates the winter dry season region of the Atlas of Australian Resources (Plumb 1977). Cape York has a shorter dry season along the east coast and is perhaps not directly comparable with the other two areas. Nevertheless, comparison of the three areas is a useful exercise in that the similarity between the Northern Territory/Western Australian regions becomes apparent. Of the total number of species recorded for the Northern Territory and the Kimberleys,

approximately 70% are common to both regions. The Northern Territory and Cape York, on the other hand, share only twelve species or 18%. Added to this, Cape York has members in the Idiogenes and Monocalyptus, subgenera which are absent from the other regions (Table 7). The Northern Territory tropics has relatively few eucalypt species, concentrated mainly in Blakella, Corymbia and Eudesmia. Subgenus Blakella, which is the only subgenus with a predominantly tropical distribution, is singled out here for further discussion.

Subgenus Blakella

Of the nine species mapped by Chippendale & Wolf (1981) only one (unnamed) species does not occur north of 16°S latitude. The group is a natural one, possessing several characteristics which are common to all or most of the species and which distinguish it from other subgenera. Flowering time is a character which has rarely, if ever, been used in eucalypt classification above the species level. In Blakella, flowering time is decidedly uniform, the wide-ranging *E. tessellaris* (which does not occur in the monsoonal tropics) being the exception. Flowering time, which just precedes the wet season (Fig. 27), is a clue to the possible origins of the group. The behaviour of species in Blakella corresponds closely to the deciduous species of other families which characterise the monsoonal tropics, i.e. flowering coincides with the emergence of new leaves at the end of the dry season. In some species (e.g. *E. confertiflora*, *E. grandifolia*), the flowers may precede the flush of new leaves while in

Table 7. Distribution of subgeneric groups of *Eucalyptus* in Northern Australia.

Taxonomic group (Pryor & Johnson 1971)	Blakella	Corymbia sect. Rufaria	Corymbia sect. Ochraria	Eudesmia	Symphyomyrtus sect. Equatoria	Symphyomyrtus sect. Exsertaria	Symphyomyrtus sect. Adnataria	Symphyomyrtus sect. Umbrawarria	Symphyomyrtus sect. Transversa	Monocalyptus	Idiogenes	% of species in common with Kimberleys	% of subgeneric groups in common with Kimberleys	% of species in common with Cape York	% of subgeneric groups in common with Cape York
Total species	9	25	8	15	3	29	66	1	14	115	1	–	–	–	–
Kimberleys	5	14	–	4	1	8	7	1	–	–	–	100	100	–	45
NT	6	14	–	3	1	7	10	1	–	–	–	70	100	18	45
Cape York	4	6	1	2	–	6	11	–	2	2	1	–	45	100	100

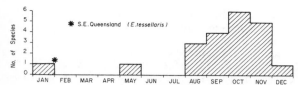

Figure 27. Flowering calendar *Eucalyptus* Blakella group.

Figure 28. Flowering calendar *Eucalyptus* species other than Blakella group.

others (e.g. *E. papuana, E. aspera*), flowerbuds and leaves are produced together. Not all species are deciduous (see Blake 1953a, Carr 1972, Walter 1971) but two which are conspicuously so are *E. grandifolia* and *E. clavigera*. These display canopies of wine-red foliage when in new leaf. The value of flowering in the late dry season is in the timing of seed shed which will coincide with the wet season – the relatively short period (four months) in which the seed can germinate and the seedling become established. The period from flowering to seed shed is quite short, observed to be four weeks in a tree of *E. clavigera* near Darwin. None of the other groups of eucalypts in the region (Fig. 28) are synchronised in this way with the climate. Their seeds will be lying on the forest floor for at least part of the dry season and thus be more susceptible to predation. Perhaps the keystone to the facility for shedding seed at the appropriate time is the nature of the Blakella capsule. The capsule walls and valves are thin and appear to dry and shrink immediately the capsule is mature. Seed release from the capsule is very effective and is completed for the whole tree within a few days.

ACKNOWLEDGEMENT

We are grateful to the Director of the Queensland Herbarium, Brisbane, for information about *Eucalyptus tessellaris*.

REFERENCES

Barlow, B.A. 1981. The Australian Flora: its origin and evolution. In A.S.George (ed.), *Flora of Australia* 1: 25-75. Canberra: Aust. Govt. Printing Service.

Beard, J.S. 1967. Some vegetation types of tropical Australia in relation to those of Africa and America. *J. Ecol.* 55: 271-290.

Beard, J.S. 1976. The monsoon forests of the Admiralty Gulf, Western Australia. *Vegetatio* 31: 177-192.

Beard, J.S. 1977. Tertiary evolution of the Australian flora in the light of latitudinal movement of the continent. *J. Biogeog.* 4: 111-118.

Beadle, N.C.W. 1981. *The Vegetation of Australia.* Cambridge: Cambridge University Press.

Begg, R.J., K.C.Martin & N.F.Price 1981. The small mammals of Little Nourlangie Rock, N.T. V. The effects of fire. *Aust. Wildl. Res.* 8: 515-527.

Blake, S.T. 1953a. Botanical contributions of the northern Australian regional survey 1. Studies on northern Australian species of Eucalyptus. *Aust. J. Bot.* 1: 185-352.

Blake, S.T. 1953b. Plant communities in Survey of Katherine-Darwin region, 1946. In C.S.Christian & G.A.Stewart, *Land Research Series* 1: 58-68. Melbourne: CSIRO.

Bourlière, F. & M.Hadley 1983. Present-day savannas: an overview. In F.Bourlière (ed.), *Ecosystems of the World 13, Tropical Savannas*: 1-15. Amsterdam: Elsevier.

Burbidge, N.T. 1960. The phytogeography of the Australian Region. *Aust. J. Bot.* 8: 75-209.

Burgman, M.A. & E.J.Thompson 1982. Cluster analysis ordination and dominance structural classification applied to diverse tropical vegetation at Jabiluka, Northern Territory. *Aust. J. Ecol.* 7: 375-387.

Carlquist, S. 1979. *Stylidium* in Arnhem Land: new species modes of speciation on the sandstone plateau, and comments on floral mimicry. *Aliso* 9: 461-511.

Carr, S.G.M. 1972. Problems of the geography of the tropical eucalypts. In D.Walker (ed.), *Bridge and Barrier: The natural and cultural history of Torres Strait*: 153-181. Canberra: Australian National University Press.

Carr, S.G.M. & D.J.Carr 1969. Oil glands and ducts in *Eucalyptus* L'Herit. I. The phloem and the pith. *Aust. J. Bot.* 17: 471-513.

Carr, S.G.M. & D.J.Carr 1970. Oil glands in the bark of Victorian Eucalyptus. *Vic. Nat.* 87: 120-126.

Chappell, J. & B.G.Thom 1977. Sea level and coasts. In J.Allen, J.Golson & R.Jones (eds.), *Sunda and Sahul: prehistoric studies in south-east Asia, Melanesia and Australia.* London: Academic Press.

Chippendale, G.M. & L.Wolf 1981. *The Natural Distribution of Eucalyptus in Australia.* Special Pub. No. 6. Canberra: Australian National Parks and Wildlife Service.

Christian, C.S. & G.A.Stewart 1953. General report on survey of Katherine-Darwin region, 1946. *Land Research Series* 1. Melbourne: CSIRO.

Clifford, H.T. 1964. The systematic position of the grass genus *Micraira* F.Muell. *Univ. Queensl. Papers* 4: 87-94.

Dansereau, P. 1957. *Biogeography: An Ecological Perspective.* New York: Ronald Press.

Dunlop, C.R. & R.J.Begg 1981. The small mammals of Little Nourlangie Rock, N.T. I. Description of Study Site. *Aust. Wildl. Res.* 8: 51-56.

Fox, R.W., G.G.Kelleher & C.B.Kerr 1977. *Ranger Uranium Environmental Inquiry Second Report.* Canberra: Australian Government Publishing Service.

Gaff, D.F. & P.K.Latz 1978. The occurrence of resurrection plants in the Australian flora. *Aust. J. Bot.* 26: 485-492.

Galloway, R.W. & E.M.Kemp 1981. Late Cainozoic environments in Australia. In A.Keast (ed.), *Ecological Biogeography of Australia*: 51-80. The Hague: Junk.

Gardner, C.A. 1942. The vegetation of Western Australia with special reference to the climate and soil. *J. Roy. Soc. Western Aust.* 28: 11-87.

George, A.S. 1981. The genus Banksia L.f. (Proteaceae). *Nuytsia* 3: 239-473.

Gill, A.M. 1975. Fire and the Australian flora: A review. *Aust. For.* 38: 4-25.

Gill, A.M. 1981. Adaptive responses of Australian vascular plant species to fires. In A.M.Gill, R.H.Groves & I.R.Noble (eds.), *Fire and the Australian Biota*: 243-271. Canberra: Australian Academy of Science.

Gillison, A.N. 1983. Tropical savannas of Australia and the southwest Pacific. In F.Bourlière (ed.), *Tropical Savannas*: 183-243. Amsterdam: Elsevier.

Good, R. 1974. *The geography of the flowering plants*. London: Longman.

Green, J.W. 1981. *Census of the vascular plants of Western Australia*. Perth: W.A. Department of Agriculture.

Hill, R. & G.Webb 1982. Floating grass mats of the Northern Territory Wetlands – an endangered habitat? *Wetlands* 2: 45-50.

Hnatiuk, R.J. & K.F.Kenneally 1981. A survey of the vegetation and flora of Mitchell Plateau, Kimberley, Western Australia. *Biological Survey of Mitchell Plateau and Admiralty Gulf, Kimberley, Western Australia*: 13-94. Perth: Western Australian Museum.

Hoare, J.R.L., R.J.Hooper, N.P.Cheney & K.L.S. Jacobsen 1980. Unpublished report to Australian National Parks and Wildlife Service, Canberra.

Hughes, R.J. 1978. *The geology and mineral occurrences of Bathurst Island, Melville Island and Cobourg Peninsula, Northern Territory.* Bureau of Mineral Resources Bull. No. 177.

Huntley, B.J. & B.H.Walker 1982. Editorial. In B.J.Huntley & B.H.Walker (eds.), *Ecology of Tropical Savannas*: 657-660. New York: Springer.

Hutchinson, J. 1964. *The Genera of Flowering Plants*. London: Oxford University Press.

Johnson, L.A.S. & B.G.Briggs 1975. On the Proteaceae – The evolution and classification of a southern family. *Bot. J. Linn. Soc.* 70: 83-182.

Johnston, R.D. & C.J.Lacey 1983. Multi-stemmed trees in rainforest. *Aust. J. Bot.* 31: 189-195.

Kemp, E.M. 1981. Pre-Quaternary fire in Australia. In A.M.Gill, R.H.Groves & I.R.Noble (eds.), *Fire and the Australian Biota*: 3-21. Canberra: Australian Academy of Science.

Kenneally, K.F. 1983. Flora. In N.L.McKenzie (ed.), *Wildlife of the Dampier Peninsula, South-West Kimberley, Western Australia*. Wildl. Res. Bull. West. Aust. No. 11: 27-39.

Kern, J.H. 1974. Cyperaceae. In C.G.G.J.van Steenis (ed.), *Flora Malesiana* 7: 435-753. The Hague: Nijhoff/Junk.

Kershaw, A.P. 1978. The analysis of aquatic vegetation on the Atherton Tableland, north-east Queensland, Australia. *Aust. J. Bot.* 3: 23-42.

Kershaw, A.P. 1985. An extended late Quaternary vegetation record from north-eastern Queensland and its implications for the seasonal tropics in Australia. *Proc. Ecol. Soc. Aust.* 13: 179-189.

Kikkawa, J. & G.B.Monteith 1980. Animal ecology of monsoon forests of the Kakadu region, Northern Territory. Unpublished report to Australian National Parks and Wildlife Service.

Kikkawa, J., L.J.Webb, M.B.Dale, G.B.Monteith, J.G.Tracey & W.T.Williams 1981. Gradients and boundaries of monsoon forests in Australia. *Proc. Ecol. Soc. Aust.* 11: 39-52.

Lacey, C.J. 1974. Rhizomes in tropical eucalyptus and their role in recovery from fire damage. *Aust. J. Bot.* 22: 29-38.

Lacey, C.J. & P.I.Whelan 1976. Observations on the ecological significance of vegetative reproduction in the Katherine-Darwin Region of the Northern Territory. *Aust. For.* 39: 131-139.

Lacey, C.J., J.Walker & I.R.Noble 1982. Fire in Australian Tropical Savannas. In B.J.Huntley & B.H.Walker (eds.), *Ecology of Tropical Savannas*: 246-272. New York: Springer.

Lazarides, M. 1979. Micraira F.Muell. (Poaceae, Micrairoideae). *Brunonia* 2: 67-84.

Lazarides, M. 1984. New taxa of tropical Australian grasses (Poaceae). *Nuytsia* 5: 273-303.

Love, L.D. 1981. Mangrove swamps and salt marshes. In R.H.Groves (ed.), *Australian Vegetation*: 319-334. Cambridge: Cambridge University Press.

Luke, R.H. & A.G.McArthur 1978. *Bushfires in Australia.* Canberra: Aust. Govt. Printing Service.

McCusker, A. 1984. Rhizophoraceae. In A.S.George (ed.), *Flora of Australia* 22: 1-10. Canberra: Aust. Govt. Publishing Service.

McKenzie, N.L. & K.F.Kenneally 1983. Background and Environment. In N.L.McKenzie (ed.), *Wildlife of the Dampier Peninsula, South-West Kimberley, Western Australia*. Wildl. Res. Bull. West. Aust. No. 11: 5-23.

Michael, P.W. 1980. A new perennial species of Echinochloa from New Guinea. *Telopea* 2: 31-33.

Mutch, R.W. 1970. Wildland fires and Ecosystems – a hypothesis. *Ecology* 51: 1046-1051.

Paijmans, K. 1976. Vegetation. In K.Paijmans (ed.), *New Guinea Vegetation*: 23-104. Canberra: Australian National University Press.

Perry, R.A. & C.S.Christian 1954. Vegetation of the Barkly Region. In Survey of the Barkly Region, 1947-48. *Land Research Series* 3: 78-112. Melbourne: CSIRO.

Plumb, T.W. 1977. *Atlas of Australian Resources*. Canberra: Division of National Mapping.

Pryor, L.D. & L.A.S.Johnson 1971. *A classification of the Eucalypts*. Canberra: Australian National University Press.

Purdie, R.W. & R.O.Slatyer 1976. Vegetation succession after fire in sclerophyll woodland communities in south-eastern Australia. *Aust. J. Ecol.* 1: 223-236.

Russell-Smith, J. & C.R.Dunlop 1985. The status of monsoon vine-forests in the Northern Territory: a perspective. In G.L.Werren & A.P.Kershaw (eds.), *Australian National Rainforest Study Report to the World Wildlife Fund (Australia)* 1. Proceedings of a workshop on the past present and future of Australian rainforests, Griffith University, December 1983. Melbourne: Monash University.

Sarmiento, G. & M.Monasterio 1975. A critical consideration of the environmental conditions associated with the occurrence of savanna ecosystems in Tropical America. In F.B.Golley & E.Medina (eds.), *Tropical ecological systems*: 223-250. New York: Springer.

Simon, B.K. 1981. An analysis of the Australian grass flora. *Austrobaileya* 1: 356-371.

Smitinand, T., L.J.Webb, T.Santisuk & J.G.Tracey 1982. A co-operative attempt to compare the habitats of primary forest in Thailand and northern Australia. In Kamis

Awang et al. (eds.), *Ecological Basis for Rational Resource Utilisation in the Humid Tropics of Southeast Asia*: 77-107. Serdang: Universiti Pertanian Malaysia.

Specht, R.L. 1958a. The climate, geology, soils and plant ecology of the northern portion of Arnhem Land. In R.L.Specht & C.P.Mountford (eds.), *Records of the American-Australian Scientific Expedition to Arnhem Land* 3: 333-413. Melbourne: Melbourne University Press.

Specht, R.L. 1958b. The geographical relationships of the flora of Arnhem Land. In R.L.Specht & C.P.Mountford (eds.), *Records of the American-Australian Scientific Expedition to Arnhem Land* 3: 415-478. Melbourne: Melbourne University Press.

Specht, R.L. 1958c. The history of botanical exploration in the Arnhem Land Aboriginal Reserve. In R.L.Specht & C.P.Mountford (eds.), *Records of the American-Australian Scientific Expedition to Arnhem Land* 3: 1-7. Melbourne: Melbourne University Press.

Specht, R.L. 1981a. Biogeography of halophytic angiosperms (salt-marsh, mangrove and sea-grass). In A.Keast (ed.), *Ecological biogeography of Australia*: 575-589. The Hague: Junk.

Specht, R.L. 1981b. Responses to fires in heathlands and related shrublands. In A.M.Gill, R.H.Groves & I.R.Noble (eds.), *Fire and the Australian Biota*: 395-415. Canberra: Australian Academy of Science.

Specht, R.L. 1981c. Major vegetation formations in Australia. In A. Keast (ed.), *Ecological biogeography of Australia*: 163-197. The Hague: Junk.

Specht, R.L., R.B.Salt & S.T.Reynolds 1977. Vegetation in the vicinity of Weipa, North Queensland. *Proc. Roy. Soc. Qld.* 88: 17-38.

Speck, N.H. 1960. Vegetation of the North Kimberley area, W.A. Lands and Pastoral Resources of the North Kimberley Area, W.A. *Land Research Series* 4: 41-63. Melbourne: CSIRO.

Speck, N.H. 1965. Vegetation and pastures of the Tipperary area. General Report on Lands of the Tipperary area, northern Australia, 1961. *Land Research Series* 13: 81-98. Melbourne: CSIRO.

Speck, N.H. & M.Lazarides 1964. Vegetation and pastures of the West Kimberley area. In General Report on Lands of the West Kimberley Area, W.A. *Land Research Series* 9: 140-174. Melbourne: CSIRO.

Stocker, G.C. 1968. The plant communities of Karslake Peninsula, Melville Island, Northern Territory. Unpublished MSc. thesis, University of New England.

Stocker, G.C. 1971. The age of charcoal from old Jungle Fowl nests and vegetation changes on Melville Island. *Search* 2: 28-30.

Stocker, G.C. & B.P.Hyland 1981. Semi-evergreen vine forests (monsoon forests). In N.C.W. Beadle (ed.), *The Vegetation of Australia*: 179-195. Cambridge: Cambridge University Press.

Story, R. 1969. Vegetation of the Adelaide-Alligator Area. In Lands of the Adelaide-Alligator Area, Northern Territory. *Land Research Series* 25: 114-130. Melbourne: CSIRO.

Story, R. 1976. Vegetation of the Alligator Rivers area. In Lands of the Alligator Rivers Area, Northern Territory. *Land Research Series* 38: 89-111. Melbourne: CSIRO.

Taylor, J.A. & C.R.Dunlop 1985. Plant communities of the wet-dry tropics of Australia: The Alligator Rivers region of the Northern Territory. *Proc. Ecol. Soc. Aust.* 13: 83-127.

Tinley, K.L. 1982. The influence of soil moisture balance on ecosystem patterns in southern Africa. In B.J.Huntley & B.H.Walker (eds.), *Ecology of Tropical Savannas*: 175-192. New York: Springer.

van Steenis, C.G.G.J. 1979. Plant geography of east Malesia. *Bot. J. Linn. Soc.* 79: 97-178.

Walker, J. & A.N.Gillison 1982. Australian savannas. In B.J.Huntley & B.H.Walker (eds.), *Ecology of Tropical Savannas*: 5-24. New York: Springer.

Walter, H. 1971. *Ecology of tropical and subtropical vegetation.* Edinburgh: Oliver & Boyd.

Webb, L.J. 1959. A physiognomic classification of Australian rainforests. *J. Ecol.* 47: 551-570.

Webb, L.J. 1968. Environmental relationships of the structural types of Australian rainforest vegetation. *Ecology* 49: 296-311.

Webb, L.J. 1973. Eat die or learn – the botany of the Australian Aborigines. *Aust. Nat. Hist.* 17: 290-293.

Webb, L.J. 1978. A general classification of Australian rainforests. *Aust. Plants* 9: 349-363.

Webb, L.J. & D.M.Smyth 1984. Ecological guidelines and traditional empiricism in Rural development. In J.Hanks (ed.), *Traditional Life-Styles, Conservation and Rural Development*. Gland: Int. Union Conservation Nature.·

Webb, L.J. & J.G.Tracey 1979. An ecological survey of the monsoon forests of the north-western region of the Northern Territory. Unpublished report to Australian National Parks and Wildlife Service.

Webb, L.J. & J.G.Tracey 1981a. Australian rainforests: patterns and change. In A.Keast (ed.), *Ecological Biogeography of Australia*: 607-694. The Hague: Junk.

Webb, L.J. & J.G.Tracey 1981b. The rainforests of Australia. In R.H.Groves (ed.), *Australian Vegetation*: 67-101. Cambridge: Cambridge University Press.

Webb, L.J., J.G.Tracey & L.W.Jessup 1986. Recent evidence for autochthony of Australian tropical and subtropical rainforest floristic elements. *Telopea* 2: 575-589.

Webb, L.J., J.G.Tracey & W.T.Williams 1976. The value of structural features in tropical forest typology. *Aust. J. Ecol.* 1: 3-28.

Webb, L.J., J.G.Tracey & W.T.Williams 1984. A floristic framework of Australian rainforests. *Aust. J. Ecol.* 9: 169-198.

Wells, A.G. 1982. Mangrove vegetation of northern Australia. In B.F.Clough (ed.), *Mangrove ecosystems in Australia*: 57-78. Canberra: Australian National University Press.

Wells, A.G. 1983. Distribution of mangrove species in Australia. In H.J.Teas (ed.), *Tasks for vegetation science* 8: 57-76. The Hague: Junk.

Whitmore, T.C. 1975. *Tropical rainforests of the Far East.* Oxford: Clarendon Press.

APPENDIX 1: FIELD KEY TO STRUCTURAL TYPES OF RAINFOREST AND MONSOON FOREST VEGETATION IN THE NORTHERN TERRITORY (FROM WEBB 1978)

1 Mesophylls* and notophylls most common
  2 Robust lianes, vascular epiphytes, plank buttresses, macrophylls and compound mesophylls prominent; trunk spaces generally obscured by hemi-epiphytic aroids and climbing palms; stem diameters irregular, many av. 60-80 cm; canopy level av. 20-30 m
    3 Deciduous emergent and top canopy trees rare or occasional
      4 Palm trees not prominent in canopy
        Complex mesophyll vine forest (CMVF)
        5 Feather palm trees prominent in canopy
          CMVF, + feather palms
    3 Deciduous emergent and top canopy trees conspicuous
      4 Mostly mesophylls . . . . Semi-deciduous mesophyll vine forest (SDMVF)
      4 Mostly notophylls . . . . . Semi-deciduous notophyll vine forest (SDNVF)
  2 Robust lianes and vascular epiphytes not conspicuous in upper tree layers which are simplified; spur rather than plank buttresses prominent; trunk spaces open, stem diameters (except for evergreen emergents) generally regular, av. 60 cm; canopy level av. 15-25 m
  Sclerophylls (e.g. *Acacia*) may be scattered in canopy
    3 Deciduous emergent and top canopy trees rare or absent
      4 Mostly mesophylls. Palm trees not prominent in canopy Mesophyll vine forest (MVF)
        5 Fan palm trees prominent in canopy
          MVF, + fan-palms

*For leaves of roughly lanceolate or elliptical shape the leaf-size class can be determined by length of leaf blades: nanophyll (<2.5 cm), microphyll (2.5-7.5 cm), notophyll (7.5-12.5 cm), mesophyll (>12.5 cm).

1 Notophylls and microphylls most common
  2 Robust and slender woody lianes, vascular epiphytes, plank buttresses, and compound entire leaves prominent; trunk spaces generally obscured by epiphytes; stem diameters irregular, many av. 60-80 cm
    3 Canopy level uneven, av. 20-30 m, occasional deciduous species; emergents if present mostly evergreen and umbrageous . . . . . Complex notophyll vine forest (CNVF)
      4 Patches of feather palm trees prominent in canopy . . . . . . CNVF + feather palms
  2 Robust and slender lianes generally present, wiry lianes (climbing ferns) generally conspicuous in understorey; vascular epiphytes and plank buttresses inconspicuous; feather palms generally conspicuous; tree crowns evergreen; canopy level av. 20-25 m . . Evergreen notophyll vine forest (ENVF) ± feather palms
  2 Robust, slender wiry lianes generally inconspicuous; fleshy vascular epiphytes may be prominent on trunks; plank buttresses inconspicuous; simple entire leaves prominent; deciduous species generally absent but many tree crowns become sparse during the dry season, i.e. semi-evergreen; typically mixed with sclerophyllous emergents and co-dominants
    3 Canopy level av. 10-20 m . . Simple evergreen notophyll vine forest (SSENVF)
    3 Canopy level av. 3-9 m, generally even, and canopy trees often branched low down (shrub-like) . . . . Simple semi-evergreen notophyll vine thicket (SSENVT)

1 Microphylls most common
  2 Mossy and vascular epiphytes inconspicuous in top tree layers; robust lianes generally prominent; plank buttresses absent; prickly and thorny species frequent in usually dense shrub understorey; ground layer sparse; compound leaves and entire leaf margins common
    3 Canopy level uneven, av. 9-15 m with mostly evergreen emergent and upper tree layer species; deciduous emergents rare or absent Low microphyll vine forest (LMVF)
    3 Canopy level uneven and discontinuous, av. 4-9 m; practically all emergents are deciduous, and many understorey species are deciduous or semi-evergreen; swollen stems ('Bottle Trees' and other species) may be common
      Deciduous microphyll vine thicket (DVT)

Plate 4. The Red Water, or Lotus, Lily *Nelumbo nucifera* (Photograph: G.J.W.Webb).

Plate 5. Mixed paperbark *Melaleuca* and monsoon forests, with drier elements, beside East Alligator River during floods, February 1984. The sandstone escarpment of Arnhem Land can be seen in the background (Photograph: I. Morris).

Plate 6. Floating vegetation mats along a water body unaffected by buffaloes at Bullcoin Finniss River, February 1983 (Photograph: G.J.W.Webb).

Plate 7. Tall open forest, with the palm *Livingstonia humilis*, August 1974 (Photograph: M.G. Ridpath).

Plate 8. Burning in the late dry season at Meeaguinea in north central Arnhem Land, October 1979 (Photograph: G.J.W.Webb).

CHAPTER 5

Use and impact of fire

C.D.HAYNES

INTRODUCTION

In Chapter 4 Dunlop & Webb discussed the role of fire as a determinant of vegetation, with particular reference to plant morphology and adaptive traits. In this chapter I wish to outline in more detail the evolution of the occurrence of fire in the monsoonal lowlands and to pay particular attention to the position taken by humans in its use. We shall see that before the introduction of modern fire suppression equipment it was possible for people to use their knowledge of fire behaviour to manipulate the spread and intensity of fire and hence to influence both immediate and longer term effects on the biota.

Throughout the world it has long been recognised that there is a conundrum in the interaction of fire and vegetation: each can be a determinant of the other. The question is often posed as to which has the more powerful effect. Another quite valid question is whether there is some kind of natural equilibrium, or whether current practices, whatever they might be, produce a disequilibrium. Such a disequilibrium may lead to impoverishment of the flora and/or fauna. These questions will also be examined here.

In the most fundamental sense, spread of bushfires is determined by the factors influencing any combustion: fuel, heat and oxygen. In the case of bushfires the parameters which have the most important bearing on fire behaviour can be considered in two broad groups: weather conditions and the nature and quantity of fuel. Weather conditions determine availability of oxygen and direct warmth, and can be related quite directly to the index of severity of a bushfire, which is called fire intensity. Intensity is directly related to such parameters as air temperature, wind speed and more indirectly related to relative humidity and atmospheric stability. These factors also influence the moisture content of fuel.

Within the seasonally wet tropics, many conditions for the spread of fire are quite different from those occurring elsewhere in Australia.

1. Much of the fuel is sufficiently cured to carry fire for eight months (April-November) in most years. Where no fire occurred in the previous dry season it is possible for fire to be sustained during the dry spells of the wet season. This will be considered in greater detail below, but it should be noted at this point that a fire season lasting 7-8 months is unusual at higher latitudes.

2. Temperature throughout the fire season is consistently high. Maximum temperatures are rarely less than 25°C and frequently 34°C or more. Whilst dry season evening temperatures are sufficiently low for condensation of water vapour to cause fires to extinguish themselves in most situations, it should be noted that they rarely go below 10°C. This temperature is adequate for spread of fire where fuel has been allowed to accumulate, but that is a rare situation.

3. Compared with other regions there is very little relief through precipitation during the 'fire season'. Some rains might fall after the first fires of the season, and some fires still occur after the first rains of the wet season. Very rarely rain falls during the mid-dry season months, and such falls moisten fuel for a few days at the most.

4. Much available fuel is herbaceous even within forests. In some places annual grass growth can contribute 1000 kg/ha per year of well aerated fuel (Rowell & Cheney 1979). This is common in other tropical savannas such as Africa (Komarek 1972) but rare in temperate forests.

These conditions indicate why fire is so common in the tropical north. Visitors are surprised to see so many fires, with little effort directed towards suppression. Elsewhere in Australia attempts are made to extinguish all fires but those lit under controlled conditions. In some places in the region, policies of suppression have been instituted with limited success (Haynes 1978), but efforts are now being directed towards much stricter control in some areas such as National Parks and managed forests (Commonwealth of Australia 1980, Fox 1976).

A second striking feature of the north is that it is one of very few remaining parts of Australia in which Aboriginal people still remember the technology of using fire, the fire rituals, the totemic and religious significance associated with fire and the language of the fire ecology of their culture. In many places Aboriginal people still use fire as an integrated part of their lives, carefully, deliberately and pragmatically (Plate 8). Often contemporary pragmatism has seen the replacement of the hunter's fire stick by matches flicked from the back of a utility. This has in some ways obscured the tradition of detailed knowledge which has only recently started to become clear. Some of this knowledge will be described by reference to a case study in this chapter, but first some aspects of the long fire history of the region should be examined.

FROM NO MAN'S LAND TO FIRE STICK FARMING

As indicated earlier fire and vegetation can be determinants of each other. This has led to considerable debate, particularly over the role that humans may have played in determining vegetation type through their use of fire. Monsoon forests, for example, can be replaced by eucalypt dominated open forest where certain fire regimes prevail (Stocker & Mott 1981, Russell-Smith 1985, Haynes 1985). It is a relatively straightforward conclusion that humans, on their arrival in north Australia, were responsible for the decline of rainforests and monsoon forests (Stocker & Mott 1981). Whilst such a conclusion may appeal to many students in this area, it presents several problems. The most obvious problem is that monsoon forest itself does not ignite easily, nor does it carry intense fire without prior disturbance. Further, disturbance must lead to invasion by xeric vegetation which carries an intense fire (Stocker & Mott 1981).

To look more carefully at this question it is necessary to outline the history of the vegetation over a period which extends well beyond human occupancy of Australia. Since the Middle Eocene epoch the lands which are now part of the wet-dry tropics have undergone a geographic shift which must not be underestimated. Kemp (1981) traces movement from about 25°S in the Middle Eocene (45 million years BP), through 16°S in the Middle Miocene (15 million years BP) to the present day latitude of 12-13°S. The shift in the continent was accompanied by a general increase in temperature and aridity, and coincided with the evolution of many taxa which comprise the xeric vegetation of the wet-dry tropics, e.g. Myrtaceae, Mimosaceae, Poaceae.

The evolution of these taxa, established well before the Pleistocene (Kemp 1981), produced the ingredient so necessary for the spread of fire: fuel which does not break down quickly under moist conditions, and which often contains fire promoting essential oils. The combination of such fuels and the more arid climate

contributed to periodic fire occurrence, evidence of which is contained in burnt fossil coal deposits (Kemp 1981).

Until the arrival of humans in the region, ignition was caused by lightning. Causes which have been postulated for other regions (e.g. exposed coal seams, volcanoes) could not have been significant here. Lightning fires were probably very extensive but would have been limited by factors such as whether storms produced enough rain to put fires out, and as indicated earlier, the probability of fires entering an area of closed forest. The probability of ignition by lightning increases with fuel quantity. It is possible to see this demonstrated in those parts of Australia where fuel is allowed to accumulate through the absence of regular burning: lightning is the most common cause of fires in many regions (Luke & McArthur 1978). It has been an almost unknown source in the contemporary Australian tropics (Stocker & Mott 1981), but only because fires which are ignited by humans have already occurred. In recent years many observations have been made in forestry project areas (at Murgenella and Maningrida in Arnhem Land) and in Kakadu National Park, of lightning ignition in fuels which have been allowed to accumulate for as few as 2-3 years.

The consistent and often unrelenting pressure on the mesic vegetation caused by gradual warming and drying of the continent through the Eocene and Miocene epochs had diminished by the start of the Pleistocene, some 2 million years ago. The Pleistocene showed great climatic extremes which in fact allowed periods of expansion of mesic vegetation. The expression of these extremes is still unclear but palynological data from the Atherton Tableland (Kershaw 1975, Singh et al. 1981) indicate wide variation in vegetation during the Late Pleistocene and Holocene epochs (125,000 years BP to present). The pollen cores also show wide variation in charcoal which is taken to be an indication of the presence of fire.

At some stage during the Late Pleistocene humans started to make their presence felt. It has been argued by Jones (1968) and other writers (e.g. Hallam 1975) that, at the time of their arrival in Australia, humans had already at their disposal the technology of the firestick. If they did not, the environment dictated that they learnt the technology rapidly or lose their living space beyond the closed forest, coasts and rivers. In fact humans had colonized the principal ecological zones of Australia by at least 20,000 years BP (Jones 1973). In the north, White (1964) has published evidence of occupation by 23,000 years BP. Humans have therefore been present in the region throughout the terminal Pleistocene and Holocene changes in climate and resultant changes in vegetation. All evidence points to the fires of this period being predominantly, if not absolutely, within the control of humans.

FIRE MADE BY HUMANS

The determination of how Aboriginal people used fire throughout Australia has been widely debated among specialists for at least two decades. Historical records from the time of first contact (Hallam 1975, Nicholson 1981) establish that fire was widely used in the landscape. It is, however, difficult to ascertain from those early writers many of those things about fire which we now wish to know: e.g. season of burning, frequency of occurrence, pattern of ignition. Indeed, as will be discussed below, it is often difficult to pose the questions that we wish to have answered.

Some journalists of contact times have provided richer information than others. For example colonists and explorers like George Fletcher Moore and Sir George Grey in south-western Australia were able to learn much of fire and its effects through the vocabulary of the resident Aboriginal people (Hallam 1975). As Hallam points out this vocabulary is as useful to prehistoric research as much of what can be dug from the sand. Moreover it illustrates discernment, discrimination and order in the culture of Aboriginal burning: it provides us with clues for investigating further the process of Rhys Jones' now well-known phrase 'fire-stick farming' (Jones 1969).

As indicated previously, a striking feature of northern Australia is that its Aboriginal people still remember, and often still practice, the complex technology of fire management as it was practiced at the time of contact with Europeans. A great deal of such technology has been difficult to obtain and interpret in western terms, and it seems likely that much of the technology is still to be discovered by Europeans. Several reasons for failure to understand it are worth mentioning here.

1. Contact has frequently involved European style economic activity which comes into conflict with Aboriginal land use. Fires have been used by Aborigines in retaliation both in this region (Haynes 1977, 1978) and elsewhere in Australia (Hallam 1975, Nicholson 1981).

2. Contact has almost always caused redistribution of resident populations away from living areas towards European-built town areas. The process of movement is quite complex (Haynes 1978, Altman 1982) but causes immediate changes in fire use because the people are no longer at the right place at the right time to continue customary burning.

3. Resettlement caused substantial economic changes. There was no longer the need to use the land for complete sustenance. Much, and in some cases, all of the Aboriginal diet came from external sources. Correspondingly there was less need to fire the country for food gathering.

4. Substantial cultural and linguistic barriers produce a variety of problems, not the least of which is the formulation of ecological questions which are meaningful to Aboriginal informants. Because of specialisation in disciplines among European investigators, ecologists frequently have lacked knowledge that anthropologists take for granted, whereas anthropologists frequently miss a wealth of ecological data that could be readily obtained (Stocker & Mott 1981). Webb (1977) has outlined an approach to this problem by interdisciplinary research teams working together in the field.

5. Successful enquiry rests on the assumption that informants are willing to divulge information. There are many good reasons for informants to be unready to do so. For example they want to be satisfied that the enquiry is not to be used against their interests, particularly where religious/ceremonial information may fall into the hands of uninitiated people.

A CASE STUDY OF ABORIGINAL FIRE USE

The study emerged from my own enquiries arising from fires being lit within a forestry project near Maningrida in Arnhem Land (Haynes 1985). Figure 29 shows the general location and the approximate core locations for the languages spoken in the area. The area was remote from centres of European settlement, e.g. towns and former missions, until Maningrida was established as a permanent centre in 1957 (Hiatt 1965). Many outstations, that is centres comprising only Aboriginal people, were established in the area in the early 1970s on former traditional camping places. My main informants had left their traditional domains for little more than a decade, and even during this time had let no year go by without returning at some time during the dry season.

At the time I was concerned with seeing how a small forestry project might be undertaken based on the stands of cypress pine (*Callitris intratropica*) scattered through the area (Haynes 1978), and did not realise what an unusually good opportunity would be presented to work with people who had maintained such close contact with their traditional country and associated use of fire. Eventually it was possible to look in detail at the burning of an area of about 9000 ha in the Upper Cadell Valley through one complete fire season (1976).

Several years previously (1971-74) I had been concerned about Aboriginal people lighting fires in the Government forest project areas, especially those near the newly created outstations, and I started to contact

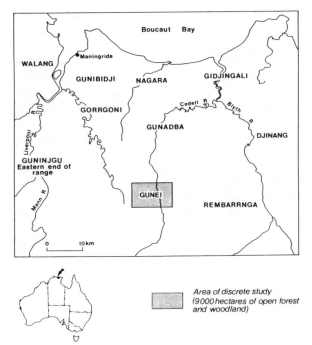

Figure 29. An area of approximately 5700 km² in the Maningrida locality of Arnhem Land showing languages spoken (language names are indicated at approximate core locations).

the residents to see how they might cooperate in maintaining *their* forestry project which was essentially based on the exclusion of fire to enable cypress pine regeneration to develop. The people were offended at the intrusion of the project on to their traditional lands (Haynes 1977) but to their great credit they were prepared to discuss the forestry project, and give polite answers to my question 'why do you use fire?'.

The answers to this question were relatively unproductive. Answers were similar to those postulated by Jones (1969), i.e. for fun, signalling, to clear the ground, hunting, regeneration of plant food, and extending human habitat. An unexpected emphasis was given to 'cleaning the ground'. Many people said things like 'We must clean this country up because that is a very important custom for us'. There was a sense of urgency and compulsion which anthropologists who had frequently seen and recorded use of fire in Arnhem Land (Thomson 1949, Warner 1937, Berndt & Berndt 1970) had not reported. Quite possibly this was unrecorded not only because the anthropologists were not in a position to pursue ecological enquiry (indeed Thomson made particularly good ecological records), but because they did not represent challenge to established order as I did with a project which sought to exclude fire absolutely.

The line of enquiry was not otherwise productive. The question, after all, implied that in my opinion fires should not be lit and that brought immediate conflict with people who unanimously thought they should. As a result I was forced to take another avenue which was: 'You say that it is your custom to have bushfires; but I know that some things get burnt up in bushfires too. There must be some special way of burning off. How do you do it?' The answer to that from the Guneispeaking people of the Upper Cadell Valley was 'At the right time we will show you the way we do it in our custom'. And so I learnt primarily by recording what they did and noting their commentary. I was also assisted by information from people of other language groups. This was necessary because each group has different resources and adopted burning strategies to suit. These differences will be discussed below.

Climate and seasonal perception

The study area lies in one of the drier parts of the tropical north, receiving average precipitation of less than 1200 mm with most rain falling between December and March. There is considerable variation between years (see Chapter 2), both in total rainfall and time of occurrence. Recognition of this among Aboriginal people of this region is widespread (Thomson 1949, Jones 1980, Stevenson 1985, Morris 1978, Alderson et al. 1980).

Morris (1978) devised an excellent presentation for north-east Arnhem Land showing how each identified season was distinguished from another by changes in weather and behaviour of the biota: this was adapted by Alderson et al. (1980) for use in Kakadu National Park. Figure 30 shows the seasonal divisions made by the Gunei people and demonstrates how their six seasons in the year can be equated to changes in some climatic parameters. In this case daily records of temperature range, 9 am and 3 pm relative humidity, and rainfall are traced through a sixteen month period. Within the study area (Fig. 29) there are some significant differences in the way the seasons are perceived. For example the Gidjingali people from the lower Blyth River area determine only three major seasons: djambirr, the rain time, followed by midawarr, the time between the rain and the beginning of the hot dry weather, and barparanga, the hot period from about August until the succeeding wet season (Jones 1980). Both Jones' informants and my own were resolutely defensive about their seasonal divisions: they each know of the others' concepts and insisted that their own were more valid. The differences in perception can be at least partly explained by differences in the resources available to each group and, significantly, in the way that fire is managed.

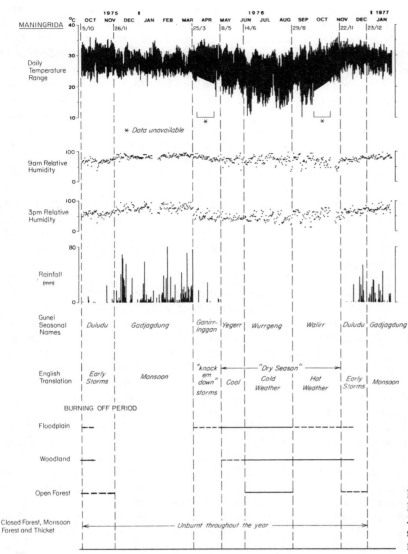

Figure 30. Some parameters of climate at Maningrida between October 1975 and January 1977 and corresponding Gunei season names and times for burning in four biomes.

Physical landscape, population, and resources

The area included within Figure 29 is about 5,700 km² in extent. Apart from a small section of escarpment in the south it is otherwise similar to much of the wet-dry tropics: most is vegetated by woodland and, less frequently, open forest which is developed on deeper soils. Towards the coast, the large rivers are surrounded by thousands of hectares of open floodplain and associated fringe communities including closed forest and thicket (Jones 1980).

The population at time of contact was distributed over the whole area, usually in bands of 15-30 people except where ceremonies were being conducted. Each band (see Peterson 1974), much smaller than a language group, moved over a limited range according to

season and utilised the resources of their range. Unless they were involved in a ceremony or trading mission near the floodplain, Gunei and Rembarrnga people had very little access to floodplain country and even less access to the sea. Conversely, the Gidjingali, Nagara and Gorrgoni people of the plains had relatively little access to the open forest. In all cases they had limited areas of open forest within their normal range, but to them it assumed much less importance by comparison with the language groups inland.

Lighting up the bush

Informants from all language groups agreed that no burning took place during the monsoon season. As pointed out earlier, it is possible to ignite fires during this period only where fuel has accumulated for several years, and all informants agreed that this was not allowed to happen except in areas of closed forest. The first fires of the season were lit on the floodplains in the season the Gunei call ganirringgan, the beginning of midawarr of the Gidjingali, before the final storms had run their course (see Chapter 12). At this time fires often burn only a few square metres before being extinguished as they run into adjacent green fuel.

By the time the last rains had fallen, the beginning of the season the Gunei call yegerr, burning on the floodplains was well under way and some burning in the woodland had started, with particular emphasis on the creation of stands of fresh grass to which antilopine wallaroo and agile wallaby (*Macropus antilopinus* and *M. agilis*) were attracted. At about this time burning around the edges of closed forests and thickets was conducted with the express purpose of excluding the subsequent fires of the season from such areas. The protective fires were small and went out on the verges of the closed forest. At about this time also the Gunei people burnt off very small areas around their camps, but that was all. As yegerr went by, the grass in the tall forest cured and could be burnt with ease. My Gunei informants resisted my suggestions that it was time to start: 'Those other people can burn off now, but it is still too soon for us: we must wait for the strong winds and dew of the proper cold weather, wurrgeng' (i.e. about June-August).

The beginning of wurrgeng is usually quite distinct, being marked by the passage of the first strong anti-cyclone of the southern Australian winter. This produces strong dry south-easterly winds and a drop in temperature. Such a distinct beginning occurred during the study in the Upper Cadell Valley (14 June) and my informants immediately took up fire sticks and matches: 'we will show you our old custom burning now'. During the ensuing ten week period I was able to record the areas burnt (Fig. 31). A most important feature of the fires was that they all went out during the evening of the day of ignition. As the season progressed, each fire became larger and extended further into the evening. Eventually the point was reached where a halt was called: 'it's too late now for this country' (the open forest): 'if we keep on lighting now we have lost (control of) the fires'. Thereafter the only deliberate lighting took place within designated hunting areas, in woodland, which had already been burnt around.

Another important feature of the fires of wurrgeng was their response to wind. Once burning began, my informants commented that burning earlier, i.e. in yegerr (about April/May) makes the fires slow-moving and hot, often scorching the complete canopy of the forest. The flames of the wurrgeng fires, on the other hand, bent well forward and caused scorch of rarely more than 3 m and usually less than 1.5 m. The implications for protection of the biota are discussed below.

The end of the burning in the open forest marked the beginning of the hot dry season, walirr (usually late August-late October). During this season an escaped hunting fire from further east burnt into the woodland but was precluded from entering the open forest because of the previous burning. My informants commented: 'It's all right there. That is only desert (i.e. has no important food resources). We have protected what we use'. The fires of this season spread rapidly with complete scorch during the day, continuing more slowly through the evenings, and being extinguished only when they meet a natural barrier or previously burnt ground.

The end of walirr is marked by a rise in the humidity, the build-up of cloud, and the first storms. My informants used this time to visit neighbours to the south and demonstrated burning some areas of recently germinated annual grasses (*Sorghum* spp.) which carried the unburnt thatch from the previous season. They said it was a good way of 'keeping the grass down', and again 'cleaning up before the rain'.

The pattern of burning formed a mosaic and was used by all groups (see Chapter 12). That shown in Figure 31 was produced partly in open forest and partly in woodland. The woodland mosaic was not as 'fine grained' as that for the open forest. Woodland fuel cures earlier in the year, and fuel temperature and wind speed are greater: this allows greater rate of spread and individual fires generally burn much larger areas. The floodplain mosaic was comparable in most aspects to that created in the open forest because burning started so early in the year and fires encountered frequent patches which were insufficiently cured to allow combustion. These were burnt later in the year. One important difference between floodplain and open forest was the amount of area burnt. Table 8 shows the area burnt after each two weeks during the Gunei burning programme. At the end of the season nearly 40% of the area was left unburnt. The Gunei people were adamant that adequate area had been burnt: 'those patches will be good for burning next year'. By contrast no unburnt areas of floodplains could be seen by the time of the first rains. Because of regrowth after the first fires, some areas had even been burnt twice (Jones 1980).

Ecological perception of fire effects

The literature on Aboriginal use of fire has been limited almost entirely to the way in which it is used exploitatively. The 'fire-drives' in which groups of

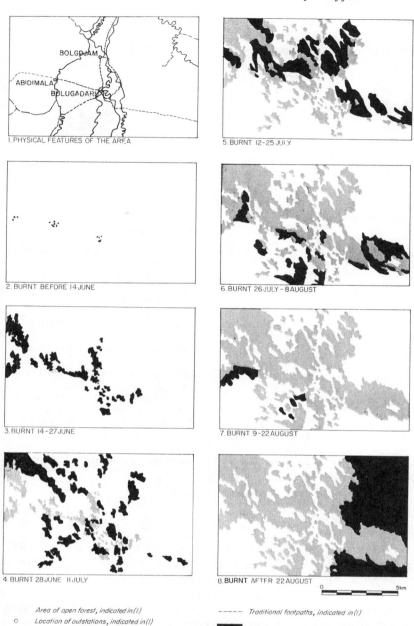

1. PHYSICAL FEATURES OF THE AREA

2. BURNT BEFORE 14 JUNE

3. BURNT 14–27 JUNE

4. BURNT 28 JUNE – 11 JULY

5. BURNT 12–25 JULY

6. BURNT 26 JULY – 8 AUGUST

7. BURNT 9–22 AUGUST

8. BURNT AFTER 22 AUGUST

0   5km

Area of open forest, indicated in (1)

○   Location of outstations, indicated in (1)

〜   Watercourses, including the main channels
      of the Cadell River, indicated in (1)

- - - - -   Traditional footpaths, indicated in (1)

▮   Area burnt in current period, indicated in (2–8)

░   Area burnt in previous periods, indicated in (3–8)

Figure 31. Sequence of burning in 9,000 hectares of open forest and woodland during 1976.

Aborigines lit fires to drive large macropods into an ambush appear to be referred to more often than other fire uses (Cleland 1957, Gould 1970, Hallam 1975). The technique of using fire to flush out animals is still used in Arnhem Land (Altman 1982). All informants spoke of being able to set a fire and stalk the macropods, gonobolo (agile wallaby) and gandagij (antilopine wallaroo) on their own.

During the intensive recording in the Upper Cadell Valley in 1976 I was able to witness hunting of yok (bandicoot, *Isoodon macrourus*) and galawan (sand goanna, *Varanus gouldii*) both of which were deliberate and specific. In neither case did there appear to be any complex plan as the quarry and its behaviour was predictable. It was simply necessary to set the fires and chase the small animals as they ran away.

In some cases fires are set in the knowledge that they will overcome a proportion of the small animals. Jones

Table 8. Areas (hectares) burnt under Aboriginal management of a 9059 ha block with 5562 ha of open woodland and 3497 ha of open forest.

Period	Open woodland	Open forest	Total
1 Before 14 June	5 (0.08%)	24 (0.69%)	29 (0.32%)
2 14-27 June	312 (5.61%)	274 (7.84%)	586 (6.47%)
3 28 June-11 July	539 (9.69%)	581 (16.60%)	1120 (12.36%)
4 12-25 July	766 (13.76%)	671 (19.19%)	1437 (15.86%)
5 26 July-8 August	440 (7.90%)	331 (9.47%)	771 (8.51%)
6 9-22 August	137 (2.46%)	95 (2.72%)	232 (2.56%)
7 After 22 August	2230 (40.11%)	28 (0.80%)	2258 (24.93%)
Unburnt	1133 (20.39%)	1493 (42.69%)	2626 (28.99%)

Table 9. Protection of economic plant resources by manipulation of fire.

Species	Part of plant	Ecosystem	Means of protection
Dioscorea spp. (yams)	Tubers (location indicated by dry stems)	Closed forest margins	Burning around margins of closed forest very early in year
Carpentaria acuminata (palm)	Terminal bud	Closed forest	Exclusion of fire (protection of whole plant)
Gronophyllum ramsayi (palm)	Terminal bud	Open forest	Burning in early cold weather (protection of plant)
Callitris intratropica (cypress pine)	Rosin	Open forest	Burning in early cold weather (protection of whole plant)
Terminalia carpentariae (almond tree)	Seed	Open forest, closed forest margins	Avoid burning until after fruit is ripe
Buchanania obovata (plum tree)	Fruit	Open forest	Burning in early cold weather to avoid scorch of developing flower buds
Persoonia falcata (plum tree)	Fruit	Open forest	Burning in early cold weather to avoid scorch of developing flower buds
Syzigium suborbicularis (red apple tree)	Fruit	Wetter areas of open forest up to closed forest margins	Burning in early cold weather to avoid scorch of developing flower buds
Vitex glabrata (plum tree)	Fruit	Closed forest margins	Burning around margins very early in year

(1980) saw a large number of dusky rats (*Rattus colletti*) die in the fast moving fires which bore down on them, and then be collected, cooked and eaten shortly afterwards.

In all situations described above, success in the hunt required not only skill with the spear, or sharp reactions and fleetness of foot, but also knowledge of where the animals might be, their likely reaction to fire, and the behaviour of fire under the prevailing conditions. Fire always assisted the hunter, and in some situations fire was a necessary part of exploitation. The same can be said of the creation of small patches of regrowth early in the dry season, described above. The removal of dry stems and leaves stimulates fresh growth on perennial grasses and shrubs, the former having particular attraction to macropods which the Aborigines will hunt.

The creation of such new growth is one form of manipulation of fire, followed by response of vegetation, which has received attention in the literature (e.g.

Cleland 1957, Jones 1969, Stocker & Mott 1981). During the course of the case study I became aware of many other examples of manipulation of fire to achieve a particular response of vegetation, some of which are shown in Table 9. Fire management was aimed at protecting the plants for later exploitation. The fruit trees which grow in the open forest or on the closed forest margins invoked particular comments on flames, and scorch height. One informant told me 'We must burn the country for these trees in the wurrgeng time. If we burn too early, in yegerr like on the gapalg (floodplain) the flames go up and kill the tucker (flower buds in this case). If we burn too late, the fire does the same. It burns up everything'.

The species which had prompted my original question, cypress pine, attracted considerable voluntary comment. When a patch of cypress, killed by some late season hot fire of several years earlier, was encountered, many informants expressed their disapproval of the owners of that country who were accused of 'walking out' and not burning at the 'right' time. This had happened because people who were living at Maningrida for a decade or more were returning at irregular intervals, often in the hot dry weather, and lighting fires at the 'wrong' time. The effects on the more sensitive species such as cypress were often severe.

A fire culture

The recognition that cypress pine was being killed by late season fires created a problem for some informants. What else could they have done as a result of returning late in the dry season? They still had to 'look after' their country, and by 'look after' they meant burning it. This was not burning for an expressed ecological reason, but rather for the need to 'clean the country' which appeared to take precedence over the destruction it caused. What was the need to treat mandalgurug, the state of country carrying thick, 'dirty' grass, or mulurr, where the leaf litter had grown thick? The rationale could be, and would eventually have to be, that each year made the inevitable fire worse. Part of the fire culture of all groups was simple protection of oneself and one's resources from the fires of other people, as the Gunei people had done in the Upper Cadell Valley.

It is possible to look at the part of fire in cultural life in a number of ways. Most white Australians view bushfire as a nuisance, or something to be feared, regardless of intensity. At even the most superficial level, my informants had a radically different view. In driving or walking through freshly burnt country, with some logs still smoking, my companions frequently commented on how good and again, 'clean', the country looked, whilst to me it appeared bad: black, dry and lifeless.

A clue to understanding the significance of cleanliness, in that sense, came from the ritual approach to a dangerous totemic site, djalag (Haynes 1977). The site had to be approached by particular individuals (and is forbidden to all others) who sing its totemic song, burning the grass as they continuously approach the site from the north-west. This can only be done while the south-east winds of the cold weather are blowing. Smoke blows back over them as they approach. 'The smoke keeps us clean, and our country too. If everything is clean, the dreaming will be quiet'.

There are still many parts of Aboriginal fire culture to be understood by Europeans, and since parts of it are contained within the religious ceremonies, it is probable that some parts will never be told. Most anthropologists have concentrated their attention on fire as an abstract matter (e.g. the fire myth, Maddock 1970; fire totemism, Warner 1937), its use in ceremonies (Warner 1937, Hiatt 1966) and in brief descriptions of

its pragmatic use, discussed above. For the present we are assured by Aboriginal people that the country must be cleaned, regularly, and if possible, in the 'old custom way which came from the time of the first men'.

FROM ABORIGINAL/EUROPEAN CONTACT TO PRESENT DAY

First contact between Aborigines and the European settlers has been complex and diffuse. By the time many Aboriginal people of this region had seen a white person for the first time they had already heard about them from their compatriots.

The settlement of Darwin, subsequent gold mining at Pine Creek, and the development of the railway line south of Darwin in the late 1880s provided a focus for Aboriginal curiosity. Many Aborigines looked and then went back to their homelands, but many never returned and depopulation of many parts of the region occurred. If the example presented to me by the Gunei people, among the last to be affected by European settlement, is true, depopulation would have had a significant effect on the pre-contact fire regimes. The production of the mosaic shown in Figure 31 required extensive walking within the relatively few weeks of the cold weather. Depopulation could then be critical, especially as the fit and active people who lit the fires were the most common absentees. Progressively more fires would have been later in the year and injured the more sensitive species, the most obvious of which is cypress pine.

This is indeed what happened. The demise of cypress has been assisted by systematic logging of many areas (Bateman 1955). Logging not only removed the best trees, but also left fallen tree crowns and opened up the stands, making way for intense fires. Nevertheless, many stands which were never logged now show all the signs of decline. A healthy stand is surrounded by regeneration, the mature trees having no or few fire scars, with all trees alive. The model of the regime presented by the Gunei people, described above, actually maintained such stands. By contrast almost all stands in Kakadu National Park, many of which were remote from logging, appear to have been declining for several decades. In some cases, the only evidence of a stand is a few charred and very dead stumps, burnt up in a succession of wide ranging, uncontrolled, late season fires which have been the common pattern since the Aboriginal population dispersed and declined, and de facto European usage began.

The account of change, recorded above, represents the erosion of the pragmatic fire policy that existed under stable Aboriginal occupation of the land. Over much of the land, depopulation has had a significant effect; but in many notable areas within the region, land has been put to use by the newcomers, who had fire policies of their own. Most of the European settlers, small in numbers as they have been until relatively recently, recognised that they could use fire to their own ends. In almost all cases they had to rely on an Aboriginal workforce and often allow the Aborigines to burn at their own discretion. In other cases they directed their employees to burn early firebreaks for protection of pastures through the dry season (H.T. Lewis, pers. comm.).

In yet other cases, perhaps the majority, Europeans failed to recognise the importance of timing and placement, and copied what they perceived to be a 'scorched earth' policy. As a result, there are signifiicant areas, particularly around closely settled places such as Darwin, which are burnt relentlessly year after year.

Some European-initiated policy has been based on the other extreme of fire exclusion where possible (e.g. the forestry projects discussed earlier). This form of management produces a build-up of dense fuel and an inevitably increased intensity when fire occurs. As a result it is essential to have well organised vigilance and effective fire suppression forces. These are expensive and so such a policy can be implemented only where economic benefits justify the expense.

During the period of closer settlement, which has intensified only within recent decades, more specialised land use has developed. With this has developed a variety of fire policies, and it seems likely that further variations of policy and practice will arise with further differentiation.

THE FUTURE: RESEARCH AND POLICY

It is always difficult to predict the future of land use in a developing region but it appears likely that fire policy will continue to be directed towards two major purposes: rural production and conservation. Research directed towards more fruitful fire policies in either area is likely to benefit the other. To date most research has been directed towards the former; but the recent recognition of the value of conservation and the benefits it imparts to the development of tourism should promote interest in fire research in national parks and other nature conservation areas (Fox 1976, Bowman 1987, Bowman et al. 1988).

Interest in the effects of fire regimes on animals is growing, and it has been postulated that some endangered species have declined because of the adverse effects of fire. Forshaw (1981) has discussed this with respect to the hooded parrot which occurs in the region: the birds' diet consists of seeds of grasses which may have been severely affected by the changes in fire regimes discussed earlier. Many other species, or

groups of species, of sensitive animals and plants warrant further examination. The situation of cypress pine and some other trees is now reasonably well understood, but the circumstances of some grass finches, and rare macropods (e.g. the spectacled hare wallaby and the northern nail-tailed wallaby) and of the grasses on which these animals are thought to depend, are obscure. The status of these groups warrants special examination and investigation into the effects of fire on these groups should follow. The significance of fire in determining the status of species which are now abundant also needs investigation since any alteration in fire regimes has effects on these species, which we cannot predict from present knowledge.

Meanwhile it appears that the basis for a conservative strategy of fire use lies in the approach used by the Aboriginal people of this region (Press 1987). Modern methods, which incorporate the use of aircraft, have already been used in conjunction with traditional Aboriginal knowledge to produce a variety of fire regimes. In conclusion it should be remembered that the biota of this region has survived throughout millennia of Aboriginal fire use and is to some extent a product of it. A suitable strategy for future policy could arise from understanding its mechanisms and where necessary adapting Aboriginal techniques to new and specialised requirements.

REFERENCES

Alderson, M., T. Gangali & C. Haynes 1980. Seasonal Calendar for the Kakadu Region in Gundjeidmi (Maiili) Language. *Kakadu National Park. Plan of Management*: 41. Canberra: Commonwealth of Australia. Australian National Parks and Wildlife Service.

Altman, J.C. 1982. Hunter-gatherers and the State: The Economic Anthropology of the Gunwinggu of North Australia. Unpublished Ph.D. Thesis. Canberra: Australian National University.

Bateman, W. 1955. Forestry in the Northern Territory. *Forestry and Timber Bureau Leaflet* 72: 11. Canberra: Commonwealth of Australia.

Berndt, R.M. & C.H. Berndt 1970. *Man, Land and Myth in North Australia: The Gunwinggu People*: 212-236. Sydney: Ure Smith.

Bowman, D.M.J.S. 1987. Stability amid turmoil?: Towards an ecology of north Australian Eucalypt forests. *Proc. Ecol. Soc. Aust.* 15: 149-158.

Bowman, D.M.J.S., B.A. Wilson & G.W. Davis 1988. Response of *Callitris intratropica* to fire protection, Murganella, northern Australia. *Aust. J. Ecol.* 13: 147-159.

Cleland, J.B. 1957. Our Natives and the Vegetation of Southern Australia. *Mankind* 5: 149-162.

Commonwealth of Australia 1980. *Kakadu National Park. Plan of Management*: 312-313. Canberra: Commonwealth of Australia. Australian National Parks and Wildlife Service.

Forshaw, J.M. 1981. *Australian Parrots*: 231-232 (second ed.). Melbourne: Lansdowne.

Fox, R.E. 1976. Report on the use of fire in National Parks and Reserves. Darwin: Commonwealth of Australia. Department of the Northern Territory.

Gould, R.A. 1970. Uses and Effects of Fire among the Western Desert Aborigines of Australia. *Mankind* 8: 14-24.

Hallam, S.J. 1975. *Fire and Hearth*. Canberra: Commonwealth of Australia. Australian Institute of Aboriginal Studies.

Haynes, C.D. 1977. Evidence. *Joint Select Committee on Aboriginal Land Rights in the Northern Territory*: 753-780. Canberra: Commonwealth of Australia. Hansard.

Haynes, C.D. 1978. Land, Trees and Man. *Commonwealth Forestry Review* 57: 99-106.

Haynes, C.D. 1985. The Pattern and Ecology of *Munwag*: Traditional Aboriginal Fire Regimes in North Central Arnhemland. In M.G.Ridpath & L.K.Corbett (eds.), *Ecology of the Wet-Dry Tropics, Proc. Ecol. Soc. Aust.* 13: 203-214.

Hiatt, L.R. 1965. *Kinship and Conflict*: 10-11. Canberra: Australian National University Press.

Hiatt, L.R. 1966. Mystery at Port Hacking. *Mankind* 6: 313-317.

Jones, R. 1968. The Geographical Background to the Arrival of Man in Australia and Tasmania. *Archeology and Physical Anthropology in Oceania* III: 186-215.

Jones, R. 1969. Fire-Stick Farming. *Australian Natural History* 16: 224-228.

Jones, R. 1973. Emerging Picture of Pleistocene Australians. *Nature* 246: 278-281.

Jones, R. 1980. Hunters in the Australian Coastal Savanna. In D.R.Harris (ed.), *Human Ecology in Savanna Environments*: 107-146. New York: Academic Press.

Kemp, E.M. 1981. Pre-Quaternary fire in Australia. In A.M.Gill, R.H.Groves & I.R.Noble (eds.), *Fire and the Australia Biota*: 3-21. Canberra: Australian Academy of Science.

Kershaw, A.P. 1975. Late Quaternary Vegetation and Climate in Northeastern Australia. In R.P.Suggate & M.M.Cresswell (eds.), *Quaternary Studies*: 181-187. Wellington: Royal Society of New Zealand.

Komarek, E.V. 1972. Lightning and Fire Ecology in Africa. In E.V.Komarek (ed.), *Fire in Africa: Proc. Tall Timbers Fire Ecology Conference* 11: 473-511.

Luke, R.H. & A.G.McArthur 1978. *Bushfires in Australia*. Canberra: Commonwealth of Australia. Australian Government Publishing Service.

Maddock, K.J. 1970. Myths of the acquisition of fire in Northern and Eastern Australia. In R.M.Berndt (ed.), *Australian Aboriginal Anthropology*: 174-199. Perth: University of Western Australia Press.

Morris, I. 1978. Seasonal Calendar of N.E. Arnhemland. Unpublished manuscript. Canberra: Australian National Parks and Wildlife Service.

Nicholson, P.H. 1981. Fire and the Australian Aborigine: An Enigma. In A.M.Gill, R.H.Groves & I.R.Noble (eds.), *Fire and the Australian Biota*: 55-76. Canberra: Australian Academy of Science.

Peterson, N. 1974. The importance of women in determining the composition of residential groups in Aboriginal Australia. *Woman's Role in Aboriginal Society*: 16-27. Canberra: Commonwealth of Australia. Australian Institute of Aboriginal Studies.

Press, A.J. 1987. Comparisons of the extent of fire in different land management systems in the Top End of the Northern Territory. *Proc. Ecol. Soc. Aust.* 15: 167-175.

Rowell, M.N. & N.P.Cheney 1979. Firebreak preparation in tropical areas by rolling and burning. *Australian Forestry* 42: 8-12.

Russell-Smith, J.J. 1985. A Record of Change: Studies of Holocene Vegetation History in the South Alligator Region, Northern Territory. In M.G.Ridpath & L.K.Corbett (eds.), *Ecology of the Wet-Dry Tropics. Proc. Ecol. Soc. Aust.*, 13: 191-202.

Singh, G., A.P.Kershaw & R.Clark 1981. Quaternary vegetation and fire history in Australia. In A.M.Gill, R.H.Groves & I.R.Noble (eds.), *Fire and the Australian Biota*: 23-54. Canberra: Australian Academy of Science.

Stevenson, P.M. 1985. Traditional Aboriginal resource management in the wet-dry tropics (Tiwi Case Study). In M.G.Ridpath & L.K.Corbett (eds.), *Ecology of the Wet-Dry Tropics. Proc. Ecol. Soc. Aust.*, 13: 309-315.

Stocker, C.G. & J.J.Mott 1981. Fire in the tropical forests and woodlands of northern Australia. In A.M.Gill, R.H.Groves & I.R.Noble (eds.), *Fire and the Australian Biota*: 425-439. Canberra: Australian Academy of Science.

Thomson, D.F. 1949. Arnhem Land: Explorations among an unknown people. *Geographical Journal* 113: 1-8.

Warner, W.L. 1937. *A Black Civilization*. New York: Harper.

Webb, L.J. 1977. Ethnobotany: The co-operative approach to research. *Australian Institute of Aboriginal Studies Newsletter* 7: 43-45.

White, C. 1964. Plateau and Plain. Unpublished Ph.D. Thesis. Canberra: Australian National University.

Part 3:
The vertebrate animals

Introduction

This part comprises six chapters which deal with the ecology of the freshwater fishes, the crocodiles, the remaining reptiles with the frogs, the birds, the native mammals, and the feral mammals. In spite of its great importance in tropical ecosystems (Ridpath 1985), the ecology of invertebrate animals is not treated here. That is because it is a vast but still very poorly explored realm in the monsoonal lowlands. It presents a major challenge to ecologists for we shall never understand tropical ecosystems properly till we can fill that gap better. This is illustrated by the large number of species of fishes, frogs, reptiles and birds which depend on invertebrates, especially insects, for food.

Since the Pliocene epoch the fauna of the region has been moulded by a series of pronounced climatic changes, in parallel with a succession of connections to and severances from New Guinea (see Chapters 1 and 13). The processes are explained by Morton in Chapter 9, and illustrated with reference to birds. Habitats dominated by lowland tropical rainforest (Irian), highland and once temperate rainforest (Tumbunan), a relict from much earlier and generally cooler times), and arid-adapted *Eucalyptus* forest/woodland (Torresian) have all contracted and expanded repeatedly in response to these environmental changes in northern Australia. The present mix of Irian and Tumbunan elements in a Torresian-dominated fauna is the consequence of this history. Fragmentation of these faunal elements by such changes has been a major cause of speciation.

Endemism of freshwater fishes (Chapter 6), and frogs and lizards (Chapter 7) is high in the monsoonal lowlands; though generally not quite as high as it is in the adjacent more rugged and slightly drier Kimberley region to the west. Whereas the freshwater fishes, frogs and lizards of Cape York show quite close relationships to those of New Guinea, only 200 km away across the Torres Straits, they are mainly less closely related to those of the drier monsoonal lowlands or the Kimberley. This suggests that in these classes the problems of adaptation to the wet-dry tropics are very different from those faced in the more humid tropics of northeast Queensland and New Guinea. In birds, which are a more mobile group of animals, the level of endemism is less marked. About 90% of the avifauna of the monsoonal lowlands is shared not only with the Kimberley but also with Cape York (Chapter 9). Nonetheless, as in the other three classes, the relationships of the avifauna of the monsoonal lowlands, and of the Kimberley, are much less close to the avifauna of New Guinea than are those of the avifauna of Cape York to New Guinea. This higher endemism among the vertebrates of northwest Australia, other than birds, springs not only from its being somewhat less humid but also from its frequent and long periods of isolation by both marine and arid barriers (Chapters 1 and 13).

The most recent change experienced by the vertebrate fauna of the monsoonal lowlands has come about as a result of the introduction of domesticated mammals after European colonization. Nine forms of these have since escaped and established feral populations (Chapter 11). The most successful have been those with tropical origins, although the absence of large competing native herbivores may also have played a part.

The dictates of tropical latitude impose upon its biota permanent warmth and a dramatic annual alternation between a very wet and a very dry season; the onset and duration of both being characteristically variable. The six chapters deal largely with the responses of vertebrate animals to these conditions, especially to marked seasonality. The permanent tropical warmth generally advantages reptiles and amphibia which, as ectotherms, must severely curtail their activity during winter at higher latitudes. For amphibia this advantage is offset by the aridity of the tropical winter but that is a much less serious drawback for reptiles, which withstand dry conditions well (Chapter 7). Even in the tropics the winter temperatures at night, though relatively warm, are sufficiently lower than in summer for some vertebrates to show special behavioural and physiological adaptations to cope with them. Examples are given in crocodiles (Chapter 8), native and feral mammals. For homeotherms such as mammals the greater problem appears to be to keep cool enough by day in the hot humid summer. Some species, such as the tiny native *Planigale maculata* or the large feral Brahman cattle, avoid overheating by having a low basal metabolic rate which keeps down their production of heat. However, in the tropics mammals as a class may be close to their effective upper limit as regards physiological regulation of body temperature. Thus various behavioural strategies for avoidance or loss of heat are common (Chapters 10 and 11).

The effects of seasonality dominate the ecology of vertebrate animals of the monsoonal lowlands. Most of this seasonal influence relates to rainfall in one way or another. The aquatic habitat available to freshwater fishes expands enormously in the wet season by flooding across the lowlying black-soil plains and swamps; at which time water quality becomes uniform throughout (Chapter 6). Fishes breed in watercourses early in the wet season, later on the flooded plains, then move back to the deeper pools in watercourses as the dry season progresses. As this happens these residual water bodies begin to show increasing differences between each other in water quality, which affects their characteristics as refuges in the dry season. In the wet season the distribution and feeding of crocodiles is affected, rather similarly to the fishes, by the filling and interconnection of freshwater swamps and billabongs and by the decreasing salinity in tidal rivers (Chapter 8). However, as reptiles, crocodiles can withstand the conditions of the dry season quite well, accommodating partly by moving and partly by a reduction in their metabolism. In fact the Freshwater Crocodile actually breeds in the dry season, even though it ceases to feed and grow then. As amphibia, frogs are well adapted to the conditions of the wet season, when they breed prolifically (Chapter 7). In the dry season there is a dramatic decline in their abundance and activity.

The use and/or mitigation of the effects of alternating wet and dry seasons also plays a preponderant part in the lives of the birds, native and feral mammals of the monsoonal lowlands (Chapters 9, 10 and 11). Mobility is a very conspicuous response among birds. Among feral mammals, the Buffaloes at the peak of the wet season graze the wetlands less and the higher forested lands more, when the former are flooded and the latter verdantly grassed. As the dry season arrives the forests dry out more quickly and so grazing increases on the wetlands and their low-lying forested margins. Eventually most grazing is concentrated in these margins to which most green herbage and free water become restricted. This example illustrates a widespread strategy among mammals and birds (and some reptiles and fishes), which depend on their mobility to track the best environmental conditions and resources. This often includes escape to local refuges from the tropical vicissitudes of floods or

droughts. This also occurs on a much larger, continental, scale among the highly mobile birds. Thus the monsoonal lowlands provide an important refuge area for migrants from summer drought in inland and southern Australia, and for winter migrants from northern Asia (Chapter 9).

Seasonality also strongly influences when birds and mammals breed. Resources such as fruits, nectar and insects peak in the middle of the wet season, as does the breeding of those birds dependent on them. On the other hand seed-eaters, waterbirds and the swamp-dwelling dusky rat generally breed late in the wet season when the vegetation on which they depend has completed its cycle of growth (Chapters 9 and 10).

Although the dominating alternation between wet and dry seasons is annual, its timing within the year is quite variable. Thus the transition from dry season to wet, in terms of plant growth, occurs any time between the beginning of October and the end of November or even later; and that from the wet season to dry any time between the beginning of April and the end of May, or even early June (Ridpath 1985). The responses of animals to such a high degree of climatic variability in the tropics, and thus of resources, are well illustrated by Williams and Newsome with respect to native mammals (Chapter 10). Their strategies include forms of opportunistic breeding normally regarded as more typical of the arid zone. This also helps to explain the dramatic oscillations of the population densities of certain species.

The diversity in the monsoonal lowlands of the five classes of vertebrates varies according to habitat. Structural diversity of habitats beneath the surface of the water is greatest in the deeper channels and billabongs of the flood-plains, where freshwater fishes are richest in species (Chapter 6). On land, the greatest structural diversity in the habitats of terrestrial vertebrates occurs in the open forests and woodlands, and these habitats are richest in species of reptiles, birds and mammals. However, the semi-aquatic frogs have most species in the wet, low-lying forested margin between the open forest and the seasonally flooded wetlands (Chapter 7). The seasonally flooded wetlands are structurally the simplest habitat and poorest in species of frogs, reptiles, birds and mammals. This is in spite of the fact that these fertile wetlands have a much higher productivity of vegetation and, for most of the year, support a much greater total biomass of terrestrial vertebrates than do the open forest and woodlands.

The monsoonal lowlands have not yet been so much affected by European settlement as southern and eastern Australia. Nonetheless the fauna still poses a number of conservation and management issues. Most concern introduced plants and animals or fire, but some concern pollution and human-induced barriers to

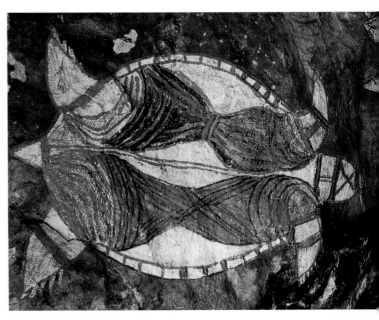

Plate 9a. The uncommon Pig-nosed Turtle *Carretochelys insculpta* which inhabits freshwater and estuaries (Photograph: G. Miles).

Plate 9b. Aboriginal rock painting of the Pig-nosed Turtle (Photograph: G. Miles).

Plate 10a. Northern Snake-necked Turtle *Chelodina rugosa*, a common aquatic reptile and favoured food of Aborigines and also of the White-bellied Sea Eagle (Photograph: G. Miles).

Plate 10b. Aboriginal rock painting of the Northern Snake-necked Turtle (Photograph: G. Miles).

Plate 11a. Darter *Anhinga melanogaster*, a common fish-eater, drying its plumage (Photograph: G. Miles).

Plate 11b. Aboriginal rock painting of a Darter being hit by a spear propelled from the woomera (spear-thrower) of a hunter, to right (Photograph: G. Miles).

Plate 12a. Agile wallaby *Macropus agilis* (Photograph: I. Morris).

Plate 12b. Aboriginal rock painting of agile wallaby, a common item of diet (Photograph: I. Morris).

the vital movements of aquatic fauna (Chapter 6). The dense, prickly, introduced *Mimosa pigra* shrub threatens the wetland habitat of several vertebrates, especially waterbirds. The highly invasive, mat-forming water hyacinth has been held so far to a single infestation, where it remains a serious potential menace to the habitat of the aquatic fauna of the billabongs and channels of the entire monsoonal lowlands. Likewise, the mosquito fish is another escape, from local aquaria with careless owners, which may affect native fishes. However it is the escaped domesticated mammals now living feral, especially Asian Water Buffaloes, that have had the biggest overall impact on the ecosystems of the monsoonal lowlands so far (Chapter 11, also 6 and 9). Heavy grazing by buffaloes on vegetation unadapted to it has had several important effects, the most pronounced and longest lasting being associated with the grazing of longest standing. This mobile, amphibious ungulate's habitat extends from the wetlands up to higher drier open forests therefore its influence is widespread. There is direct and/or indirect evidence for the reduction of several species of plants, and for significant effects on the habitats, among others of saltwater crocodiles, magpie geese and the fauna of the very small relict patches of monsoon forest, including their largely endemic avifauna. Progressive reduction of buffalo populations in the 1980s have seen a corresponding return of several elements of the vulnerable biota (Braithwaite & Werner 1987).

The eight species of feral mammals are of further concern because they are important reservoirs of several serious diseases of stock and humans, including bovine tuberculosis, and potentially of others in nearby Asia. One remedy to all these problems is to try to reduce feral populations permanently. That is often costly in time and effort, but one possible alternative is redomestication leading to disease-free utilization (Chapter 11).

The indigenous vertebrate animals of the monsoonal lowlands are one of tropical Australia's most important natural resources (Plates 9-12). From the standpoint of science, national parks and tourism this resource is coming under new pressures, as illustrated above. Conservation cannot be left to chance.

REFERENCES

Braithwaite, R.W. & P.A. Werner 1987. The biological value of Kakadu National Park. *Search* 18: 296-301.

Ridpath, M.G. 1985. Ecology in the wet-dry tropics – how different? In M.G. Ridpath & L.K. Corbett (eds.), *The Ecology of the Wet-Dry Tropics, Proc. Ecol. Soc. Aust.* 13: 3-20.

CHAPTER 6

The freshwater fishes of northern Australia

KEITH A. BISHOP & MALCOLM A. FORBES

INTRODUCTION

Fishes are the most numerous of the vertebrates. Best estimates range from 17,000 to 20,000 recent species. Approximately 8,000 species inhabit fresh water and 6,700 of these are derived from purely freshwater ancestors.

The incredible diversity of fishes reflects their antiquity and the extent and variety of their habitats. The earliest known fishes have been found in deposits of Early Ordovician age, about 400 to 450 million years ago. Speciation since then has been considered to be largely a function of the development of diversity of living conditions in water (Lagler et al. 1962). Fish now occupy Antarctic waters below freezing to hot springs more than 40°C, and from soft fresh waters to hypersaline lake waters. They may be found from approximately five kilometres above sea level in turbulent sunlit mountain streams, to nearly twelve kilometres beneath sea level in quiet, dark, oceanic waters.

To focus better on the chapter theme, we start by specifying some of the unique features of the origins of the Australian continent and the nature of its freshwater habitats. This background will in turn provide an appreciation of the origins and nature of Australia's freshwater fish fauna. This appreciation is necessary before we document in more detail the type of freshwater habitats available to fishes in the northern rivers of the continent.

ORIGINS OF THE AUSTRALIAN FRESHWATER FISH FAUNA

During glacial maxima, sea levels were up to 150 m lower than today (see Chapter 1). The Australian continent was periodically connected by land to New Guinea (Jennings 1972), forming the Australian region, a region probably isolated geographically since it split away from Antarctica during the Late Cretaceous (some 90 million years ago). The nature of the region's contact with the parent continent (Gondwanaland)

prior to this period, whether to points in India, Africa or South-east Asia, is not agreed upon. The effect of this early contact with Gondwanaland, followed by long isolation, interacting with considerable long-term climatic variability, including extreme aridity and extensive marine transgressions, would be expected to have had a strong limiting influence on the development and hence the character of the present day Australian freshwater fish fauna.

The most obvious feature of the fauna attributable to the above historical factors is the lack of true 'primary' freshwater fish. These fish are of unique significance in zoogeography because they are closely bound to land masses. Like their ancestral forms, they do not possess the physiological ability to survive in saline waters. They can pass from one isolated stream to another only through the slow physiographic change of landscapes. Hence their dispersal over the world is slow and strongly reflects old distribution patterns. Some significant primary freshwater fish families found in Africa or Asia are missing in Australia. The cyprinids (carps), cyprinodontids (minnows and killifishes), nandids (leaf fishes), cichlids, anabantids (labyrinth fishes) and chanids (milk fishes) are examples. These freshwater fish families have widespread tropical distributions in Africa and Asia and their absence from Australia, particularly from the northern tropics, may be explained by Australia's geographical isolation. These families possibly spread into South-east Asia from Africa and Asia after Gondwanaland split up. However, a small complement of primary freshwater fishes had probably colonised the rivers of the Australian landmasses before Gondwanaland fragmented. Descendants of these species may include *Neoceratodus fosteri* (the Queensland lungfish, originally limited to the Mary and Burnett Rivers), *Scleropages leichhardti* (spotted barramundi, originally limited to the Fitzroy River, Queensland), *S. jardini* (northern spotted barramundi or saratoga, rivers of northern Australia and southern New Guinea) and *Lepidogalaxis salamandrioides* (salamander fish, southern Western Australia). It is possible that the synbranchid eels, *Ophisternon* spp., of northern Australia and New Guinea also have Gondwanian origins (McDowall 1981). The plotosid and ariid catfishes of Australia's northern fresh waters are descendants of the primary freshwater group Ostariophysi. These catfishes are however of secondary marine derivation, i.e. their ancestral stocks evolved in fresh waters, moved to the sea, and then dispersed into Australian fresh waters.

The primary freshwater fishes only represent a very small proportion of Australia's freshwater fish fauna. The remainder represent invaders from the sea which may have replaced old fauna and evolved into endemic families, genera and species. McDowall (1981) indicated that, on an overall basis, Australia has 110 endemic freshwater species, a further 35 shared with New Guinea, and only about 20 widespread species. In another perspective recent estimates (Hoese 1977, unpubl.) put the number of species which can be found in Australian fresh waters at approximately 235, of which 65 are essentially marine forms. The remaining 140 to 150 species may complete their entire life cycles in fresh waters. Australia has thus been viewed as having a depauperate freshwater fish fauna when compared with the faunas of similarly sized continents or divisions of larger continents. Comparisons of this nature are not strictly valid as no consideration is given to the comparative amounts of surface water resources on the continents being compared. This depauperate state has been related to the fact that Australia, as a whole, is a huge, flat, dry island, having a past history and characteristics scarcely conducive to the development of a diverse freshwater fish fauna.

Australia is of course not homogeneously hot, dry and flat. Rainfall is higher on the northern, eastern, south-eastern and south-western coastal fringes. This geographic variability is compounded by considerable temporal variability in rainfall. As a consequence, there is marked seasonal and year-to-year variation in river and stream flows. Seasonal and secular (year to year) variations in flow rates are most pronounced in continental Australia, less so towards Tasmania. Great seasonal variations in flow rates are displayed in the tropical northern Australian rivers where monsoonal weather patterns strongly influence climate, as described in Chapters 2 and 3. Considerable secular variation occurs in the time distribution of flows in the 'wet' season as well as total annual discharge in these northern rivers, however, the seasonal occurrence of flow is generally a reliable phenomenon. Great secular variation exists in arid inland drainage basins.

Our attention now focuses on the former drainage systems where rivers may carry huge volumes of water in summer, the 'wet' (November to March), yet may almost cease to flow in winter, the 'dry' (May to September). We shall proceed further to characterise these northern Australian catchments as well as their fish fauna. Particular attention will be placed on how this fauna copes with the pronounced seasonal water flows brought about by the tropical wet-dry climatic cycle.

TROPICAL SEASONAL RIVERS OF NORTHERN AUSTRALIA

Lake (1971), in his book on the freshwater fishes and rivers of Australia, recognises fifteen separate catchment systems draining Australia. Four of these systems actively drain the northern tropical coastal margins of the continent. The contributions of total annual runoff (TAR) of these systems (Fig. 32) to Australia's total annual runoff (ATAR = 350 km^3) are shown in Table 10.

These systems, therefore, discharge 65% of the total discharged by all Australian rivers. The Murray-Darling catchment system which drains most of inland south-eastern Australia (an area over 1,000,000 km^2 or a seventh of Australia's landmass) has a remarkably small total annual discharge of 22 km^3 representing only 6% of that of all Australian rivers.

It should be noted that the north-eastern slopes and Indian Ocean catchments are not completely confined to the tropics, and that the north-eastern slopes are subject to orographic climatic effects as well as summer monsoonal effects. The orographic effects arise from the presence of the Great Dividing Range, and cause 'low flow' rather than essentially 'no flow' winter conditions in many of the catchments on these slopes. Strictly speaking, the Gulf of Carpentaria and the Timor Sea catchment systems cover Australia's true wet-dry monsoonal tropics and discharge 132 km^3, almost 40% of the total discharge by all Australian rivers. Williams (1983) appreciates this system distinction in a description of Australia's river and stream environments. He referred to the numerous rivers which drain the northern tropical margins of the continent from Cape York Peninsula (Qld) to the Kimberleys (W.A.) as northern flood-drought rivers (Fig. 32).

The major contributors in the Gulf of Carpentaria catchment system are the Mitchell, Gilbert-Staaten, Flinders-Norman, Leichhardt, Gregory-Nicholson and Roper Rivers. Some of the rivers in the north-west of Cape York have continuous flow as their upper reaches receive orographic rains. The Gregory River has continuous flow as it is fed by springs in the Barkly Tablelands. There is considerable variety in the Gulf rivers. Most, however, flow through a flat coastal belt which extends about 150 km from the sea. In places numerous anabranches and distributaries link many of the rivers together.

The major rivers in the Timor Sea catchment system

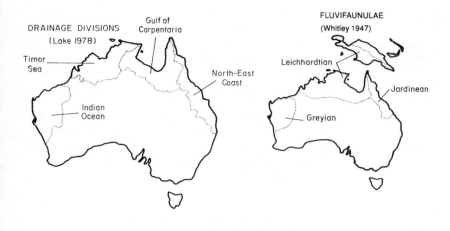

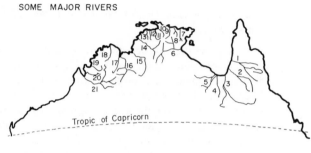

I	Mitchell	8	Blyth	15	Victoria
2	Gilbert - Staaten	9	Liverpool	16	Ord
3	Flinders - Norman	10	East Alligator	17	Pentacost- Chamberlain
4	Leichhardt	11	South Alligator	18	Drysdale
5	Gregory - Nicholson	12	Mary	19	Isdell
6	Roper	13	Adelaide	20	Meda-Lennard
7	Goyder	14	Daly-Katherine	21	Fitzroy - Hann

Figure 32. Drainage divisions, fluvifaunulae regions and some major rivers of northern Australia.

Table 10. Total annual runoff (TAR) of the four drainage systems which drain northern Australia.

Catchment system	TAR (km³)	ATAR* (%)
North-eastern slopes	84	24
Gulf of Carpentaria	64	18
Timor Sea	75	21
Indian Ocean	6	2
Total	229	65

*ATAR, Australian Total Annual Runoff.

are the Daly, Victoria, Ord and Fitzroy. The East and South Alligator, Adelaide and Mary Rivers drain into Van Diemen Gulf, while to the east in Arnhem Land, the largest rivers include the Liverpool, Blyth and Goyder. In the Kimberleys, some sizable rivers include the Pentecost, Drysdale, Isdell and Meda. All of them are similar in several respects to the Gulf rivers in their lower reaches in that they possess wide coastal floodplains which reduce to chains of billabongs in the winter dry. However, in contrast to the Gulf rivers, many of the Timor Sea rivers arise in rocky plateau country and often drain through deep escarpment gorges before entering lowland areas. All of the rivers of both catchment systems are small (in length terms) when compared to the Murray-Darling river system of south-eastern Australia (Table 11).

Williams (1983) highlighted how little is really known about our rivers and the fauna they contain. Certainly, very little work has been published on the ecology of our freshwater systems. The tropical fresh waters of Australia have been investigated in even less detail than the inadequately researched temperate fresh waters. This is particularly the case for the northern flood-drought rivers whose catchments are sparsely populated and seasonally inaccessible. However, in the past five years the rivers and creeks of one region, the Alligator Rivers region some 200 km east of Darwin, have been extensively investigated in parallel with the development of nearby uranium mining. It is from these ongoing investigations that case

Table 11. Approximate lengths of major rivers in the Gulf of Carpentaria and Timor Sea drainage systems.

Drainage and river	Approximate length from mouth (km)
Gulf of Carpentaria	
Mitchell	510
Gilbert-Staaten	650
Flinders-Norman	620
Leichhardt	550
Gregory-Nicholson	320
Roper	440
Timor Sea	
Goyder	150
Blyth	150
Liverpool	170
East Alligator	230
South Alligator	260
Mary	200
Adelaide	180
Daly-Katherine	690
Victoria	880
Ord	500
Pentacost-Chamberlain	320
Drysdale	325
Isdell	250
Meda-Lennard	250
Fitzroy-Hann	650
Murray-Darling	
Murray-Darling (combined)	5270

studies will be drawn later to characterise physiographic features of a typical catchment, and exemplify structural and limnological changes in the aquatic habitats brought about by the wet-dry climatic cycle.

Tropical, seasonally-flooding rivers also occur in Africa, South America and Asia. Such rivers, like those of the northern coastlands of Australia, inundate immense areas seasonally on a scale not known in temperate regions. The scale of inundations in these other continents are many magnitudes higher than that experienced in Australia's northern rivers. Some overseas tropical seasonal rivers inundate areas greater than that of the entire catchment areas of some of Australia's northern rivers! The northern Australian rivers did not even rate a mention in a recent worldwide review (Welcomme 1979) of floodplain rivers and their fisheries – the Fly and Sepik rivers of Papua New Guinea were the geographically closest floodplain areas mentioned in that text.

Fish have been investigated more than any other vertebrate group in the tropics as they are so important as a food source (a considerable proportion, perhaps the majority, of the world's freshwater fish catch is taken from running waters and their lateral seasonal flood zones). Tropical studies are still in their infancy, and theories about tropical communities are based on

few hard data (Lowe-McConnell 1975). Overseas results have generally been published in specialist reports not readily available to the general ecologist and the wider public. With this in mind, let us consider what information is available on the freshwater fishes found in Australian northern rivers, and, to begin with, the relationships of the northern fauna within the considerable longitudinal range and with fauna from elsewhere in the Australian zoogeographic region (including New Guinea). The systematics of most major fish groups have been studied, or are currently being reviewed.

ZOOGEOGRAPHY OF THE NORTHERN AUSTRALIAN FISHES

Faunal relationships within the tropical coastland drainages

Iredale & Whitley (1938) divided Australia and New Guinea into ten zoogeographic regions which had distinctive mammals, birds, fishes and molluscs. These faunal elements were termed fluvifaunulae. Whitley (1947) modified the original fluvifaunulae regions (Fig. 32) after the study of more material. While these regions may be helpful abstractions of distributions they give only a hazy picture of the dynamics of fish distribution and dispersal. The regions were named in honour of naturalists and explorers who were associated with the respective regions. The region (Fig. 32) encompassing most of Australia's northern flood-drought rivers, as well as southern New Guinea, is the Leichhardtian region (coincident with Lake's Timor Sea and Gulf of Carpentaria catchment systems). Ludwig Leichhardt was the ill-fated explorer who in 1844 made the first crossing of the northern Australian rivers from Queensland to Port Essington (see Chapter 12).

The Leichhardtian has the most diverse inland fish fauna in Australia. Williams (1983) indicated that 40 to 49 species may be found in the Timor Sea catchment system while 50 to 59 may be found in the Gulf of Carpentaria and north-eastern slopes (Whitley's 1947 Jardinean) systems. Only a few catchments in northern Australia have been extensively surveyed for fish. The results from surveys of five northern Australian regions, the Kimberley Province, W.A. (Allen 1982), Victoria, N.T. (Australian Museum Records), Alligator Rivers, N.T. (see later), Jardine River, Qld (Allen & Hoese 1980) and Black-Alice River, Qld (Beumer 1980) (Fig. 33), will be used to describe the faunal relationships within the tropical coastland drainages. It must be noted that the species lists used from these regions may be subject to changes as a result of further survey work and taxonomic investigations. In order to make valid regional comparisons of freshwater fauna, a number of criteria must be kept in mind. The most important of these concern uniformity of sampling effort, area sampled and proximity of sampling sites to estuarine reaches. All of these criteria may not necessarily have been met in some of the comparisons which follow, nevertheless strict adherence to these criteria is not necessary in this section as the depiction of coarse rather than fine faunal patterns is intended.

All the fishes found in fresh waters from these five regions do not necessarily live in fresh waters throughout their lives. Each species has its own level of association with fresh waters; but, for convenience in this section, three associations are recognised. Marine vagrants are those which are mainly found in marine or estuarine waters and which move into fresh water for short times and distances (e.g. river whaler sharks, saw sharks, stingrays, some herrings, nursery fish, whiting, mangrove jacks, some gobies, soles and puffer fish). Diadromous species are those which undertake extensive migrations into fresh waters from estuarine reaches for considerable times. Most of the diadromous species (e.g. ox-eye herring, anguillid eels, some fork-tailed catfish, silver barramundi and most mullet) in the regions appear to be catadromous, i.e. moving into fresh waters for feeding and refuge opportunities and then returning to estuarine or marine reaches for breeding purposes. Principal freshwater species are those which can live in fresh waters for their entire lives (e.g. eel-tailed catfish, rainbow fish, hardyheads, perchlets, grunters, gudgeons and some gobies). Many of these principal species are capable of entering estuarine waters for brief periods of time. The saratoga (*Scleropages jardini*), a primary freshwater species, is considered to be intolerant of estuarine waters. Principal freshwater species are dominant in terms of number of species in four of the five regions under examination (Table 12A). The Black-Alice River system was dominated by marine vagrants probably as a result of the proximity of sampling sites to estuarine

DRAINAGE KEY

1 Davenport	6 Fly (P.N.G)
2 Lake Eyre	7 Laloki
3 Murray-Darling	8 Purari
4 Tweed	9 Sepik
5 Shoalhaven	

Figure 33. Location of drainages used in species composition and number comparisons. Other drainages mentioned in the text are also shown.

Table 12A. Percentage composition of fish associations in five northern Australian coastal regions (number of species in parenthesis).

Freshwater association	Overall (131)	Northern Australian coastal drainages				
		Kimberley Province (53)	Victoria River (39)	Alligator Rivers (53)	Jardine Rivers (31)	Black-Alice River (40)
Marine vagrants	34	17	31	23	3	53
Diadromous species	13	11	8	15	13	20
Principal freshwater species	53	72	69	62	84	28

Table 12B. Percentage of fish fauna found in common between five well surveyed northern Australian coastal drainages. Note that the rows are expressed as a percentage of the columns (number of species in parenthesis).

Coastal drainages	Kimberley Province (53)	Victoria River (39)	Alligator Rivers (53)	Jardine River (31)	Black-Alice River (40)
Kimberley Province	100	56	40	35	20
Victoria River	42	100	47	29	20
Alligator Rivers	40	64	100	61	30
Jardine River	21	39	36	100	23
Black-Alice River	15	21	23	29	100

reaches. Diadromous species generally made up the minority of species found in fresh waters of the regions.

The similarity of the five regional faunas was examined in terms of the percentage number of species in common between regions. The similarity matrix for the five regions (Table 12B) generally shows that geographic proximity is usually positively related to the similarity of faunas. The exception is again the estuarine-biassed Black-Alice River fauna which had low similarities with all other regions. The pairs of regions which had over half of their fauna in common were the Victoria River-Alligator Rivers, Jardine River-Alligator Rivers and Victoria River-Kimberley Province. Similar trends followed at the generic and family levels.

The Kimberley Province had relatively low similarities with other regions. This low similarity reflects the uniqueness of the fauna (sixteen endemic species and one endemic genus) in the Province. The fauna of the Kimberley, like that of the Pilbara (a province in Whitley's Greyian region or Lake's Indian Ocean catchment system), is dominated by teraponids (grunters) and eleotrids (gudgeons) which collectively comprise close to half of the species. The high endemism can be attributed to the rugged topography of the province which has meant the effective long-term isolation of rivers and streams.

Recent widespread fish survey work in the Timor Sea and Gulf of Carpentaria catchment systems (Midgley, unpubl. reports) has indicated disjunct distributions of many of the widely found species, particularly the ariid catfishes. A good example of this is to be found in the Alligator Rivers region of the Northern Territory where *Arius australis, A. proximus, Anodontiglanis dahli* (toothless eel-tailed catfish) and *Toxotes lorentzi* (primitive archerfish) are found in the fresh waters of the South Alligator River, but not in the East Alligator – two adjacent catchments. This disjunction has also been noted in the Gulf of Carpentaria by Allen & Hoese (1980). These authors considered that temperature clines were responsible for such distributions; however, in areas of similar latitudes and distances from the coast, geological history and physiographic factors are likely to be very important. Further work is required in the compilation and integration of existing fish survey data in the Northern Territory before well-based explanations of such distributional anomalies can be presented.

The Alligator Rivers region of the Northern Territory has been subjected to intensive fish surveying for many years. In 1948, R. R. Miller conducted the first significant survey of fishes of the region during the American-Australian Scientific Expedition to Arnhem Land (Taylor 1964). The discovery of uranium mineralization in the region in the late sixties initiated further fish survey work with progressively more specialised aims. The net result of all this work is that there is a clearer understanding of the factors which control the distribution and abundance of fishes within this one large region. This information can also be readily used in sorting-out distribution patterns of closely related species in areas outside the region (see later). About 50 species have so far been collected in the fresh waters of the region. Magela Creek catchment has yielded about 35 species. Only a few species (e.g. the black anal-finned grunter, *Pingalla midgleyi* nov. sp., and the Arnhem Land blue-eye, *Pseudomugil tenellus*) are endemic to the region. The small-bodied fish fauna of the region is in overall abundance dominated by centropomids (perchlets), melanotaeniids (rainbow fish) and atherinids; the larger-bodied fish fauna is dominated by ariids and plotosids (fork and eel-tailed catfish), clupeids (boney bream and tarpon) and, to a lesser extent, teraponids. Considerable seasonal and longitudinal catchment abundance variation is, of course, involved.

Allen & Hoese (1980) made collections of freshwater fishes from the Jardine River at the northern tip of Cape York Peninsula. This catchment is interesting because of its proximity to New Guinea and its intermediate position between the Gulf of Carpentaria and north-eastern slopes catchment systems. The latter interest can hardly be followed through without more systematic surveys in all the mentioned catchment systems. The affinities of the Jardine River fauna with the southern New Guinea fauna will be discussed in the next section. Allen & Hoese (1980) produced an annotated checklist for the Jardine River, which included 30 species in 24 genera and 16 families. No endemic species were found. The collection was dominated in number by eleotrids, centropomids and melanotaeniids. Allen & Hoese listed 3 plotosid, 2 melanotaeniid, 2 atherinid, 2 centropomid, 1 eleotrid and 1 synbranchid (one gilled eel) species which display disjunct distributions to the south of the Jardine River in Gulf of Carpentaria catchments. The authors also noted that *Scleropages jardini* and *Hephaestus carbo* were absent from portions of the Gulf drainage between the Gregory and Mitchell Rivers. Another pattern of disjunction was also noted by the authors for 3 Jardine River fishes in the north-eastern slopes catchment system.

Faunal relationships outside the Australian tropical coastlands

Munro (1964) listed 24 principal freshwater species common to fresh waters of Australia and southern New Guinea. Allen & Hoese (1980) in sampling the Jardine River on the tip of Cape York were particularly interested in seeing the extent of relationships with the

Table 13. Percentage of northern Australian coastal fish fauna found in common with outside drainage systems. Note that the rows are expressed as percentages of the columns (number of species given in parenthesis).

Rivers outside coastal northern Australia	Ref.*	Northern Australian coastal drainage basins				
		Kimberley Province (53)	Victoria River (39)	Alligator Rivers (53)	Jardine River (31)	Black-Alice River (40)
Papua New Guinea						
Laloki River (43)	1	15	21	19	26	18
Fly River (103)	2	23	31	40	52	23
Australian continent Internal:						
Davenport Ranges (7)	3	13	18	13	6	2
Lake Eyre (17)	4	11	18	13	6	7
South-east coastal:						
Murray-Darling (24)	4	9	15	11	3	10
Shoalhaven River (18)	5	2	3	2	7	8

*1. Berra et al. (1975), 2. Roberts (1978), 3. Bishop & Larson (1984), 4. Lake (1978), 5. Bishop (1979).

fauna of southern New Guinea, just 205 km distant across the geologically recently inundated Torres Straits. Not surprisingly, they found that there was a strong resemblance (63% of the Jardine species) with the coastal lowlands of southern New Guinea opposite Torres Straits. They considered that this similarity would probably be increased with further collections in both areas and clarifications of taxonomic problems.

The north-west extreme of the Australian continent, the Kimberley Province of W.A., shows a weaker resemblance (18% of Kimberley species) of fish fauna to southern New Guinea than does Cape York. The closer-proximity/greater-similarity relationship which appears to hold within Australia appears to be valid when extended to include New Guinea. This relationship is better depicted when similarities are calculated with some rivers in New Guinea and the five northern Australian regions already mentioned (Table 13).

Moving in a north-westerly direction from New Guinea along the Malay-Indonesian Archipelago, a dramatic difference in freshwater fauna occurs across a deep sea channel running between the Molucca Islands and the Celebes. This apparently impassable barrier (Weber's Line) has prevented the interchange of the rich primary freshwater fauna from the south-east Asian to the Australian region. The Laloki River of New Guinea shares in excess of 20% of its fauna with northern Australia yet only 8% with Thailand (Berra et al. 1975); this 8% represents only 2% of Thailand's fauna! All of these common species are marine vagrants or diadromous species with either Indo-Pacific or circum-tropical distributions.

Again on the Australian continent, there are similarities between the northern Australian faunas and those of internal and south-eastern coastal drainages (Table 13). Once more the closer-proximity/greater-similarity relationship appears to hold. The internal drainage

systems of the Davenports and Lake Eyre (and to a lesser extent the south-east coastal Murray-Darling system) resemble the fauna of the Timor Sea drainage systems rather than those found at the margins or within the north-eastern slopes catchment system. The north-eastern slopes fauna more closely resembled the south-eastern slopes and Murray-Darling fauna.

Invasions of continental Australia by northern warm water fishes and southern cold water fishes are evident (McDowall 1981). Grunters, archerfish, eel and fork-tailed catfish, rainbow fish, some herrings, perchlets, atherinids and eleotrids are characteristic of northern waters and decline in diversity to the south. The complement of southern cold water invading fishes is less diverse and as a result the northern Australian freshwater fish fauna is usually most diverse showing a decline in diversity going south. The invasion south by northern fishes can be depicted when faunal similarities are examined (Table 14) between northern rivers and internal drainage basins as well as south-east coastal rivers. The closer-proximity/greater-similarity relationship is also valid in this con-

Table 14. Percentages of internal and south-east coastal fish fauna found in common with five northern Australian drainages. Note that the rows are expressed as a percentage of the columns.

Northern drainages	Internal		South-east coastal	
	Davenports (7)	Lake Eyre (17)	Murray-Darling (27)	Shoalhaven (18)
Kimberley Province	100	35	21	6
Victoria River	100	41	25	6
Alligator Rivers	100	41	25	6
Jardine River	86	12	4	11
Black-Alice System	29	18	17	17

text and seems to reflect an 'ease of invasion' factor. This invasion phenomenon is very evident in the Davenport internal drainage basin where all the species present are found in adjacent northern coastal catchments. It is interesting to note here that the combination of species found in the Davenport basin most strongly resembles plateau and upper escarpment communities in the Alligator Rivers region. These communities are made up of species which have strong upstream migratory abilities, i.e. species which are most likely to invade adjacent catchments by crossing watersheds in wet periods.

RELATIVE DIVERSITIES OF FRESHWATER FISH COMMUNITIES

High diversity and complexity of plant and animal communities in the tropics is a well-documented phenomenon. Explanations have usually involved concepts relating to geological time, climatic stability, spatial heterogeneity, competition, predation and productivity (Lowe-McConnell 1969). To avoid delving into these considerations too far, and too soon, let us first investigate some simple questions.

Are the freshwater fish communities found in the tropical coastlands of northern Australia really more diverse than their temperate counterparts? Further, does the widely published comment that Australia has a depauperate fish fauna really hold when the tropical northern coastlands are considered? Diversity in this section relates to the total number of species found in a community (i.e. species richness), and is further considered by Braithwaite et al. and by Morton & Brennan (Chapters 7 and 9). Any consideration of diversity must be related to some frame of reference when comparisons are to be drawn. Diversity in natural streams has been related by previous workers to position in catchment, depth, gradient, stream length, discharge and area of drainage basin (Lake 1982). All of these parameters relate to stream size and possible habitat diversity. In terrestrial environments, species-area relationships (MacArthur & Wilson 1967) have been used to compare patterns of species diversity. The same approach will be used in this section where diversities are compared relative to drainage basin area of the systems in question.

Welcomme (1979) examined the above diversity-drainage area relationships with a log-log scale in 45 overseas floodplain rivers. A hand fitted representation of Welcomme's relationship ('Welcomme's line' in this chapter) is shown in Figure 34. Superimposed on this figure are data points for some tropical and temperate Australian region rivers (including New Guinea). The most obvious feature of the resulting pattern

is the dual positioning of the tropical and temperate data points; all of the tropical systems are on the upper side of Welcomme's line, while most temperate systems are below the line. The exception for the temperate systems is the Tweed River which is the northernmost temperate example and which could perhaps be better classified as subtropical. The lowest outlier from Welcomme's line was the Murray-Darling system. The highest outliers from the line were the Black-Alice system (probably because of the strong estuarine species component of the communities) and a number of small catchments in the Alligator Rivers region. However, Welcomme noted that the log-log species-drainage area relationship was probably not a reliable index of diversity for extremely small basins. A remarkable example of such an anomalous situation is the comparison between the Murray-Darling system (27 species: 1,000,000 km² catchment) and an escarpment perennial spring pool (Baroalba Springs) in the South Alligator River system (22 species: pool area 100 m² and catchment area of 10 km²).

In summary:

1. Diversities (relative to drainage basin areas) of freshwater fish communities in Australia's northern tropical coastlands appear higher than in temperate counterparts;

2. Diversities in the northern tropical coastlands appear to be higher than those in overseas floodplain rivers; conversely, diversities in temperate Australia appear to be lower than in overseas rivers;

3. Other factors seem to be influencing diversities when small basins or subsections of basins are examined.

The first conclusion may have been expected. Its explanation may lie amongst historical factors relating to the invasion of the Australian region by northern and southern marine fauna. Studies on coral reef fish communities have indicated that the central Indo-West Pacific region contains the highest number of species; the global pattern of variation in diversity is one of a pronounced cline of decreasing diversity away from the region (Sale 1980). It could be speculated on these grounds that the northern invading marine communities are more diverse than their southern counterparts which results in greater diversity in Australian tropical fresh waters. This explanation may answer questions within the Australian region; however the underlying question still remains unanswered as to why the Indo-West Pacific has the highest diversities. The causes may relate to geological history and be outside the immediate scope of ecological investigation.

The second conclusion was not expected. It appears that the depauperism of the Australian fauna is over-emphasised, and in the tropical coastlands the reverse may be true. This over-emphasis could well be related

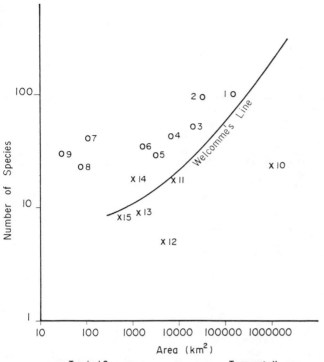

Tropical:O			Temperate:X		
Code	System	Source	Code	System	Source
1	Fly R.	Roberts (1978)	9	Baroalba Ck.	Bishop & Harland (1981)
2	Purari R.	Haines (1979)	10	Murray-Darling	Lake (1979)
3	Alligator Rvs.	Bishop *et al* (1982)	11	Shoalhaven R.	Bishop (1979)
4	Laloki R.	Berra *et al* (1975)	12	Ironstone Ck.	Lake (1982)
5	Jardine R.	Allen & Hoese (1980)	13	Parson's Ck.	Lake (1982)
6	Magela Ck.	Bishop *et al* (1982)	14	Tweed R.	Richardson (1984)
7	Black-Alice R.	Buemer (1980)	15	Seven Creeks	Cadwallader (1979)
8	Gulungul Ck.	Bishop & Harland (1981)			

Figure 34. Number of species present in different river systems plotted according to their basin areas. The points are derived from Australian systems while the line is an overseas summary produced from Welcomme (1979).

to the fact that many Australian freshwater ichthyologists have been restricted to temperate latitudes, and in particular the depauperate Murray-Darling system.

IMPORTANT FACTORS INFLUENCING FISH COMMUNITIES: THE ALLIGATOR RIVERS REGION AS A CASE STUDY

Within the tropical northern coastlands many factors appear to influence diversities and other community parameters within catchment systems. Some of these factors can be most readily described from studies on fish communities under way in the Alligator Rivers region. Some details of the region's physiography and aquatic habitats are first required.

Physiography and aquatic habitats of the region

The region is made up of five distinct subregions: the plateau, lowlands, floodplains, tidal flats and southern hills and basins. Escarpment and corridor areas inter-face the lowlands towards the plateau and floodplain subregions respectively. This section will describe the first three subregions and corresponding aquatic habitat subdivisions (Fig. 35).

Plateau habitats
The plateau is a massive sandstone formation towering up to 250 m above the adjoining lowlands. The plateau is mainly comprised of bare rock and thin soil with a scanty vegetation cover. All major streams in the region have their headwaters on the plateau. Most of the plateau is made inaccessible to aquatic fauna from the lowlands by many high vertical drops at the escarpment in the courses of the rivers and streams.

Escarpment area habitats
The irregular edges of the plateau, and the gorges intruding into it, sometimes along fault lines, form an escarpment area which is abrupt and scenically striking. Most of the channels and streams in the escarpment are scoured by floodwaters, and some banks are frequently inundated during the wet season. Waters in

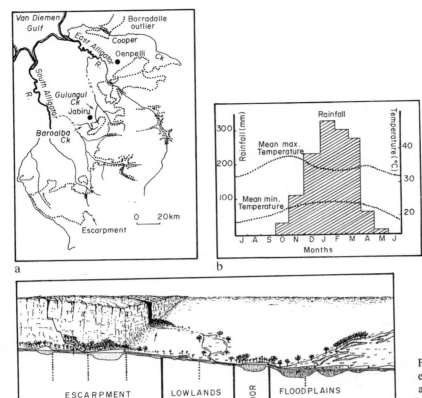

Figure 35. (a) Map of the Alligator Rivers region; (b) Annual rainfall regime and mean maximum and minimum temperatures at Oenpelli; (c) Long-section of a typical creek in the region showing major aquatic habitats.

this zone are generally cool and clear. Three major escarpment habitats can be defined.

Main channel water holes. Within the deep gorges many large waterbodies are found which are up to 1.5 km long and up to 7 m deep. The gorges usually terminate in rocky plunge-pools less than 100 m in diameter and up to 15 m deep, in which waterfalls and cascades enter from the plateau in the wet season. At the lowland outlets of the gorges there are alluvial fans of sand and gravel. The banks of these waterbodies are in places steep and generally covered by thick vegetation. Submerged fallen timber is often associated with rainforest vegetation.

Small feeder streams. Small feeder streams, less than 4 m wide, enter the main channels in the wet season. These streams are often surrounded by rainforest vegetation and are transformed into a series of isolated rock pools during the dry season.

Spring-fed perennial streams. In the dry season in these small streams less than 4 m wide flow arises from groundwater trapped in aquifers underneath the pla-

teau. The flow persists a few kilometres onto the lowlands during the dry season. These streams flow over bedrock and alluvial gravels and through the extensive exposed root systems of dense rainforest vegetation which encloses most of these habitats.

Lowland habitats
The present land surface of this zone has been formed by erosion and deposition of materials from the escarpment and plateau. Two major aquatic habitat types are found.

Sandy creekbed. All creekbeds in this zone are scoured by wet season flows, have shifting sandy substrates within anastomosing channels, and are surrounded by paperbarks and pandanus or rainforest flora. Aquatic plants abound around submerged clayey banks during the wet season. Creekbeds are without water flow for about 5 to 6 months during the dry season. Isolated sandy pools diminish and frequently disappear during the dry season.

Backflow billabongs. Backflow billabongs with shelving clay banks exist on tributaries and on some anabranches of the main creek channels. Most of these billabongs are near the mouths of the tributaries or in depressions behind levee banks beside the main channels. Water either flows through these billabongs to the main stream, or flows back from the main channel up the tributaries to the billabong in the wet season. Backflow tends to deposit fine sediments and organic matter in these billabongs. Thick plant growth occurs in most of these billabongs in the wet season during the time when their wooded banks are submerged by floodwaters.

Corridor habitats

A series of large stream channels and waterholes up to 1 km long and up to 6 m deep effectively funnel waters between the lowlands and floodplains. These channels and waterholes are well flushed each wet season by main creek flow and remain relatively cool and clear for most of the year. The banks are steep and thickly covered by pandanus, paperbarks and grasses. The substrates are generally very sandy, but aquatic plants are found in muddy substrates in the slow flowing shallow edges of the waterbodies during the wet season. These habitats may receive seepage input from sandy aquifers for some time after creek flow stops.

Floodplain habitats

The floodplain portion of Magela Creek, a well studied tributary of the East Alligator River, covers 150 km² of the 1600 km² catchment. The floodplain consists of poorly drained paperbark swamps, open perennial and annual swamps, billabongs and grass/sedge herb fields. The billabongs on the floodplains are up to 6 m deep and have muddy substrates with shelving clay banks. The substrate in these areas was probably formed when clays and muds were deposited in river estuaries. The floodplains emerged from the estuaries probably 3,000 to 4,000 years ago. An unusual feature of the Magela Creek floodplain is that it has no continuous channels, and so the manner of water movement across it is not readily discernible. Depressions around the margins of the floodplains may first receive water in the early wet season from run-off from adjacent catchments or from spillover from the mainstream. The capacity of the Magela floodplain is approximately 60 × 10⁶ m³ when the once dry plain is covered with up to 2 m of water. The average daily input during the wet season is 3% of its capacity. During the dry season the floodplain dries out to approximately 8% of its capacity (Williams 1979).

Seasonality in the Alligator Rivers region

Climate

The region, in common with much of northern Austra-lia, has a monsoonal climate; the dry season lasts from about May to September, and virtually all the rain falls in the wet season, which varies in starting and ending time and in intensity distribution, but is generally confined to the period from November to March (Fig. 35) (see also Chapters 2 and 13). A prominent feature of the region's climate is the high temperature sustained throughout the year. The average annual rainfall is 1560 mm while the evaporation is 2200 mm. A small range in day length occurs at this latitude (12°S). The mean hours of cloud-free sunshine range between 9.7 to 10.2 hours in the dry season and between 4.5 and 7.5 in the wet season.

Hydrological regime

The seasonal rainfall pattern causes large variations in the hydrological regime of the region's creeks and rivers. The size of the whole aquatic environment thus fluctuates seasonally and considerably from year to year. For example, in the floodplains of Magela Creek about 150 to 200 km² are flooded in most years and it is estimated that, on average, the area flooded may extend to more than 300 km² once in ten years and to nearly 500 km² once in 100 years (Galloway 1976). Flow in the region's waterways usually occurs in the wet season (except in some escarpment habitats) when it comprises a series of floodpeaks superimposed on a base flow which begins on average during mid-December and ceases about the end of July (Table 15). The flow increases the diversity of the aquatic environment; large areas, notably floodplains and lowlands, take on quite different appearances during the wet season compared to the dry. Upstream escarpment habitats are relinked via waterways with all downstream habitats in their respective catchments. The rising water runs first down the lowland channels and creeks from the escarpment in a distinct 'flow tongue'. Once it reaches the corridor and floodplain habitats it connects billabongs and swampland. The number of waterholes which persist throughout the dry season varies from year to year as does the amount and depth of water left in permanent billabongs.

Table 15. Flow characteristics of Magela Creek lowlands for seven years.

Wet year	Flow starts	Total discharge (m³ × 10⁶)	Flow stops
1972/73	December 8	274	July 17
1973/74	November 18	597	August 10
1974/75	November 3	499	August 7
1975/76	November 2	793	July 23
1976/77	December 1	349	July 17
1977/78	November 2	395	July 15
1978/79	November 26	282	June 30

Limnological changes in permanent billabongs
The marked dichotomy in climate and hydrology of monsoonal Australia is matched by a marked seasonal limnology which influences aquatic community dynamics. Whilst the variations in physico-chemical state of surface waters discussed below are of great consequence to fish communities they are of relevance to other aquatic flora and fauna. In billabongs there is a generalised dichotomy between the uniformity of physico-chemical properties at periods of high flow over the wet season and the progressive solute concentration accompanying their hydrological isolation during the dry season. This generalised feature is common to other drainage basins in monsoonal Australia.

Wet season. The immense volume of water from intense rainfall and rapid run-off over the wet season ensures concomitant relative uniformity in water quality from the plateau to the floodplain. General features of wet season surface water characteristics are illustrated in Table 16. The surface waters of the wet season are amongst the softest in Australia with total dissolved solids being typically below 20 mg/l and hardness and alkalinity consistently below 10 mg/l. Buffering capacity is therefore exceptionally low and as a result these waters are subject to marked pH shifts.

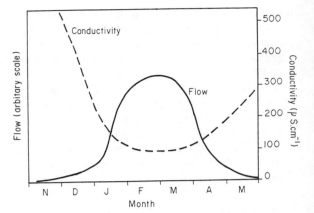

Figure 36. Idealised diagram of the relationship in water flow over the wet season with conductivity (solute concentration).

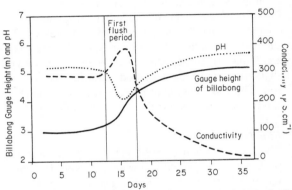

Figure 37. Idealised figure of variation in conductivity and pH in waters on flushing of a billabong as evidenced by gauge height at arrival of first floodwaters during the onset of the wet season (adapted from Koontz et al. 1983).

Table 16. Average wet season surface water quality in the Alligator Rivers region (source: Dames & Moore 1978).

Parameter (mg/l except where stated)	Value
Temperature °C	31.0
pH	6.2
Conductivity (μ S/cm)	16
TDS (res. on evap. 180°C)	11
Suspended solids	16
Turbidity (NTU)	1
Hardness (as $CaCO_3$)	
Total	3.29
Carbonate	2.50
Non-carbonate	0.78
Total alkalinity	2.67
Cations	
Calcium (Ca)	3.27
Magnesium (Mg)	0.59
Sodium (Na)	1.38
Potassium (K)	0.22
Iron (total) (Fe)	0.05
Anions	
Bicarbonate (HCO_3)	3.27
Sulphate (SO_4)	0.95
Chloride (Cl)	1.95
Nitrate (NO_3)	0.15
Phosphate (PO_4)	0.02
Nitrate as N	0.02
Nitrite as N	< 0.01
Ammonia as N	< 0.01
Silica (SiO_2)	3.38

Ionic composition is generally influenced by an oceanic type rainfall with orders of dominance of Na^+ > Mg^{2+} >> Ca^{2+} and Cl^- >> SO_4^{2-} > HCO_3^- in plateau areas whereas in the lowlands anions are dominated by bicarbonate. Nutrient levels are at their lowest but, on a world scale, total phosphorus levels are high and billabongs would be regarded as mesoeutrophic to hypereutrophic (Walker & Tyler, in press). Although rainfall is acidic with pH values frequently below 5.0 (Noller, pers. comm.), surface waters tend towards neutral.

Concentrations of solutes are greater at the beginning of the wet season during initial flows of streams when material accumulated in catchments over the dry season is washed downstream (Fig. 36). Indeed, at the first flush or inundation by waters entering the corridor and floodplain billabongs marked transient variations in pH and conductivity have been noted occasionally

Table 17. Comparison of fish habitats on the basis of physico-chemical properties at the end of the dry season (after Walker & Tyler, in press).

Fish habitat	Ionic classification	Other physico-chemical properties
Channel water holes, springs, corridor billabongs	Na/Mg HCO_3 group	Maintain ionic composition of wet season throughout the year. Solute concentration factor 3-5; $K_{18} < 80$ μScm^{-1} (all year). $5.8 < pH < 6.5$ (late dry season). Nutrient status < 70g PL^{-1}. Turbidity < 50 NTU
Backflow billabongs	NaCl group	Progression during dry season towards mean world seawater ionic proportions. Solute concentration factor > 10; $30 < K_{18} < 550 \mu g$ PL^{-1}. Turbidity > 100 NTU (late dry season)
Floodplain billabongs	SO_4^{2-} group	Selective concentration of SO_4^{2-} from ground water ingress in addition to progression towards mean world seawater ionic proportions. Solute concentration factor > 10; $50 < K_{18} < 1000$ μScm^{-1} (late dry season). Nutrient status $100 <$ Total P $< 300 \mu g$ PL^{-1}. Turbidity > 100 NTU (late dry season)

(Koontz et al. 1983). The period of the first flush and the presence of an acidic 'slug' (Fig. 37) has occasionally coincided with fish kills (see later).

Following the first flush, water quality is relatively uniform throughout, although, where billabongs are not subject to strong flows and where extensive stands of macrophytes are present, marked diurnal variations in thermal and dissolved oxygen profiles may occur well into the wet season (Walker & Tyler, in press).

Dry season. Flow ceases with the onset of the dry season, with waters receding to remnant habitats as described above. As emphasised in Chapters 2, 3, 10, 11 and 13, the pattern of rainfall within seasons is highly variable with concominant variation in the amount of water remaining over the dry season from year to year. This variation is reflected in the year to year variation in physico-chemical properties.

The relative uniformity of the wet season provides a marked contrast to the progressive deterioration in water quality throughout the dry season. Habitats over the dry season are hydrologically closed systems except for spring habitats. Although individual billabongs exhibit discrete characteristics, by the end of the dry season three generalised groups can be distinguished (Walker & Tyler, in press) by the nature and extent of ionic change and nutrient status (Table 17). Least changed are the Na/Mg HCO_3 group, largely owing to their characteristic low surface area to volume ratio. The ionic composition of the other two groups moves to an NaCl dominance, and where groundwater ingress occurs, as with many of the floodplain billabongs, the NaCl dominance is accompanied by sulphate enrichment. Evaporative loss causes a progressive increase in conductivity with a greater than tenfold solute concentration factor. A decline in pH is also observed.

With an essentially closed drainage system coupled with evaporative concentration, inputs from biotic sources, ingress of groundwaters in some situations, and internal loadings from resuspended sediments and the die-off of extensive macrophyte stands, backflow and floodplain billabongs exhibit a dramatic change in nutrient status. Indeed, phosphorus levels indicate hypereutrophy. Table 18 indicates the effect which biotic inputs, in this case caused by the feral water buffalo, can have on the nutrient status and other physico-chemical characteristics of a small water body (see also Ridpath, Chapter 11). Although nutrient concentrations rise throughout the dry season in the NaCl and SO_4 groups, productivity is dampened with a severe deterioration of the underwater light

Table 18. Water quality of a water body with and without influence from buffalo.

	With buffalo influence Year 1; 29 September	Without buffalo influence Year 2; 12 September
pH	6.9	7.1
Conductivity (μScm^{-1})	352	72
Total dissolved solids	244	52
Suspended solids	168	4.2
Hardness (as $CaCO_3$) (total)	120	31.2
Cations		
Calcium	9.5	1.54
Magnesium	23.0	6.60
Sodium	8.6	3.15
Potassium	45.0	0.78
Anions		
Bicarbonate	158	37.2
Sulphate	8.6	1.4
Chloride	42.4	3.0
Nitrate as N	0.01	<0.01
Nitrite as N	n.d.	<0.001
Phosphate (total)	1.91	0.04
Ammonia as N	n.d.	0.009
Silica	27.0	11.3

n.d.: not determined; concentrations in mg/L except where stated.

climate caused by high loads of triptonic turbidity (Walker et al., in press). Turbidity in these permanent surface waters is induced when falling water levels permit resuspension of sediments from wind action and aquatic bird and feral water buffalo activity.

In temperate climates, dissolved oxygen in surface waters is employed as a criterion for safeguarding fish habitat. As flow ceases with the progression of the dry season, permanent billabongs in monsoonal Australia display thermal and dissolved oxygen stratification in their profiles. Walker & Tyler (in press), in their extensive studies of the thermal behaviour of billabongs in the Alligator Rivers region had demonstrated stratification, but this is not comparable to the enduring hydraulic stratification of temperate latitudes. In contrast, billabongs over the dry season are 'characterised by considerable spatial heterogeneity of water masses and considerable dynamism, with frequent limited mixing but infrequent holomixis' (Walker & Tyler, in press). Whilst persistent hydraulic stratification does not occur, there is generally a sufficient barrier to mixing to provide for progressive oxygen depletion in the hypolimnion. Each billabong has its individual hydraulic behaviour, depending on its morphometry, orientation and degree of shelter with respect to wind action. A spectrum of hydraulic behaviour therefore exists in billabongs ranging from stratification and sustained hypolimnetic anoxia for several months (one unique record reported by Walker & Tyler (in press)), to billabongs exhibiting sustained periods of hypolimnetic anoxia separated by periods of mixing, to those exhibiting regular alternation of holomixis and brief episodes of stratification (Walker & Tyler, in press). Most billabongs are of the latter type (Fig. 38).

Spatial and temporal changes in fish communities of the region

The diversity of fish communities in the region varies greatly between seasons and habitat types. Insurmountable obstacles in escarpment cascade areas are responsible for the greatest reduction of community diversities along creeks. This reduction is most apparent on the plateau where only seven species have been recorded. Usually only two or three species (most frequently, purple spotted gudgeons and rainbow fishes) are found within each plateau stream. Moving upstream from the lowlands into progressively more rugged escarpment country, species diminish in abundance according to their swimming abilities and endurance to high-flow. Temperate freshwater fish in Australia are also prone to such distribution restrictions as evidenced by a reduction of species diversity in the upper Shoalhaven catchment (Bishop 1979) caused by the presence of steep escarpments downstream. The manner in which some species have colonised the Arnhem Land plateau from lower reaches is intriguing, especially when the presence of 150 m high waterfalls is considered. Possible explanations for such distributions include extraordinary out of water climbing ability of some species, previously easier access, and aerial transport by birds or whirlwinds.

There are, of course, many other factors which influence diversity of communities in the region. Escarpment main channel waterbodies, corridor and floodplain billabongs of the region contain fish communities more diverse than communities in other habitat types. This diversity, as well as similar community biomass trends, can be positively related to the maximum depth of waterbodies, i.e. shallow waterbodies have fewer species and less biomass than deep waterbodies. Deep water habitats appear to be very important dry season refuge areas for fish. Floodplain billabong diversities usually fall to a minimum and biomass rises to a maximum towards the end of the dry season as water levels drop. In the wet season floodplain billabong communities have much reduced biomasses but increased diversities.

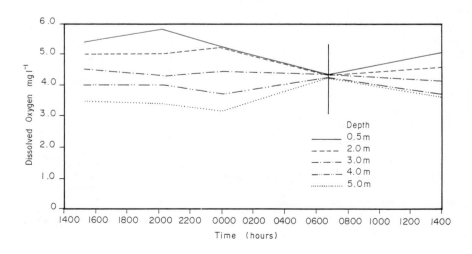

Figure 38. Diurnal behaviour of a billabong in the Alligator Rivers region.

Lowland habitats contain constantly changing fish communities with fluctuating lower diversities. Backflow billabongs are least diverse towards the end of the dry season and even into the early wet season, depending on the timing of backfilling and the rate of fish colonization via the inundated sandy creekbed channels. The diversity and biomass in backflow billabongs is highest in the late-wet and mid-dry seasons when the waterbodies become isolated and start to contract in size thereby 'concentrating' fish. Like backflow billabongs, sandy creekbed habitats are least diverse in the late-dry and most diverse in the late-wet to mid-dry. In the early-wet season the diversity, biomass and composition of creekbed communities increases and changes markedly as isolated pools are transformed into rapidly flowing channels which are used as both migration routes and living space by many fish species.

Communities in escarpment perennial streams appear to be relatively unstable because their diversities may fluctuate markedly. These communities are most diverse and have high biomass in the dry season. In the early-wet season diversities and biomass plummet as many members of the community disperse downstream to the lowlands with initial flows in the creeks. These habitats, like the escarpment main channels, corridor and floodplain billabongs appear to be dry season refuges for fish. Perennial streams may be sought as refuges due to their characteristic cool, clear, permanent waters rather than as a response to their overall depth.

Explanations for many of the above changes in community diversity, biomass and composition are lacking as they may involve a complex of factors such as predation, competition, fish migration, environ-mental tolerance and habitat preferences. However, one of these factors, fish migration, apparently induces considerable community change. This was alluded to above and will be expanded on in the next section.

Studies on the relationship between environmental characteristics and community diversity were pioneered by R. H. MacArthur (MacArthur & MacArthur 1961) when he demonstrated that bird community diversity increased with increasing complexity of vegetation. More recently, attempts have been made to apply the same principles in aquatic environments. Spatial structure of habitats has repeatedly been shown to be important in determining the distribution and abundance of animals, acting as a factor in guiding habitat selection and, presumably, as an ultimate factor by its associations with critical resources such as food, spawning sites, or cover from predators. Do these concepts apply to the Alligator Rivers fish fauna? That is, are more species found in the more complex aquatic habitats in the region? This question was investigated by Bishop & Harland (in press) as a precursor to ongoing studies of fish habitat preference. The dependent variable (number of species) was generated by standard netting in thirteen waterbodies during a late-wet season (a season when all fish had had an opportunity to move to preferred microhabitat types). The independent variable (habitat complexity) was measured at 100 randomly chosen points within the netted area of the above water bodies. The dimensions of the habitat complexity measure (levels in parentheses) were depths (4), substrate (7), aquatic vegetation (4), and bank cover (4). A significant relationship (Fig. 39) was shown to occur between the number of species and the habitat complexity of the waterbodies. However, other factors were influencing fish diversity,

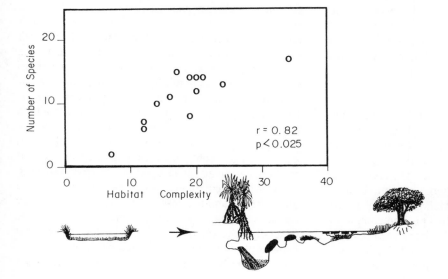

Figure 39. Scatter plot of number of fish species against habitat complexity in the Alligator Rivers region during the 1981 late-wet season.

as 40% of the variance was unaccounted for by the relationship. As mentioned earlier in the chapter, Lowe-McConnell (1969) considered that spatial heterogeneity or complexity of environments in the tropics could be responsible for higher fish community diversities relative to temperate fresh waters. This hypothesis can only be tested for Australia once similar investigations of habitat complexity are undertaken in temperate fresh waters.

Many studies have shown increased community diversity in stream fishes from upstream to downstream reaches (Larimore & Smith 1963, Sheldon 1968). In Australian tropical fresh waters, this phenomenon was shown by Beumer (1980) in the Black-Alice River system. However, Beumer's example was not well controlled, as the proximity of his sampling sites to estuarine reaches introduced another diversity factor, i.e. diversity enhancement from estuarine vagrant species penetration. Diversity reduction going upstream in the Alligator Rivers region lowlands zone is not so clear-cut due to a number of factors. An interesting example is the Cooper Creek system which drains western Arnhem Land towards the estuary of the East Alligator River. Sampling 140 km upstream from estuarine reaches yielded only 11 fish species. This is half the number of species found in equivalent habitats in the Magela system, some 50 to 60 km upstream from its estuary. Was this a species-reduction with distance-upstream phenomenon? The Cooper Creek samples contained no diadromous species and local Aborigines indicated that a very obvious and edible diadromous species, the silver barramundi, had never been captured in that section of Cooper Creek. The apparent lack of diadromous species suggested that there was a migration barrier to these species in the lower reaches of Cooper Creek. Recently such a barrier (in the form of a 5 m waterfall) was located in the Mount Borradaile outlier area in the lower reaches of Cooper Creek some 10 km upstream of its estuarine reaches. From this example the species-reduction/distance-upstream phenomenon appears to be most relevant to the distribution of diadromous fishes around migration barriers in the region.

Stream order
Stream order is another factor which is considered to affect fish community diversity. This phenomenon has been demonstrated mainly in temperate latitudes, for instance in North America (Kuehne 1962) and in the Seven Creeks system of the upper Murray-Darling system (Cadwallader 1979), though some tropical examples can be found, as in Malaya (Bishop 1973). Harrel et al. (1967) attributed this phenomenon to an increase in available habitat and a decrease in environmental fluctuations as stream order and size increase. This relationship is not so straightforward in the Al-

ligator Rivers region, probably because the stream order concept is inappropriate to the region's physiography. Problems are encountered defining stream order in the region from season to season and from habitat to habitat. The fact that streams dry up into a series of isolated waterbodies for most of the year, and in the wet season inundate the entire floodplain and lowland areas adjacent to the sandy creek makes it hard to define stream order. Many streams do not originate as well-defined channels, but rather as sheet runoff in the wet season when the upper reaches of once isolated catchments can be inter-connected. In the lower reaches the channels become better defined but start to anastomose into complex networks in the lower corridor and upper floodplain areas. Fish communities which are retained in water bodies in the dry season are usually subsets of a well mixed, widespread, wet season community (see later) rather than subsets strongly orientated to stream order parameters. There are however some species which are characteristically found in first order streams in the mid- to late-wet season. These are usually small bodied species (e.g. black striped rainbow fish, blue-eyes and occasionally the purple spotted gudgeon) which are generally suited to live in shallow waterbodies. The larger fish species (e.g. forktailed catfish, tarpon, and silver barramundi) tend to be captured mainly in deep waterbodies and usually in high stream-order channels. An inference from this which is supported by data is that there is generally a stronger positive relation between biomass and stream order than with community diversity in the region.

Fish moving upstream frequently orientate to channels with greatest discharge, i.e. high stream-ordered channels. Within this generalization there may lie an explanation for some of the association apparent between fish species and stream order.

Characteristic escarpment communities
A degree of reservation has already been implied with regard to the general concept that the number of fish species increases in a downstream direction. If this concept applies at all in the region it relates mainly to species addition, particularly at migration barriers, of part-time members of estuarine communities. Overlapping this pattern is a species loss sub-pattern which usually operates downstream of escarpment habitats in the dry season and downstream of corridor habitats in the wet season. The species lost (e.g. the black-striped rainbow, gertrudae's blue-eye, three teraponid species: black anal-finned grunter, sharp-nosed grunter and the black bream, and to a lesser extent, archerfish) below the above habitats at various times of the year are generally characteristic of headwater escarpment habitats throughout most of the region. A similar situation exists in New Guinea in respect of the headwater

distribution of teraponids (Mees & Kailola 1977).

The larger species of fish in overseas floodplain systems can be divided into two fairly distinct groups on the basis of their behaviour in response to the peculiar conditions of flood-drought rivers (Welcomme 1979):

1. The first group avoids severe conditions on the floodplains by migration to the main river channel. A few species are confined to the river channel at all times and never penetrate the plain. Species of Cyprinidae in Asia and Africa and Characoidei in Africa and South America are conspicuously members of this group. In the Mekong system species of this type are termed '*whitefish*'.

2. The second group has considerable resistance to deoxygenated conditions. They frequently remain in standing waters of flood plains in the dry season and if they move to the river they remain in vegetated fringes or in pools in the riverbed. Most silurids, chanids, anabantids, osteoglossids and lungfish belong to this category. In South-east Asia these fish are termed '*blackfish*'.

This white and black classification scheme appears relevant to the Alligator Rivers region. The headwater escarpment dwellers could be thought of as the '*white fish*' species and species such as the fork-tailed catfish, rendahl's eeltailed catfish, boney bream, one-gilled eel, penny fish, carp gudgeon and sleepy cod could be considered as '*black fish*' species. Exceptions are al-

ways present in natural systems when classification systems are used to simplify patterns. The exceptions in this case are the species (e.g. the longtom, spangled grunter, butter jew and the checkered rainbow fish) which are found in both escarpment and floodplain habitats for significant portions of the year. Possible explanations for such differing distribution patterns between species and within species on a seasonal basis are broached later in the chapter.

Fish movement in the region

Many tropical freshwater fishes migrate up-river as the water level rises in the wet season and undergo lateral movements out onto flooded plains when conditions permit (Lowe-McConnell 1975). Their reactions to water flow are complex and have been little investigated.

Ongoing analyses on fish community structures in waterbodies of the various habitats of the region have indicated lowest homogeneity between habitats in the dry season and highest homogeneity during the mid-wet and late-wet seasons (i.e. the 'widespread wet season community' concept). This increased homogeneity corresponds to the relinking across the lowlands of the dry season refuge areas of the escarpment, corridor and floodplain zones. The main channels of the creek in the wet season facilitate considerable upstream and downstream fish movement. This is

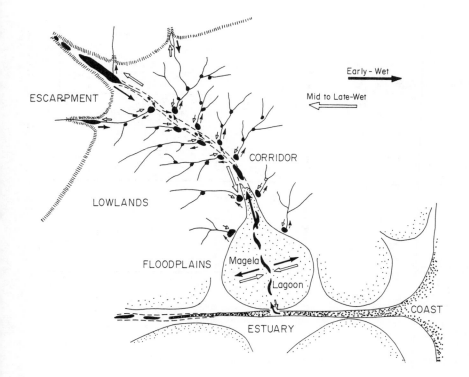

Figure 40. Schematic representations of likely fish movement patterns in Magela Creek.

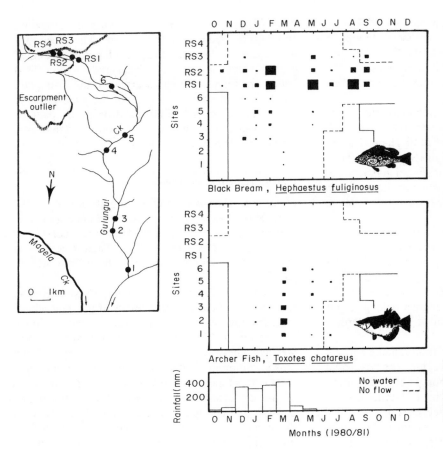

Figure 41. Distribution and abundance (8 fish = 1 mm^2) of black bream and archerfish in Gulungul Creek showing two types of movement strategies.

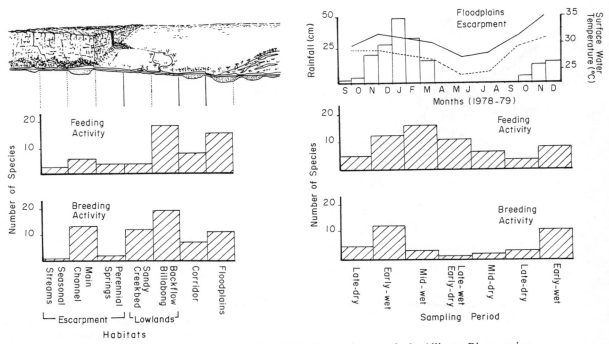

Figure 42. Variation of fish feeding and breeding activity with habitats and seasons in the Alligator Rivers region.

evidenced by the fact that communities present in lowland habitats (which were previously relatively barren as regards species), by the wet season contained many species characteristic of the escarpment and/or the corridor and floodplain zones in the dry season. Communities in the lowlands thus represent a species flux between communities in the upper and lower reaches of the catchments. A schematic representation of likely fish movement patterns in Magela Creek is presented in Figure 40.

Recolonization in the lowlands

Recolonization in lowland sandy creeks and backflow billabongs in the early-wet is the most obvious effect of fish movement in the region. During this season standard fishing methods of Bishop et al. (1986) showed a vast increase in biomass in these habitats once creek flows and backflows had commenced. Biomass by standard methods in Gulungul billabong (a backflow billabong off Magela Creek) relative to the late-dry season catches can increase from 40 to 400 fold. This increase in biomass was attributed mainly to the capture of sexually mature fork-tailed catfish, bony bream, butter-jews and other fish. Correspondingly there is a decrease in biomass of these species in floodplain, corridor and escarpment zones.

The relative biomass contribution to the lowland billabongs from either upstream (escarpment) or downstream refuge areas is difficult to determine. Occasionally however it is possible to describe communities arriving in the lowland from upstream reaches by sampling the first 'flow tongue' arriving in the early-wet season. Sexually mature archerfish, butter-jews, banded grunters, spangled grunters and checkered rainbow fish are usually found to be the most important by biomass in such frontier communities.

Underwater fish monitoring in clear escarpment streams in the region provides an excellent opportunity for migration characterization. The potential of this technique was first realised when multi-species congregations of fish species were observed below escarpment cascades in perennial streams. Investigations trying to relate the abundance and types of upstream migrants beneath these cascades to the duration and extent of wet season flows are continuing. Monitoring in Gulungul Creek, a clear and perennial escarpment tributary of Magela Creek, has indicated that the fishes therein can be grouped according to four movement strategies.

1. Distinct downstream returning movement from the escarpment perennial springs area during the wet season. This movement pattern was demonstrated by five species. The black bream (Fig. 41) demonstrated this pattern most clearly.

2. Distinct upstream returning movements from the lower reaches during the wet season. Again five species undertook such movement. The archerfish showed this pattern (Fig. 41) most clearly. However, in Magela Creek this species appears to undertake mostly Type 1 movement. A few of these species entered the creek in the early-wet season though the majority entered the system after floodpeaks in the mid-wet season.

3. Complex movement patterns. Difficulties were encountered elucidating the movements of two species as populations were present at both the springs area and lower reaches in the dry season.

4. Sporadic appearances in the creek. Ten species were sporadically found in low numbers in the creek.

In examining a small creek like Gulungul it was hoped that an understanding of larger more complex systems could be gained. An appreciation of how community diversity can change in the wet season was a side benefit of such a study. From a diversity of zero in the creek in the late dry season it rose sharply once escarpment colonisers entered with first wet season flows. Later in the wet season the diversity was further enhanced by the additional presence of lower reach colonisers, i.e. a community overlap phenomenon. Above average flood events tended to reduce diversities in the creek by literally washing colonisers down to the lower reaches of the creek. However, it takes only a few days for the creek to be recolonised. A similar finding was made by Beumer (1980) in the Black-Alice River of northern Queensland.

Movement incentives

The aim of migration for many tropical freshwater fishes is to exploit rapidly a temporary rich food source brought about by flooding and washing in of terrestrial resources or new production of food with the upsurge in supply of nutrients (Lowe-McConnell 1975). An obvious question that should be considered is why the species move from dry season refuge areas into the lowlands in the wet season to risk stranding and predation once flows subside? Do the species' movement strategies conform to the generalities above? To answer the above questions two biological parameters (feeding and breeding activity) were examined from the biological data-base of Bishop et al. (1986). The measure on these parameters was the number of species undertaking high levels of activity at particular seasons and habitats. The arbitrary definitions of 'high levels' for feeding and breeding activity were:

Feeding activity (Fig. 42): 50% of the specimens of a particular species, with stomach fullness index (Ball 1961) greater than 4 (i.e. stomach more than half full);

Breeding activity (Fig. 42): 50% of the adult specimens of a particular species, with gonad maturing stage index (Pollard 1972) equal to 4 or 5 (i.e. ready to spawn).

Feeding activity was most intense in the wet season in backflow and floodplain billabongs. A marked increase in feeding occurred in the early wet season after a minimum in the late dry season. Breeding activity was greatest in the early wet season followed by a reduction through the wet season into the dry. Breeding was most intense in the backflow billabongs followed closely by escarpment main-channel waterbodies, the sandy creek and floodplain billabongs. Recruitment of juveniles was greatest in the mid-wet season and was essentially delayed one season past maximal breeding activity (Bishop & Harland, in press). Most recruitment appeared in the more downstream billabongs (effect of downstream flow displacement of larvae), however a noticeable peak occurred in backflow billabongs.

A reason for species migration into lowlands from dry season refuge areas becomes evident from the above examination of biological parameters with respect to seasons and habitats; namely, the lowlands are valuable feeding, breeding and recruitment (nursery) areas. This situation is only temporary as many of the more mobile species must migrate back to refuge areas after the wet season as water levels fall and environmental conditions deteriorate in the lowlands. Fish movement in the catchment can therefore be seen as a response to temporal and spatial patchiness of resources, which supports generalities on tropical seasonal fish communities stated by Lowe-McConnell (1975). Environmental patchiness is recognised as pervasive and fundamental to the distribution of living organisms (Elton 1949, 1966, Andrewartha & Birch 1954). The changing distributions of the fishes and the tropical seasonal rivers of northern Australia thus provide an excellent example of animal response to environmental patchiness. The response in the Alligator Rivers region is particularly obvious in the early-wet season when the creeks flow once more.

Trophic structure of communities

In order to develop an appreciation of movements in another perspective, namely, energy and nutrient flows through freshwater fish in Australia's seasonal tropical rivers, an outline of the trophic structure of communities follows. The case example is again provided from the Alligator Rivers region.

The features of freshwater fish faunas in three tropical continents (Asia, Africa and South America) were summarised by Lowe-McConnell (1975) as (1) the importance of terrestrial plant material as a direct food; (2) the important role of insects as food; (3) the important role of mud and detritus as direct food for some species, and; (4) the presence of large numbers of individuals and kinds of piscivorous fishes. Data on the feeding habits of the freshwater fishes of the Alligator Rivers region do not support these generalities.

Terrestrial plant material is only directly important to a few fish species in the escarpment zone; however, some omnivores consumed this material in the lower reaches during the dry season. The teraponids of the escarpment community consumed some terrestrial material in the form of fallen berries from overhanging trees and bark scraping from submerged *Pandanus* trunks. Terrestrial insects and aquatic insect larvae were important fish foods in the region but crustaceans were of comparable importance in the aquatic component of the diets. Direct utilization of mud and detritus as food was partially important to only a few fish species, notably the bony bream and the mullets. Large numbers and kinds of piscivorous fish species are not present in the fresh waters of the region.

The percentage frequency and the number of non-rare fish species in four representative feeding niches (herbivore/detritivore, omnivore, carnivore and piscivorous carnivore) in a longitudinal array of habitats in the Magela Creek catchment are shown in Table 19. Similar data (some niches have been redesignated for comparison's sake) are presented in this table for fish communities from various habitats in tropical seasonal rivers of New Guinea and South America. In considering the trophic structure of the region's habitats, a number of points can be made.

1. A high proportion of herbivores/detritivores and omnivores are found in escarpment plunge pool and perennial spring communities.

2. A preponderance of carnivores exists in escarpment main channels downstream across the lowlands to the floodplains.

3. The highest proportion of piscivores is found in sandy creekbeds followed by floodplain billabongs.

In general terms it appears that trophic-structure variations within one geographic region can be just as large as intercontinental variation. In other words, concepts regarding energy and nutrient flows in such aquatic systems should be developed from the catchment zone level and should not be overly influenced by the results from geographically close or distant areas.

There is considerable seasonal variation superimposed on the above habitat variation of trophic structure in the region. This is mainly induced by community changes caused by fish movement and species' dietary changes caused by food supply fluctuations. An explanatory summary of habitat and some seasonal changes in trophic structure of communities in the region follows.

The high proportion of herbivores/detritivores and omnivores in upper escarpment habitats, particularly in terminal plunge pools (Fig. 43), generally reflects a limited food supply. A trend of increased omnivorous-

Table 19. Feeding structure of communities from various habitats found in tropical seasonal rivers (number of fish species in parentheses).

Origin of samples	Feeding niche Herbivores/ detritivores	Omnivores	Carnivores	Piscivorous carnivores	No. of species
Australia					
ARR, Magela Creek[1]					
Escarpment					
Perennial springs	22.2 (2)	55.6 (5)	22.2 (2)	–	9
Terminal main channel	50.0 (2)	50.0 (2)	–	–	4
Main channel	7.1 (1)	35.7 (5)	50.0 (7)	7.1 (1)	14
Lowlands					
Sandy creekbed	–	27.2 (3)	54.5 (6)	18.1 (2)	11
Backflow billabongs	–	31.5 (6)	57.8 (12)	10.5 (2)	19
Corridor					
Main channel	9.1 (2)	22.7 (5)	54.5 (12)	13.6 (3)	22
Floodplain billabongs	5.3 (1)	21.1 (4)	57.9 (11)	15.8 (3)	19
Papua New Guinea[2]					
Purari River					
Springs off river	44.4 (4)	–	44.4 (4)	11.1 (1)	9
Main channel above delta	34.2 (13)	7.9 (3)	42.1 (16)	15.8 (6)	38
Waterbodies in delta	19.2 (5)	3.8 (1)	57.6 (15)	19.2 (5)	26
South America					
Amazon River					
Forest stream tributaries[3]	38.4 (15)	26.6 (10)	30.7 (12)	5.0 (2)	39
Lake Redondo[4]	17.5 (7)	25.0 (10)	45.9 (18)	12.4 (5)	40
Middle Parana-Paraguai River					
Oxbow lagoons[5]	37.5 (21)	21.0 (12)	21.0 (12)	19.6 (11)	56

1. Bishop et al. (1986), 2. Haines (1979), 3. Knoppel (1970), 4. Marlier (1967), 5. Ringulet et al. (1967).

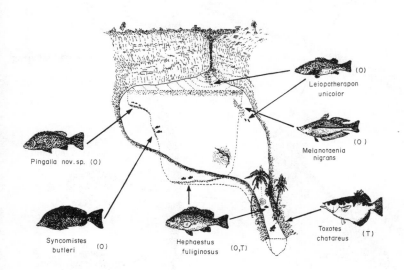

Figure 43. Feeding habits of fishes in escarpment plunge pools of the Alligator Rivers region during the dry season. Omnivores (O) browse periphyton and bark from submerged trees; terrestrial fruits and insects are utilized by some fish (T).

ness of fish communities going upstream into headwaters has already been shown by other workers. These reaches are generally oligotrophic as regards nutrients and are scoured by wet season flows. Consequently aquatic plants, zooplankton, aquatic insects and hence small carnivorous fish are virtually absent. *Crocodylus johnstoni* is generally the only piscivore present. The fish dwelling in these habitats must be able to sustain themselves on the limited amount of periphyton and washed-in and adjacent terrestrial matter. Archerfish, which are frequently abundant in escarpment habitats, have the unique ability to utilise as food terrestrial animals which are not in contact with the water. As mentioned earlier, all the fish occupying these reaches must also have extraordinary capacities for swimming and enduring high velocities. Many of these upstream habitats are vacated for the lowlands in the wet season.

Piscivorous behaviour in lowland sandy creeks is most obvious in the late-wet season (Fig. 44). This is the season when piscivores are most efficient as they can readily ambush schools of small migrants moving upstream. The communities contained in isolated pools when creek flow ceases are generally subsets of the migrating communities plus piscivores which fail to return to dry season refuges. The community structure of these pools changes noticeably through the dry season; the extreme example occurs when a multi-species community is transformed into one large, rapidly growing, piscivore. The newly arrived colon-

ists and the surviving residents in the early-wet season consume large quantities of terrestrial animals which are washed into the water from the recently inundated banks of the creek.

The slow flowing billabongs in the lowlands and floodplains are habitats for deposition and breakdown of organic matter arising from terrestrial sources. These billabongs are consequently mesotrophic as regards nutrients, have muddy substrates and hence luxuriant plant growth, and dense zooplankton and aquatic insect populations. As a result dense populations of small carnivorous fish exist which in turn support sizeable populations of piscivorous fish (Fig. 45). It is interesting to note that catadromous species most extensively utilize the lowest (i.e. the mullets feeding on periphyton and detritus) and highest (i.e. the silver barramundi and tarpon feeding on small fish) trophic food sources in the floodplains. It is also worth noting that these systems do not have any strongly herbivorous species which graze aquatic plants; this is

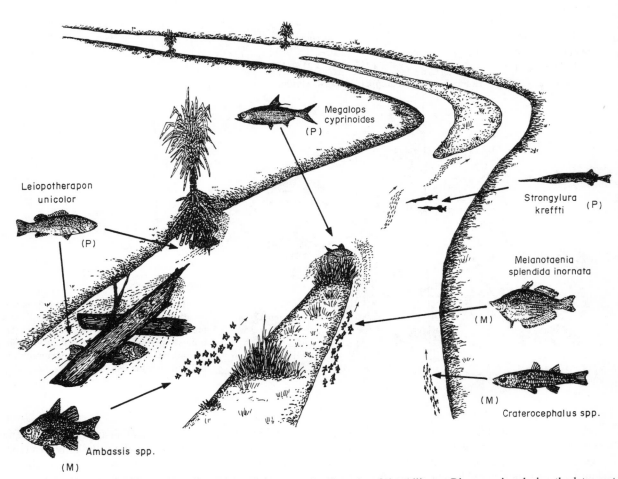

Figure 44. Piscivorous feeding habits of fishes in sandy lowland creeks of the Alligator Rivers region during the late- wet season. Ambush tactics are used by the Piscivores (P) to capture the small upstream migrants (M).

Photograph 4. Archerfish (*Toxotes chatareus*, size 150 mm) utilizes terrestrial foods close to the water's surface in upper-reach habitats and is a typical species of the escarpment area in the Alligator Rivers region (Photograph: G. Miles).

Photograph 5. Fish typical of sandy lowland creek habitats in the Alligator Rivers region: (a) The checkered rainbow fish (*Melanotaenia splendida inornata*, size 60 mm) is a widespread species which is readily ambushed by piscivores in sandy creeks during the wet season (Photograph: G. Miles).

Photograph 5 (continued). (b) The flat-headed goby (*Glossogobius giurus*, size 50 mm) is well camouflaged in sandy habitats for its predatory role, however, it can in turn be consumed by piscivores (Photograph: G. Miles).

Photograph 5 (continued). (c) The purple spotted gudgeon (*Mogurnda mogurnda*, size 40 mm) is another widespread species which has a similar trophic role as the flat-headed goby. In the wet season it is frequently washed downstream from plateau and escarpment habitats into sandy lowlands (Photograph: G. Miles).

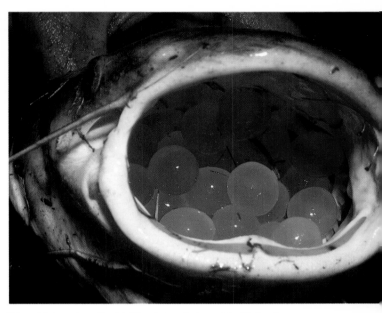

Plate 13. Fish typical of deepwater areas in lower-reach habitats in the Alligator Rivers region: (a) The bony bream (*Nematalosa erebi*, size 90 mm) consumes detritus, periphyton and some phytoplankton during the wet season.

Plate 13 (continued). (b) The fork-tailed catfish (*Arius leptaspis*, size 400 mm) is an opportunistic omnivore and its populations represent the majority of the fish biomass in lower-reach habitats. It is a buccal incubator: the male carries large yolky eggs in his mouth for many weeks in the early wet season (Photograph: G. Miles).

Plate 13 (continued). (c) The northern spotted barramundi (*Scleropages jardini*, size 600 mm) is one of Australia's few primary freshwater fish. It is a keen insectivore in the lower-reach habitats and is also a buccal incubator with the female carrying eggs in her mouth during the early-wet season (Photograph: G. Miles).

Plate 13 (continued). (d) Rendahl's eel-tailed catfish (*Porochilus rendahli*, size 120 mm) is usually a benthic carnivore which frequently enters deepwater areas from its more typical shallow water habitat (Photograph: G. Miles).

Plate 14. Fish typical of shallow water areas in lower-reach habitats in the Alligator Rivers region: (a) typical underwater scene in shallow weedy littoral zone microhabitats (Photograph: G. Miles).

Plate 14 (continued). (b) the ambassid perchlets (*Ambassis* spp., size 25 mm) are usually the most numerous fish in these habitats (Photograph: G. Miles).

Plate 14 (continued). (c) The carp gudgeon (*Hypseleotris compressus*, size 35 mm) is typically found in very weedy areas. The male (pictured) develops bright courtship colours during the wet season (Photograph: G. Miles).

Plate 14 (continued). (d) The fly-speckled hardyhead (Atherinidae, *Craterocephalus stercusmuscarum*, size 35 mm) feeds on zooplankton and some phytoplankton in weedy margins of billabongs (Photograph: G. Miles).

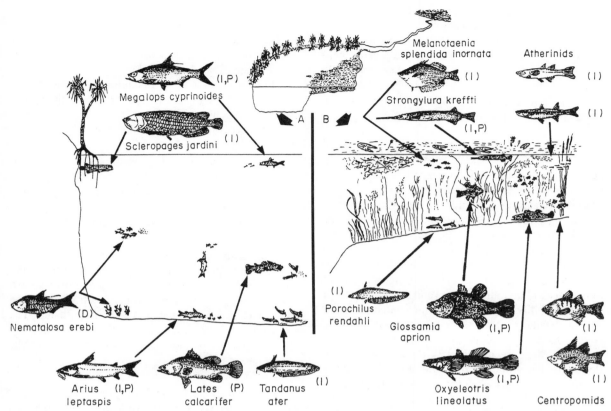

Figure 45. Feeding habits of fishes in deepwater (A) and littoral zones (B) of floodplain billabongs in the Alligator Rivers region. Invertebrates are consumed by many small fish species (I) in the littoral zone and by larger species (I) in the deepwater zone. The smaller fish are in turn consumed by piscivores (P) within the littoral zone and the deepwater zone. A few species (D) consume detritus and phytoplankton in the deepwater zone.

in contrast to overseas systems where many primary freshwater fish (e.g. *Tilapia* in the African great lakes) occupy this niche. Two main sub-habitats usually exist within these habitats, namely, the shallow weedy littoral zone and the deep open water zone. The small carnivorous fish generally feed and shelter from predators in the littoral zone while large fish occupy the open water zone. The large piscivores either make feeding incursions into the littoral zone or crop small fish which move out of the littoral zone for migration or other survival ploys.

Marchant (1982) recorded the greatest density of macroinvertebrate taxa in the littoral zone of backflow billabongs in the late-wet season in Magela Creek; the density then fell for the following dry season until rapid resurgence occurred in the early-wet season. This pattern is reflected in the density of the small carnivorous fish and consequently the feeding activity of the piscivores.

Food chains

The food chains involving the fish of the region are very complex because there is a large number of food and fish taxa, a high proportion of omnivorous fish in the communities, and high variability of diets between seasons and between habitats.

The main transfer of energy and nutrients from the primary producers to the fish fauna appears to be via the invertebrates, except in some upper reach escarpment habitats. The number of links in the chain between primary producers and the invertebrates is unknown. Marchant (1982) noted that organic detritus is probably a major source of food for macroinvertebrates within the littoral zone of muddy billabongs of the Magela Creek system. Decomposing leaves of *Pandanus, Barringtonia* and *Melaleuca* trees appear to make up much of the detritus. The trophic relationships of the variety of small fish which harvest the invertebrates are very complex as the opportunistic feeding habits of most species result in frequent overlaps in food items consumed. The piscivorous fish species are the most frequently consumed of the 20 fish species eaten by man in the region (see Meehan, Chapter 12). By following various alternative pathways of energy and nutrient transfers, the primary piscivorous fish can be classified as feeding at a range of trophic levels from third to seventh. Lowe-McConnell (1975) noted that in freshwater ecosystems, as in any other foodchains, there are rarely more than four or five links, as longer chains are expensive bioenergetically with 80 to 90% of potential energy being lost at each successive stage.

Breeding strategies of the fishes in the region

Seasonal and habitat variation in breeding activity of the fishes of the region was alluded to earlier in this chapter. Twenty-five species in the Magela Creek system appeared to have peaks in reproductive development about the early-wet season. Many small species appeared to be aseasonal spawners but they also showed peaks in development in the early-wet. The fish take advantage of the extensive flooding which often causes planktontic blooms and other increases in food availability. At this time the area and diversity of aquatic habitats increases dramatically as does the cover afforded by rapidly growing aquatic plant communities. Many species largely depend on the first floods for the initiation and induction of spawning; normally this occurs when temperatures are rising and are relatively high. Similar strategies are seen in the Murray-Darling fishes, however, their reproductive development is near complete when temperatures are in the mid to low twenties as compared with ranges in the thirties for northern Australian fishes.

Another parallel reproductive strategy with the Murray-Darling system is movement associated with spawning. Movements of five species in the Murray-Darling system have recently (Reynolds 1983) been interpreted as strategies for sustaining and expanding distribution patterns. The dry season refuge areas in the Alligator Rivers region do not appear to be ideal spawning areas. The fish move from these refuges as soon as flows commence and enter more suitable areas, namely, lowland backflow billabongs and floodplain areas which because of their recent inundation have:

1. Low densities of predators and other fish species competing for space;
2. Rapidly increasing densities of zooplankton because of recently washed-in nutrients; and
3. Rapid growth of aquatic plants which offer cover for eggs and larvae.

Nineteen species in the Magela Creek system used lowland backflow billabongs for spawning; four species apparently used these billabongs exclusively. No species apparently breed exclusively in corridor or floodplain habitats. Seven species were nearly ubiquitous spawners, apparently breeding in at least four habitat types. At least three species, the silver barramundi, tarpon and Ord river mullet, appear to migrate downstream to the estuarine or coastal areas to spawn.

The selection of breeding time and site, type and number of eggs laid, stage of development at hatching and the fate of newly hatched young have been selectively evolved to achieve maximum survival of species. Each of the more common species of the region has been classified into one of the following broad, sometimes overlapping, breeding strategies:

Live bearing species. This is the most highly developed strategy found in the evolution of fishes. This strategy is paradoxically used by the most primitive group of fishes in the region, namely, the 'estuarine vagrant' river whaler sharks, sawsharks and stingrays.

Buccal incubating species. Mouth brooding is an advantage in tropical environments as the mobility of the parent enables the eggs to avoid areas of low dissolved oxygen or other detrimental water conditions. Escape and protection from predators is also facilitated. The Saratoga, fork-tailed catfish and the mouth almighty (Glossamia aprion, an apogonid or cardinal fish) use this strategy.

Nest building species and/or species which guard eggs. This type of strategy, while affording increased protection from predation and in some cases from decreases in dissolved oxygen, is subject to destruction by changes in water level. To overcome this disadvantage the incubation time is short and numerous eggs are laid over a number of separate spawning sites. Eel-tailed catfish, some grunters, gobies and gudgeons within the region would use this strategy.

Species which attach their eggs to aquatic vegetation or other objects in the water. The eggs are attached by adhesive tendrils to submerged objects. While aquatic plants provide cover from predation for the eggs, the eggs are subject to desiccation due to receding water levels or to dispersal and damage due to high stream velocities. An advantage is gained in this and the previous strategy in that territory can be selected directly by the young without expending energy on returning upstream which would be the case if the eggs or larvae were planktonic. The rainbow fishes and the longtom use this strategy.

Species scattering demersal eggs over the substrate. This method increases the dispersal of the eggs, thus decreasing the chance of the whole brood being destroyed simultaneously. However the eggs may be subject to detrimental changes in the physico-chemical environment near the substrate. The hardyheads (atherinids) scatter a small number of relatively large eggs while the spangled grunter scatters a large number of smaller eggs.

Species with pelagic eggs. This strategy is usually adopted by 'total spawners' (Lowe-McConnell 1975). They are usually strongly seasonal (early-wet) spawners and may undergo an extensive breeding migration. Generally their fecundity is very high, eggs are small and at the end of a short incubation period, poorly developed, planktonic larvae emerge. The catadromous tarpon and silver barramundi use this strategy in estuarine and coastal waters. The archer fish and bony bream use this strategy within freshwaters of the region. Juveniles of these species are very noticeable upstream migrants in the late-wet season as they return from floodplain areas after having been effectively dispersed downstream as eggs or larvae.

A high proportion of the species examined in the region appear to exhibit some form of parental care (live bearing, buccal incubation, guarding of nests). This strategy may have evolved in response to the variable conditions imposed by the extreme wet and dry seasons. Competition for food and other 'niche parameters' due to high diversities and biomass densities in the region may also make such strategies desirable.

NATURAL FISHKILLS

Fish kills at the beginning of the wet season are frequently observed in the tropical coastlands of northern Australia. Within the Magela Creek system of the Alligator Rivers Region at least one kill has been recorded annually in recent years. The first published report on such kills was made by Bishop (1980) who examined a kill in a billabong on the Magela floodplain during the 1978 early-wet season. Bishop considered that low dissolved oxygen levels (lowered by the mixing of anoxic bottom waters by first wet season flows) may explain the kill; however, some observations indicated that low dissolved oxygen concentration may not have been the only factor which caused the mortalities. In the following year Brown et al. (1983) reported a large natural kill in another Magela floodplain billabong which involved 3,400 fish comprising eight different species. These fish deaths were assessed to be the result of a combination of natural acid water runoff (pH 2.4) into the billabong and resultant elevated aquatic biotoxic aluminium levels (500 $\mu g l^{-1}$). A year later again Brown's colleagues examined a kill in a corridor billabong and found that it could not be explained by acidity and aluminium levels. Another mechanism for fish kills was postulated by Bishop et al. (1982) in that it may be possible that the first wet season rains leach out soluble ichthyocidal compounds such as saponins from trees and shrubs surrounding billabongs and hence induce these kills. This mechanism was proposed after observations were made of Aboriginal people using plants to induce kills for food gathering purposes on the lowlands of the Alligator Rivers region. A subsequent literature review on this topic revealed that nearly 70 species of plants have been used as ichthyocides by Aboriginal people in northern Australia.

The only conclusion from all of the above is that there is no simple and universal explanation for freshwater fish kills in northern Australia. A complex of factors is surely responsible for the kills. Which factor is dominant is probably dependant on annual climatic phenomena such as the severity of the dry season and the amount of rainfall and hence the rapidity with which first flows cover the floodplains in the early-wet season. It can be readily understood why it is the early-wet season when most of the fish kills occur, for this is the season when fish are most stressed from deteriorating environmental conditions in the dry season and when they are most exposed to terrestrial toxicants which reach waterbodies in concentrated form with first flows. One thing that is particularly obvious is the scantiness of our knowledge of factors and mechanisms which affect fish survival in such waters. The ability of a species to avoid toxic waters in the first flow period of the early wet-season is likely to be the most positive survival strategy to ensure breeding opportunity later in the season.

FUTURE OF THE FRESHWATER FISHES OF NORTHERN AUSTRALIA

Environmental degradation

The future of the freshwater fishes, in terms of continued survival, is strongly dependent on the maintenance of naturally existing environmental conditions in northern Australia's aquatic systems. The consequences of thoughtless development of resources by humans in these once inaccessible lands are severe and are often focussed on the drainage systems. An example of such environmental degradation is presented by Jeffree & Williams (1975): 'The aquatic fauna of the east branch of the Finniss River (abandoned Rum Jungle Uranium mine site, Top End, N.T.) has been almost wholly destroyed in the 10 km section downstream of the mined area'. In this case, fish kills were associated with the entry of heavy metal pollutants, mainly copper and zinc, from the mining area. The kills occurred at the beginning of each wet season whenever a flush through the eastern branch of the Finniss River did not coincide with high flow in the main Finniss River channel, and provided that the fish populations had been able to recolonise from adjacent unpolluted zones.

Nowadays resource developments, at least the ones open to public attention, are subject to more effective environmental controls. Nevertheless, local damage still arises by subtle indirect routes and from developments which are deemed 'essential' for the comfortable colonization of once remote regions. So often environmental protection is considered first from a cost-benefit perspective, a perspective which usually leaves the environment a poor second. A demonstration of such developmental processes is found in the Alligator Rivers region. The resource in the region is uranium and, because uranium mining is subject to much public attention, there are tight environmental controls on its mining and processing. A town was built to accommodate the miners and the many government workers involved in environmental protection. An essential service to the town is of course a sewerage system. As it was economic to install only a primary treatment system, and because it is necessary for the system to make effluent releases into the creek system, much environmental concern and argument have been raised. If concerns are valid, the consequences of introducing sewage effluents into a normally nutrient-poor aquatic system could be dramatic – the repercussions of which would be detected all the way up foodchains to piscivorous fish. The irony of this situation is that of all the environmental attention afforded the uranium mining, little can be directed to the release of sewage because the investigation of the effects of sewage is not within the charter of agencies investigating the effects of uranium-mining! To make matters worse, the magnitude of the sewage impact could be sufficient to mask effects of releases and seepage from the uranium mine site and hence make much of the premining base line data of little value. No sewage effluent has been released in recent years.

Another resource of the region is its scenic beauty and interesting flora and fauna. This resource conflicts with the existing uranium industry but survives due to public appreciation and the tourism the resource indirectly generates. To have a tourist industry, it is essential that there are good roads from population centres. The environmental price of roads to once inaccessible regions is often high and never really seriously considered by zealous planners whose intention is to open up 'new frontiers'. Erosion, siltation and sustained turbidity levels are frequently the early impacts of such developments. A more long-term impact, particularly for raised all-weather roads with narrow high-velocity culverts over creek crossings, is retardation of movement of fish and other animals moving to feeding, breeding or refuge areas.

Movement barriers

Considerable emphasis has been placed on the impor-

tance of movement to the survival of fish in this chapter. Indeed, in monsoonal regions which are characterised by patchiness of biological resources, and temporary connections between the resources, movement as an essential survival strategy can well be expected. Consequently, man-made barriers across creeks and rivers in these regions would be expected to have considerable impacts on fish populations which have important resources upstream and downstream of these barriers. The impact of impoundment developments are first apparent immediately below the dam walls where large schools of upstream migrating fish congregate (e.g. Ord River Dam, W.A. and below Lake Moondarra, Qld on the Leichhardt River). For the first few years this phenomenon is viewed by local fishermen as leading to improved fishing; however, as time goes on, fishing deteriorates, particularly for species which have essential resources either side of the barriers. The quality of sport fishing within impoundments usually remains poor and uninteresting. Northern Australia's general lack of sport-fish which are well adapted to lacustrine life styles would be the main cause of this poor fishing. Also, the north's most sought after sport fish, the silver barramundi, cannot sustain populations in impoundments as its critical breeding resources are located in estuarine habitats. Nevertheless limited fisheries have developed in some impoundments, the most notable being a gillnet fishery for fork-tailed catfish in Lake Argyle, Ord River scheme.

Dry season refuges

The concept of dry season refuges for northern Australian fish was also introduced within this chapter. An analogous concept was introduced for Northern Territory waterbirds by Crawford (1979). Fish are, however, more restricted in their movements than birds, for they cannot move outside catchments or from lagoons once flow ceases. Crawford indicated that bird refuge areas are of critical importance in their conservation, and so similar but more emphatic conclusions can be drawn for fish refuge areas. The vulnerability of refuge populations is usually highest towards the end of the dry season. Fish at this time coexist in very high densities due to falling water levels and are subject to considerable physico-chemical stress, particularly in floodplain habitats. Recent improved access to upper reach escarpment refuges in parts of the Northern Territory has introduced another vulnerability factor, namely, overfishing of prespawning refuge populations. The black bream is most prone to this type of 'sport' as it is ridiculously easy to catch, withdraws almost totally back to refuges in the dry season, spawns in the early-wet season and moves downstream at this time in sizeable closely packed schools.

Available evidence suggests that the amateur and professional silver barramundi fishery has declined noticeably in recent years. Overfishing is thought to be mainly responsible for that decline. Effective management of biological resources usually arises from a sound knowledge of factors which influence the dynamics of the populations in question. Paucity of information about such factors is a major management obstacle in northern Australia. The situation is no different in southern Australia.

Introduced biota

Introduced animals and plants are known to have had significant effects on aquatic systems of southern Australia, and still others are presumed to have had an impact, though there is little evidence of its precise nature. The effects of introduced biota in northern Australia is just starting to be noticed and their ramifications realised. This late recognition can be attributed to the remoteness of northern Australia and its relatively recent colonization and development by white Australians. Moreover, the climate of northern Australia was inhospitable to much of the early biota introduced from temperate lands.

One early import from tropical lands that has proliferated in the coastlands of the Northern Territory is the water buffalo, *Bubalus bubalis*. The impact on aquatic systems of its wading and wallowing activity and its partiality to marsh vegetation and hydrophytes is severe and long established (see also Ridpath, Chapter 11). The direct impact of the buffalo on fishes is difficult to ascertain though some comments can be made. Firstly, the muddying of billabongs by wading and wallowing activity is likely to have an impact on fish which are sensitive to high turbidity levels or need clean water to feed. Fish species typical of clear upper reach areas may be the most sensitive to this impact and may even actively avoid affected sites. Increased turbidity levels may also aggravate stresses existing on fish which are sheltering in refuge areas towards the end of the dry season. Secondly, the reduction of hydrophyte and swamp vegetation densities by buffalo activity has probably had profound effects on the composition of fish communities. The results of ongoing studies in the Alligator Rivers region indicate that high densities of aquatic vegetation are usually associated with high densities of small fish species and low densities of large fish species. Billabongs in the region which are now weeding up due to buffalo removal contain fish communities which are undergoing changes of dominance from large to small species.

Northern Australia is relatively free of introduced fish species. However, in areas around Darwin the mosquito fish, *Gambusia affinis*, is already present. Many tropical freshwater fish, particularly the herbi-

vorous carps and cichlids, typical of Asia, Africa and South America might thrive in northern Australia if introduced. The zoogeographic boundary across the Malay-Indonesian Archipelago has effectively kept Australia free from hundreds of primary freshwater fish species for millions of years. Australia has the potential to become a 'biological cesspool' of introduced fish that the State of Florida in the USA is said to have now become (Courteney & Robins 1973). Papua New Guinea is well on the way to becoming such a cesspool with the introduction of many cichlid species (Berra et al. 1975). The Mozambique mouth-brooder, *Sarotheradon mossambica*, already infests parts of Papua New Guinea and reports exist of it and other related species in Queensland. Hubbs (1968) described the detrimental effects of *Sarotheradon* (formerly *Tilapia*) introductions in North America. We should now regard Papua New Guinea as something of an example of what northern Australia could well become without due care and attention (Tilzey 1980).

The low quality of sport fishing in northern Australian impoundments has prompted moves to have the Nile perch, *Lates niloticus*, introduced into Australia. The species of *Lates*, in contrast to Australia's silver barramundi, can complete its lifecycle within impoundments and thus self-maintain its population. There have been many objections to this introduction from the scientific community on the grounds of the impact this species would have on fish communities if it escaped into the wild. Hitherto no such proposals have received government approval.

One final note needs to be made in this chapter concerning the indirect impact of introduced plants on northern Australian fish communities. The plants of most concern are the mat-forming species such as *Salvinia molesta* and *Eichhornia crassipes* (water hyacinth) which are currently infesting river systems across northern Australia. Thick mats of these species strongly reduce available light in the water column. The ramifications of such a reduction in energy input in the northern aquatic ecosystems could well be magnified through a phytoplankton, zooplankton, macroinvertebrate, vertebrate foodchain. It is very difficult to ascertain the impact on fish communities, but their structure is likely to change markedly were new trophic routes opened up.

Opportunities for unique behavioural research

The clear waters of the upper reaches of creeks and rivers in northern Australia provide a unique opportunity for behavioural research. In these habitats, many species coexist through the dry season and then undertake well-timed evacuations and recolonizations in the wet season. The bold nature of the majority of the species makes it possible directly to observe intra- and inter-species interactions, microhabitat preferences and various behaviours associated with reproductive and feeding strategies. Many parallels can be drawn with studies being undertaken on coral reef fishes. The main difference with coral reef fish studies is that in the clear fresh waters of northern Australia there are fewer complicating variables operating which may confuse observations, since a simpler system exists as fewer species are present, the habitats are less complex and recruitment is generally more regular and timed. The clear fresh waters of northern Australia provide some of the best opportunities for such studies in comparison with tropical fresh waters elsewhere. The major advantages are the general lack of health hazard associated with its waterways, good accessibility and security.

ACKNOWLEDGEMENTS

In respect of the fish section (KAB), the willing help of Sally Allen, David Walden and Wayne Harland is greatly appreciated. Thanks also to Dr David Pollard of NSW State Fisheries for constructive discussions over the past years. One of us (MAF), acknowledges with appreciation permission from Noranda Australia Ltd to use limnological data obtained during environmental studies of their Koongarra Project, and to Dr Peter Tyler and Tim Walker for discussions on their limnological studies in the Alligator Rivers region.

REFERENCES

Andrewartha, H.G. & L.C. Birch 1954. *The Distribution and Abundance of Animals*. Chicago, Illinois: University of Chicago Press.

Allen, G.R. 1982. *A field guide to inland fishes of Western Australia*. Perth: Western Australian Museum.

Allen, G.R. & D.F. Hoese 1980. A collection of fishes from the Jardine River, Cape York Peninsula, Australia. *J. Roy. Soc. West. Aust.* 63: 53-61.

Ball, J.N. 1961. On the food of the brown trout of Llyn Tegid. *Proc. Zool. Soc. Lond.* 137: 599-622.

Berra, T.M., R. Moore & F.L. Reynolds 1975. The freshwater fishes of the Laloki River System of New Guinea. *Copeia* 1975: 316-326.

Beumer, J.P. 1980. Hydrology and fish diversity of a North Queensland tropical stream. *Aust. J. Ecol.* 5: 159-186.

Bishop, J.E. 1973. Limnology of a small Malayan River, Sungai Gombak. *Monographiae Biologicae* 22. The Hague: Junk.

Bishop, K.A. 1979. Fish and aquatic macroinvertebrate communities of a coastal river (Shoalhaven River, New South Wales) during the development of a water diversion scheme. M.Sc. thesis, Macquarie University.

Bishop, K.A. 1980. Fish kills in relation to physical and chemical changes in Magela Creek (East Alligator River system, Northern Territory) at the beginning of the tropical wet season. *Aust. Zoologist* 20: 485-500.

Bishop, K.A., S.A. Allen, D.A. Pollard & M.G. Cook 1986. *Ecological studies of the freshwater fishes of the Alligator Rivers Region, Northern Territory*. Research Report 4, Volume I, Supervising Scientist for the Alligator Rivers Region. Canberra: AGPS.

Bishop, K.A., L.M. Baker & B.N. Noller 1982. Naturally-occurring ichthyocides and a report on *Owenia vernicosa* F. Muell. (Family Meliaceae), from the Magela Creek System, Northern Territory. *Search* 13: 150-153.

Bishop, K.A. & W.G. Harland (in press). *Further ecological studies on the freshwater fishes of the Alligator Rivers Region*. Office of the Supervising Scientist, Research Institute Report Series.

Bishop, K.A. & H.E. Larson 1984. *Fish survey of the eastern drainage catchments of the Davenport and Murchison Ranges, Northern Territory, Central Australia*. Second Report to the Arid Zone Research Institute, NT Conservation Commission.

Brown, T.E., A.W. Morley, N.T. Sanderson & R.D. Tait 1983. Report of a large fish kill resulting from natural acid water conditions in Australia. *J. Fish. Biol.* 22: 335-350.

Cadwallader, P.L. 1979. Native and introduced fishes of the Seven Creeks (Goulburn River System), Victoria. *Aust. J. Ecol.* 4: 361-385.

Courteney, W.R., Jr & C.R. Robins 1973. Exotic aquatic organisms in Florida with emphasis on fishes: a review and recommendations. *Trans. Amer. Fish. Soc.* 102: 1-12

Crawford, D.N. 1979. Waterbirds: Indices and fluctuations in dry season refuge areas, Northern Territory. *Aust. Wildl. Res.* 6: 97-103.

Dames & Moore 1978. Environmental Studies, Koongarra, Northern Territory. *Noranda Koongarra Project Draft Environmental Impact Statement*: Appendices, Vol. 1. Melbourne: Noranda Australia Ltd.

Elton, C. 1949. Population interspersion: an essay on animal community patterns. *J. Ecol.* 37: 1-23.

Elton, C. 1966. *The Pattern of Animal Communities*. London: Methuen.

Galloway, R.W. 1976. Geomorphology of the Alligator Rivers area. *Lands of the Alligator Rivers area, Northern Territory*. CSIRO Land Research Series 25.

Haines, A.J. 1979. Ecology survey of the fish of the lower Purari River system, Papua New Guinea. Purari River (Wabo, Hydroelectric Scheme), *Environmental Studies* 6: 102-111.

Harrel, R.C., B.J. Davis & D.E. Dorris 1967. Stream order and species diversity of fishes in an intermittent Oklahoma stream. *Am. Midl. Nat.* 78: 428-436.

Hubbs, C. 1968. An opinion on the effects of cichlid releases in Northern America. *Trans. Amer. Fish. Soc.* 97: 197-198.

Iredale, T. & G.P. Whitley 1938. The fluvifaunulae of Australia. *S. Aust. Nat.* 18: 65-68.

Jeffree, R.A. & N.J. Williams 1975. Biological indications of pollution of the Finniss River system, especially fish diversity and abundance. In D.R. Davy (ed.), *Rum Jungle Environmental Studies*. Australian Atomic Energy Commission Report E365.

Jennings, J.N. 1972. Some attributes of Torres Strait. In D. Walker (ed.), *Bridge and barrier: the natural and cultural history of Torres Strait*: 29-38. Res. School. Pacif. Stud. Aust. Nat. Univ. Publ. BG/3.

Knoppel, M.A. 1970. Food of Central Amazonian fishes. *Amazoniana* 2: 247-352.

Koontz, D.V., A.W.Morley & N.T.Sanderson 1983. Some aspects of changes in physico-chemical parameters during the first flooding of Magela Creek. *Scientific Workshop: Environmental Protection in the Alligator Rivers Region.* Supervising Scientist for the Alligator Rivers Region, Jabiru, NT.

Kuehne, R.A. 1962. A classification of streams, illustrated by fish distribution in an Eastern Kentucky Creek. *Ecology* 43: 608-614.

Lagler, K.F., J.E.Bardach & R.R.Miller 1962. *Ichthyology: The Study of Fishes.* New York: Wiley.

Lake, J.S. 1971. *Freshwater Fishes and Rivers of Australia.* Sydney: Nelson.

Lake, J.S. 1978. *Australian Freshwater Fishes, an Illustrated Field Guide.* Melbourne: Nelson.

Lake, P.S. 1982. The relationship between fish distribution, stream drainage area and stream length in some streams of south eastern Australia. *Bull. Aust. Soc. Limnology* 8: 31-37.

Larimore, R.W. & P.W.Smith 1963. The fishes of Champaign County, Illinois, as affected by 60 years of stream changes. *Bull. Ill. St. Nat. Hist. Serv.* 28: 299-382.

Lowe-McConnell, R.H. 1969. Speciation in tropical freshwater fishes. *Biol. J. Limn. Soc.* 1: 51-75.

Lowe-McConnell, R.H. 1975. *Fish Communities in Tropical Freshwaters.* New York: Longman.

MacArthur, R.H. & J.W.MacArthur 1961. On bird species diversity. *Ecology* 42: 594-598.

MacArthur, R.H. & E.D.Wilson 1967. *The Theory of Island Biogeography.* Princeton: Princeton University Press.

Marchant, R. 1982. *The macroinvertebrates of Magela Creek, Northern Territory.* Research Report 1. Supervising Scientist for the Alligator Rivers Region, Canberra: Aust. Govt. Pubn. Service.

Marlier, G. 1967. Ecological studies on some lakes of the Amazon Valley. *Amazoniana* 1: 91-115.

McDowall, R.M. 1981. The relationships of Australian freshwater fishes. In A.Keast (ed.), *Ecological Biogeography of Australia.* The Hague: Junk.

Mees, G.F. & P.J.Kailola 1977. The freshwater Theraponidae of New Guinea. *Zoologische Verhandelingen* 153.

Munro, I.S.R. 1964. Additions to the fish fauna of New Guinea. *Papua New Guinea Agricult. J.* 16: 141-186.

Pollard, D.A. 1972. The biology of a landlocked form of the normally catadromous salmoniform fish *Galaxias macu-lata* (Jennyns). III. Structure of the gonads. *Aust. J. Mar. Freshwat. Res.* 24: 281-295.

Reynolds, F.L. 1983. Migration patterns of five fish species in the Murray-Darling River system. *Aust. J. Mar. Freshw. Res.* 34: 857-871.

Richardson, B.A. 1984. The ecology of a coastal (Tweed River, New South Wales) freshwater fish community and the management of instream flows. M.Sc. thesis, Macquarie University.

Ringuelet, R.A., R.M.Atambury & A.Alonso De Arumbura 1967. *Los Peces Argentinos de Agua Duloe.* BA, La Plata: Commision de Investigacion Gentifica.

Roberts, T.R. 1978. An ichthyological survey of the Fly River in Papua New Guinea with descriptions of new species. *Smithsonian Contr. Zool.* 281.

Sale, P.F. 1980. The ecology of fishes on coral reefs. *Oceanogr. Mar. Biol. Ann. Rev.* 18: 367-421.

Sheldon, A.L. 1968. Species diversity and longitudinal succession in stream fishes. *Ecology* 49: 193-198.

Taylor, W.R. 1964. Fishes of Arnhem Land. *Records of the American-Australian Scientific Expedition to Arnhem Land* 4.

Tilzey, R.D.J. 1980. Introduced fish. In W.D.Williams (ed.), *An Ecological Basis for Water Resource Management.* Canberra: Australian National University Press.

Walker, T.D., J.T.O.Kirk & P.A.Tyler (in press). *The underwater light climate of billabongs of the Alligator Rivers Region, Northern Territory.* Supervising Scientist for the Alligator Rivers Region, Research Report Series. Canberra: AGPS.

Walker, T.D. & P.A.Tyler (in press). *Chemical characteristics and nutrient status of billabongs of the Alligator Rivers Region, Northern Territory.* Supervising Scientist for the Alligator Rivers Region, Research Report Series. Canberra: AGPS.

Welcomme, R.L. 1979. *Fisheries Ecology of Floodplain Rivers.* London: Longman.

Whitley, G.P. 1947. The fluvifaunulae of Australia with particular reference to freshwater fishes in Western Australia. *West. Aust. Nat.* 1: 49-53.

Williams, W.D. 1979. Vegetation and stream patterns as indicators of water movement on the Magela floodplain, Northern Territory. *Aust. J Ecol.* 4: 239-247.

Williams, W.D. 1983. *Life in Inland Waters.* Melbourne: Blackwell.

CHAPTER 7

Reptiles and amphibians

RICHARD W. BRAITHWAITE, GORDON R. FRIEND & JOHN C. WOMBEY

INTRODUCTION

Reptiles and amphibians are typically air-breathing, terrestrial vertebrates with pentadactyle limbs. They differ from other terrestrial vertebrates (birds and mammals) in being ectothermic (so-called 'cold blooded') which means they have no internal mechanism for maintaining a constant body temperature and are more or less dependent on the temperature of their surroundings. However, by behavioural means (e.g. sunbasking, changes in body shape and angle) and by chemical means (e.g. changing colour to absorb heat), the thermal environment is exploited to give a degree of body temperature control. In some regions ectothermy may severely limit the extent to which these groups radiate geographically to occupy a variety of habitats. However in the monsoonal tropics, where high temperatures are sustained throughout the year and full sunlight conditions are common, reptiles and amphibians have relatively little difficulty in maintaining their body temperatures above the critical limits for activity.

Of greater importance to reptiles and amphibians in the monsoonal tropics is coping with the great contrasts between wet and dry seasons. Generally, reptiles are better adapted to arid conditions, and have a limited dependence on free water because of their relatively impermeable scaly skins, their mechanisms for reabsorbing water from faeces and excreting solid wastes (uric acid excretion instead of urine), and their characteristic of internal fertilization and laying of shelled eggs on land (oviparity) or producing live young (viviparity and ovoviviparity). Further, most reptiles are carnivorous which often supplies adequate water for physiological requirements without drinking. As their skin is relatively impervious they cannot depend on evaporation to cool the body, and this further saves water. Many reptiles, particularly the soil-dwelling forms, may have problems of excess water in the wet season in areas which are seasonally inundated such as the black-soil floodplains, and consequently there may be few reptiles in such habitats.

Frogs, on the other hand, are particularly favoured during the wet season, but may experience most problems during the tropical dry season. Their skins are very permeable and individuals can lose or gain water rapidly, depending on the moisture levels and salinity of their surroundings (so that frogs are unable to live in marine environments because of osmotic water loss). A further significant factor controlling the density and distribution of frogs is their dependence on standing freshwater for breeding, because of external fertilization, unshelled eggs and, in most, the presence in the life cycle of an aquatic larval stage, the tadpole. Consequently, amphibian species are active mainly during the wet season, when virtually all species breed, but they require effective adaptations to survive the dry season.

The constantly high ambient temperatures but extreme moisture contrast between the wet and dry seasons in monsoonal Australia would be expected to influence greatly the natural history patterns of reptiles and amphibians. In this chapter we shall describe some of the adaptations to this environment exhibited by the herpetofauna of the Kakadu area. The information used is largely drawn from on-going faunal studies in the area and is preliminary to the detailed quantitative analysis which will be published later. It is also instructive to make some broad faunal comparisons between the 'Top End' and south-eastern Australia, where the human population is greater, and the herpetofauna better known.

For details of description, geographic distribution and taxonomy of the various species mentioned, the reader is referred to Cogger (1983) as an excellent single source. However, a number of other valuable references for particular animal groups are given in the text.

HABITAT TYPES

For the purposes of this chapter, eight habitat types are recognised. These fit into three broad groupings according to the duration of inundation.

Permanently inundated habitats

Freshwater. This environment includes rivers, creeks and billabongs. Although water level fluctuates considerably throughout the year, the water body is usually perennial. Vegetation is aquatic and includes water lilies (see Chapter 4, Plate 4), floating ferns and algae. Some such water bodies become brackish during the dry season.

Estuarine mangroves. The vegetation is dominated by trees whose roots are tolerant of salt water. The ground layer consists mainly of the pneumatophores of the trees which are inundated daily by saline water. Species composition of the trees exhibits a zonation associated with the duration of exposure and the degree of salinity of the water. This habitat type is common within 20 km of the coast where it occupies about 6% of the area but scattered patches of the habitat occur up to 60 km inland.

Seasonally inundated habitats

Sedgelands (see Preface, Plates 1-3). The floodplains are extensive treeless areas of low-lying black soil clays adjacent to the lower reaches of the rivers. They are dominated by various sedge species, mainly of the genera *Eleocharis* and *Fimbristylis*. Some of the plains lie below sea level, but tidal flooding is prevented by the existence of a natural levee bank at the river's edge, so that the lower-lying areas are generally immediately adjacent to the forested low-ridge areas.

These areas are seasonally inundated for various lengths of time depending on microtopography. As water contracts to the depressions and areas adjacent to the forested ridges, the sedge vegetation browns off and is blown flat by the strong south-easterly winds, and the black clay soils crack extensively. This habitat occurs within about 60 km of the coast and occupies about 20% of the area within that zone.

The sedgeland-forest ecotone. This ecotone or 'margin' is a narrow zone (often < 200 m wide) between the treeless black soil plains and the woodlands and open forests of the more elevated land. It is a heterogeneous area, and consists of several habitat types typically arranged in bands thus: (a) paperbark (*Melaleuca* species) fringing the plains (see Chapter 4, Plate 5), (b) open 'lawn' areas of short sedges, grasses and herbs with scattered *Pandanus spiralis*, and (c) taller forest on the higher ground dominated by whitegums (*Eucalyptus papuana, E. alba*) and usually with an understorey of the introduced shrubs *Hyptis suaveolens* and *Cassia obtusifolia*. Areas of apple tree (*Syzygium* spp.) and swamp gum (*Lophostemon lactifluus*) also occur, as well as small patches of monsoon forest where soil

type is favourable (see later). During the wet season the paperbark areas become progressively flooded and the open lawn may also be inundated to shallow depths late in the season, whilst the white gums are usually only temporarily inundated by runoff after heavy rain. As the dry season progresses the water dries back until, by the late dry, only the lower-lying areas of the paperbarks retain water. The margins sustain much trampling and grazing by feral ungulates particularly buffalo (see also Ridpath, Chapter 11), and during the early dry season water may remain trapped by buffalo wallows in otherwise dry areas. This habitat mainly occurs within 80 km of the coast and occupies about 9% of the area.

Rarely inundated habitats

Lowland monsoon forest (see Chapter 4, Plate 5, and Photographs 1 and 2). These areas are characterized by a closed canopy (> 70% cover) which allows poor penetration of light to the ground. Ground cover of vegetation is generally sparse, but several strata of vegetation are usually evident, around gaps in the canopy. Although a species-poor community by comparison with north Queensland rainforest, a considerable variety of non-eucalypt tree species is present. Patches of this habitat are small in area, usually between ten and a hundred hectares, and occur on a variety of soil types in local situations where aquifers adequately support the vegetation during the dry season. Inundation is rare and brief. This habitat occurs within about 80 km of the coast and is less than 1% in area.

Woodland. The tree canopy is both low in height (< 13 m) and relatively open (10-30% cover). The habitat generally occurs on hard laterite soils, and *Eucalyptus clavigera, E. miniata, Erythrophloem chlorostachys, Planchonia careyi* and *Buchanania obovata* are the most common tree species. The ground layer is usually dominated by dense *Sorghum intrans* sp. agg., an annual grass which can grow to 3 m in the wet, but which cures quickly during the dry season. Fires are usually annual, leaving the ground almost bare except for fallen leaves, until some resprouting from below ground in the ensuing weeks. Inundation occurs only for brief periods during heavy rains. About a third of the region is covered by this vegetation.

Open forest (see Chapter 4, Plate 7). Many trees in this habitat are tall (> 13 m) and canopy cover is intermediate between that of woodland and lowland monsoon forest (30-70%). The soil is more sandy than that in woodland areas, and the ground vegetation has frequent shrubs (commonly young individuals of the canopy trees), and both annual and perennial grasses.

The most common tree species are *Eucalyptus miniata, E. tetrodonta* and *E. porrecta*, along with the sand palm, *Livistona humilis*. Fires, on average, occur in two out of three years in any one place. Within 60 km of the coast, over a third of the area is open forest but further inland the habitat is less common (< 10%).

Escarpment. Varieties of vegetation types are associated with the rocky sandstone areas known as the escarpment country. The three most common are a depauperate form of rainforest in sandy ravines, a woodland on the boulder-strewn scree slopes and a rocky spinifex grassland and woodland on the top of the escarpment plateau. Fires are much less frequent than in lowland areas. The spring-fed escarpment monsoon forest remains lush throughout the year. Several varieties of endemic plant species occur in these habitats. In Kakadu National Park (Stages 1 and 2), the escarpment habitats occupy 20% of the total area.

HERPETOFAUNA OF PERMANENTLY INUNDATED HABITATS

Although this environment does not dry up, it is subject to tidal fluctuation and considerable seasonal change in water volume, velocity, turbidity and chemical composition. The adaptations of the most spectacular inhabitants of this environment, the estuarine crocodile (*Crocodylus porosus*) and the freshwater crocodile (*C. johnstoni*), are considered in Chapter 8 as a case study. The present chapter, however, is concerned with the other less-studied herpetofauna of this environment. From the Alligator Rivers region, eight species of amphibians and eleven species of reptiles have been recorded from the freshwater habitat type. The comparable figures for the estuarine mangrove habitat are two species of frogs and twelve of reptiles (Tables 20 and 21).

Legler (1980) has described the habits of five species of freshwater turtles from the region but cautions that taxonomic revision of the group will increase this number of species. The most unusual species is the Pig-nosed or Pitted-shelled turtle (*Carettochelys insculpta*) which is the sole member of the family Carettochelyidae. Until recently, it was believed to be restricted to the south coast of New Guinea, but was recorded on the Daly River in 1969 (Cann 1978). Although rarely seen, it is now known that the species is more widespread in the Top End rivers, inhabiting both estuarine and freshwater environments. It was clearly well known to Aborigines (see Introduction to Part 3, Plates 9a, b). Legler (1980) suggests its diet includes *Pandanus* fruit, leaves and molluscs, and that it may nest more than once in the course of the dry season.

Of the other freshwater turtles, *Chelodina rugosa*

(see Introduction to Part 3, Plates 10a, b) and *Elseya dentata* lay one or more clutches of large eggs early in the dry season, whereas *Elseya latisternum* and *Emydura* sp. lay several clutches of small eggs late in the dry season (Legler 1980). While most turtles throughout the world are omnivorous, *C. rugosa* is unusual in being exclusively carnivorous while *E. dentata* adults appear to be exclusively herbivorous (Legler 1980).

Most frogs require substantial access to land and there are only a few frogs which are relatively aquatic and thus found in the freshwater habitat type. All of them are largely found in other habitat types like the sedgelands and 'margins'. They include *Litoria dahlii, Litoria rothi, Litoria wotjulumensis, Cyclorana australis, Cyclorana longipes* and *Ranidella bilingua*. The floating vegetation often provides a platform from which they call and feed, and may act as a refuge during the dry season (Hill & Webb 1982).

While many of the sedgeland frogs are also found in billabongs and backwaters of the rivers, frogs avoid brackish water and consequently are rarely observed in estuarine mangroves; Hegerl et al. (1979) recorded only two individuals, a *Litoria bicolor* and a *Limnodynastes convexiusculus*. Similarly, few geckoes and skinks have been recorded from mangroves. Hegerl et al. (1979) recorded a *Gehyra* sp. (gecko), and we have recorded the skink *Cryptoblepharus plagiocephalus*. The freshwater habitat type usually harbours none of the small lizard groups.

The mangrove monitor or goanna (*Varanus indicus*) is mainly found in the mangrove forests (Cogger 1983), whereas in freshwater habitats both Mitchell's water monitor (*Varanus mitchelli*) and Merten's water monitor (*Varanus mertensi*) are favoured. We have recorded the latter species feeding on fish and crayfish. Swanson (1976) reports that this species walks the river bottom with its eyes open in much the same manner as other monitors do on land. With its limbs pressed firmly to its sides, it uses its powerful vertically compressed tail to propel it swiftly through the water when alarmed.

A relatively large number of colubrid (solid-toothed) snakes favour the mangroves (*Cerberus rhynchops, Enhydris polylepis, E. punctata, Fordonia leucobalia, Myron richardsonii*) where they feed on crustaceans and fish (Gow 1976). Only *Styporhynchus mairii* is regularly found in the freshwater habitat type, where it mainly eats frogs. It is interesting that these colubrids and another relatively recent group to arrive from Asia, the acrochordids or file snakes, are largely inhabitants of the permanently inundated habitat types. This environment is essentially avoided by the elapid (front-fanged) and boid (python) snakes. One obvious exception is the water python (*Bothrochilus fuscus*), which is relatively common in freshwater billabongs and creeks, and their environs.

Table 20. Habitat distribution of reptile fauna in the Alligator Rivers region.

Family and species	Habitat types							
	Freshwater	Estuarine mangroves	Sedgelands	Sedgeland / forest ecotone	Lowland monsoon forest	Woodland	Open forest	Escarp-ment
Crocodylidae								
Crocodylus johnstoni	+	–	–	–	–	–	–	+
Crocodylus porosus	+	+	+	–	–	–	–	–
Carettochelyidae								
Carettochelys insculpta	+	+	–	–	–	–	–	–
Chelidae								
Chelodina rugosa	+	–	+	+	–	–	–	+
Elseya dentata	+	–	–	–	–	–	–	+
Elseya latisternum	–	–	–	–	–	–	–	+
Emydura sp.	+	–	–	–	–	–	–	+
Emydura victoriae	+	–	–	–	–	–	–	–
Gekkonidae								
Diplodactylus ciliaris	–	–	–	–	–	+	–	–
Diplodactylus occultus	–	–	–	–	–	+	–	–
Diplodactylus stenodac-tylus	–	–	–	–	–	+	–	–
Gehyra australis	–	–	–	+	+	+	+	+
Gehyr nana	–	–	–	–	–	–	–	+
Gehyra pamela	–	–	–	–	–	–	–	+
Heteronotia binoei	–	–	–	+	+	+	+	+
Nephrurus asper	–	–	–	–	–	–	–	+
Oedura gemmata	–	–	–	–	–	–	–	+
Oedura marmorata	–	–	–	–	–	+	+	–
Oedura rhombifer	–	–	–	–	+	+	+	+
Pseudothecadactylus lindneri	–	–	–	–	–	–	–	+
Pygopodidae								
Delma borea	–	–	–	–	–	+	+	+
Lialis burtonis	–	–	–	–	–	+	+	+
Pygopus nigriceps	–	–	–	–	–	+	–	–
Agamidae								
Amphibolurus caudicinc-tus	–	–	–	–	–	–	–	+
Chelosania brunnea	–	–	–	–	–	+	+	–
Chlamydosaurus kingii	–	–	–	+	+	+	+	–
Diporiphora bilineata	–	–	–	+	–	+	+	+
Diporiphora magna	–	–	–	–	–	+	–	–
Lophognathus gilberti	–	–	–	–	+	+	+	–
Lophognathus tempo-ralis	–	+	–	+	+	–	–	–
Varanidae								
Varanus acanthurus	–	–	–	–	–	–	–	+
Varanus glebopalma	–	–	–	–	–	–	–	+
Varanus gouldii	–	–	+	+	–	+	+	–
Varanus indicus	–	+	–	–	–	–	–	–
Varanus mertensi	+	–	–	–	–	–	–	–
Varanus mitchelli	–	–	–	+	–	–	+	–
Varanus panoptes	–	–	+	+	–	+	+	–
Varanus timorensis	–	–	–	+	+	+	+	+
Varanus tristis	–	–	–	–	–	+	+	+
Scincidae								
Carlia amax	–	–	–	–	–	+	+	+
Carlia foliorum	–	–	–	+	+	+	+	–
Carlia gracilis	–	–	+	+	+	+	+	–

Table 20 (Continued).

Family and species	Habitat types							
	Freshwater	Estuarine mangroves	Sedgelands	Sedgeland / forest ecotone	Lowland monsoon forest	Woodland	Open forest	Escarpment
Carlia triacantha	–	–	–	–	–	+	+	–
Cryptoblepharus plagio-cephalus	–	+	+	+	+	+	+	+
Ctenotus arnhemensis	–	–	–	–	–	+	+	–
Ctenotus borealis	–	–	+	–	–	+	+	+
Ctenotus coggeri	–	–	–	–	–	–	–	+
Ctenotus essingtonii	–	–	–	+	+	+	+	+
Ctenotus gagadju	–	–	–	–	–	+	+	–
Ctenotus kurnbodj	–	–	–	–	–	+	+	–
Ctenotus saxatilis	–	–	–	–	–	–	–	+
Ctenotus storri	–	–	–	–	–	–	+	–
Ctenotus vertebralis	–	–	–	–	–	–	–	+
Egernia frerei	–	–	–	–	–	–	–	+
Lerista karlschmidti	–	–	–	–	+	–	+	+
Menetia alanae	–	–	–	–	+	+	+	–
Menetia greyi	–	–	–	–	–	+	+	–
Menetia mainii	–	–	–	–	+	+	+	–
Morethia ruficauda	–	–	–	–	–	–	–	+
Morethia storri	–	–	–	–	+	+	+	–
Notoscincus ornatus	–	–	–	–	–	–	–	+
Proablepharus tenuis	–	–	–	–	–	+	+	+
Sphenomorphus crassi-caudus	–	–	–	+	+	+	+	–
Sphenomorphus dou-glasi	–	–	–	+	+	–	+	–
Sphenomorphus isolepis	–	–	–	+	+	+	+	+
Tiliqua scincoides	–	–	–	+	+	+	+	–
Typhlopidae								
Rhamphotyphlops guen-theri	–	–	–	–	–	+	–	–
Rhamphotyphlops tovelli	–	–	–	+	+	+	+	+
Rhamphotyphlops ungui-rostris	–	–	–	+	–	+	+	–
Rhamphotyphlops weidii	–	–	–	–	–	–	–	+
Boidae								
Aspidites melanoce-phalus	–	–	–	+	–	–	–	–
Bothrochilus childreni	–	–	–	–	–	+	–	+
Bothrochilus fuscus	+	+	+	+	+	–	–	–
Bothrochilus olivaceus	–	–	+	+	+	+	+	–
Morelia oenpelliensis	–	–	–	–	–	–	–	+
Morelia spilota	–	–	–	+	+	–	–	–
Acrochordidae								
Acrochordus arafurae	+	+	–	–	–	–	–	–
Acrochordus granulatus	–	+	–	–	–	–	–	–
Colubridae								
Boiga irregularis	–	–	–	+	+	–	+	+
Cerberus rhynchops	–	+	–	–	–	–	–	–
Dendrelaphis punctulata	–	–	–	+	+	–	–	+
Enhydris polylepis	–	+	+	–	–	–	–	–
Fordonia leucobalia	–	+	–	–	–	–	–	–
Myron richardsonii	–	+	–	–	–	–	–	–
Stegonotus cucullatus	–	–	+	+	+	–	–	+
Styporhynchus mairii	+	–	+	+	–	+	–	–

Table 20 (Continued).

Family and species	Habitat types							
	Freshwater	Estuarine mangroves	Sedgelands	Sedgeland / forest ecotone	Lowland monsoon forest	Woodland	Open forest	Escarpment
Elapidae								
Acanthophis praelongus	−	−	+	+	−	−	+	+
Cryptophis pallidiceps	−	−	−	+	−	+	−	−
Demansia atra	−	−	+	+	−	+	+	−
Demansia olivacea	−	−	−	−	−	+	+	−
Denisonia punctata	−	−	−	+	−	+	−	−
Furina diadema	−	−	−	+	−	+	+	+
Oxyuranus scutellatus	−	−	−	−	−	−	−	+
Pseudechis australis	−	−	−	+	+	−	+	+
Pseudonaja nuchalis	−	−	+	−	−	−	+	−
Simoselaps semifasciatus	−	−	−	−	−	+	−	+

HERPETOFAUNA OF SEASONALLY INUNDATED HABITATS

In a study of faunal diversity on the black soil sedgelands and the adjacent sedgeland-forest ecotone (margins), a total of 30 species of reptiles and 17 species of amphibia were recorded during systematic sampling of 70 sites near the South Alligator River between October 1979 and August 1982 (Friend & Cellier 1990). On the margins, all 30 species of reptiles were recorded and comprised nine families, whereas only nine species comprising six families were recorded on the sedgelands. The amphibian species on the other hand represented only two large families (Hylidae and Myobatrachidae), and showed little variation in the proportional representation of these families across the suite of habitats studied on the margins and sedgelands. All 17 amphibian species were recorded on the margins, but only nine were found on the sedgelands. Regionally, 34 reptilian and 18 amphibian species have been recorded from the margins, and 15 and 9 species respectively from the sedgelands (see Tables 20 and 21).

The sedgelands represent an area of great contrasts between wet and dry seasons, and in many cases the climatic influences on habitat characteristics are confounded by varying degrees of trampling and grazing by feral buffalo (see also Ridpath, Chapter 11), their effects generally being less severe towards the drier levee banks of the river. The extreme differences between seasons, as well as the simplified vegetational structure (no tree or shrub layer) which provides no shade from the sun, are probably the main factors limiting the diversity and abundance of the herpetofauna.

During the day in the dry season (and particularly the late dry season when all depressions have dried up) few if any reptiles or frogs can be found on the sedgelands, although the very occasional Sand goanna *Varanus panoptes* may be observed in the morning when temperatures are moderate. Most species take refuge in the extensive cracks of the black soils and emerge only at night. Frogs may then be readily observed but only three species are common: *Litoria nasuta*, *L. dahlii* and *L. convexiusculus*. The former two species are the most common, but *L. dahlii* is generally restricted to depressions that are extensively flooded in the wet season.

Snakes are the most conspicuous reptiles on the sedgelands at these times, hunting their main prey, frogs and rats. The Olive python *Bothrochilus olivaceus* and the Water python *Bothrochilus fuscus* are most common, but in the late dry season when nights are warm and humid and rain from thunderstorms is causing cracks to close, other species appear. Most common among these is the Death adder, *Acanthopis praelongus* (Photograph 6), which emerge from cracks at this time. Their cryptic colouration and habit of not moving when approached makes them extremely difficult to detect.

Other species of reptile are few on the sedgelands. The two skink species *Carlia gracilis* and *Cryptoblepharus plagiocephalus* may be found on the plains near the margins where some freshwater mangroves (*Barringtonia acutangula*) occur, or near the river levee bank where juvenile saltwater mangroves may grow. However these species have only been detected on a few occasions and cannot be regarded as regular sedgeland dwellers.

During the wet season when extensive flooding of the sedgelands occurs, most reptiles leave the plains for the higher ground of the margins and woodlands. However, aquatic reptiles such as *Chelodina rugosa* are then seen. At these times snakes such as the Water

Table 21. Habitat distribution of amphibian fauna in the Alligator Rivers region.

Family and species	Habitat types							
	Freshwater	Estuarine mangroves	Sedgelands	Sedgeland / forest ecotone	Lowland monsoon forest	Woodland	Open forest	Escarp- ment
Myobatrachidae								
Limnodynastes convexiu- sculus	–	+	+	+	+	+	+	+
Limnodynastes ornatus	–	–	–	+	+	+	+	–
Megistolotis lignarius	–	–	–	–	–	–	–	+
Notaden melanoscaphus	–	–	–	–	–	+	+	–
Ranidella bilingua	+	–	+	+	+	+	+	+
Uperoleia arenicola	–	–	–	–	–	–	–	+
Uperoleia inundata	–	–	–	+	–	+	+	–
Uperoleia lithomoda	–	–	–	+	–	–	–	–
Hylidae								
Cyclorana australis	+	–	+	+	+	+	+	+
Cyclorana longipes	+	–	–	+	+	+	+	+
Litoria bicolor	+	+	+	+	+	+	+	–
Litoria caerulea	–	–	–	+	+	+	+	+
Litoria coplandi	–	–	–	–	–	–	–	+
Litoria dahlii	+	–	+	+	+	–	–	+
Litoria dorsalis	+	–	–	+	+	–	+	–
Litoria inermis	–	–	+	+	+	+	+	–
Litoria meiriana	–	–	–	–	–	–	–	+
Litoria nasuta	–	–	+	+	+	+	+	+
Litoria pallida	–	–	–	+	+	+	+	+
Litoria personata	–	–	–	–	–	–	–	+
Litoria rothii	+	–	+	+	+	+	+	+
Litoria rubella	–	–	+	+	+	+	+	+
Litoria tornieri	–	–	–	+	+	+	+	+
Litoria wotjulumensis	+	–	–	+	–	–	+	+
Microhylidae								
Sphenophryne robusta	–	–	–	–	–	–	–	+

python and the King brown *Pseudechis australis* are known to prey on eggs of the Magpie goose *Anseranas semipalmata* in inundated areas (Plate 28).

Conversely, the wet conditions are ideal for frogs, and several species which are not seen during the dry season appear on the sedgelands. Even the typically tree-dwelling hylids *L. bicolor* and *L. rothii* may occur in quite large numbers amongst the thick stems of *Eleocharis sphacelata* in the deeper depressions. The tiny myobatrachid frog *R. bilingua* becomes common in most habitats on the plains, and is readily detected by its loud 'creaky' call.

The frogs appear to prefer the partially inundated habitats, or flooded areas where emergent vegetation provides resting and feeding sites. An exception is *L. dahlii* which is a most abundant inhabitant of the deeply flooded depressions. This species floats in the water with legs outstretched, and only its eyes protruding above the surface. It is a powerful swimmer, owing to extensive webbing on the long hind feet. However, this species is poorly known, and its call has only recently been reported (I. Morris, pers. comm. in Tyler et al. 1983). The species is an active predator, and as well as insects, has been observed eating individuals of *Litoria inermis*, *L. nasuta* and *L. convexiusculus*, and members of its own species (Friend 1981).

In contrast to the sedgelands, the margins are very heterogeneous with a complex spatial and temporal pattern of wet and dry habitats. Most habitats have a tree layer and some development of an understorey and litter layer, although the open lawn areas are an exception. Because of the increased habitat complexity and moderated climatic conditions, frog diversity is relatively high. However, as on the sedgelands, the reptiles and amphibians on the margins show different responses to the wet and dry seasons.

During the dry season frogs are most abundant in the paperbark areas fringing permanent swamps, and huge numbers of *L. nasuta, R. bilingua, L. convexiusculus* and *Litoria inermis* may be found, particularly where some understorey vegetation is present. There is a consistent decline in amphibian species diversity with increasing elevation (and aridity) across the ecotone, and very dry areas such as patches of *Syzygium* have few species. As expected from their different physiological adaptations, reptiles essentially show the opposite trend in species diversity. The drier habitats (e.g. *Syzygium* sp. and monsoon forest) support higher diversities of reptiles, but so too do some fringing areas of paperbark where floodwaters have receded.

The ground-dwelling skink *Carlia gracilis* may be found wherever there is a litter layer, and apart from some *Carlia foliorum* in whitegum and monsoon forest habitats, appears to be the only species of this genus on the margin. *C. gracilis* and the arboreal *Cryptoblepharus plagiocephalus* are the two most common species on the margins, particularly in the paperbark and whitegum habitats. They are most active in the morning and late afternoon when temperatures are moderate. Both these skinks eat insects from a wide range of insect families and take about 10% by volume of species with aquatic larvae (e.g. midges and mosquitoes, James et al. 1984).

Other reptile species are more cryptic and may be captured using pitfall traps, or actively searched for under logs and amongst soil and litter (e.g. *Sphenomorphus crassicaudus, Ramphotyphlops tovelli*). *Ctenotus essingtonii, Varanus timorensis* and *Lophognathus temporalis* are most usually glimpsed as they run through deep leaf-litter or up tree trunks. Still other species are mainly active at night, notably the snakes (e.g. *Stegonotus cucullatus, Bothrochilus olivaceus, Morelia spilota*) which hunt small mammals and birds (R. Shine, pers. comm. 1984), and the geckos (e.g. *Heteronotia binoei* and *Gehyra australis*) which feed on insects.

As on the plains, the onset of heavy rains in the early wet season (December-January) appears to be the cue for the emergence, renewed activity and breeding of many amphibians after a long period of aestivation or restricted activity during the dry months. A striking example of such a strategy is shown by *Cyclorana australis*. Large numbers of adults emerge during heavy rain from their dry season refuges under soil and the males congregate as breeding choruses on the open margins and around billabongs. Egg laying and development of young of this species (and virtually all of the other amphibian species) thus occurs at a time when there is ample fresh water and a bloom of the insects which constitute the major part of the diet of frogs (Tyler 1976). Depressions created by old buffalo wallows which have revegetated are favoured breeding sites on the lawn and the fringing paperbarks.

There is a dramatic increase in the richness and abundance of amphibian species on the margins in the wet, and particularly in the habitats which were completely dry in the dry season: lawn, whitegum and *Syzygium* habitats. Diversity is highest in those habitats that are temporarily or partially inundated (viz. lawn/pandanus, and the inner fringes of paperbark swamps). A wide range of species groups is present, including tree-dwellers (e.g. *Litoria caerulea*), ground-dwellers (*Litoria inermis, Litoria tornieri, Limnodynastes ornatus, Cyclorana longipes, Uperoleia inundata*) and those species which favour more deeply-flooded areas (e.g. *L. dahlii, R. bilingua*). Many species exhibit definite preferences for certain habitats or vegetation species. For example, *L. rothii* is most abundant in flooded paperbark swamps, and can be found sitting on trunks of *Melaleuca* spp. or on floating logs or sedges making its loud, bleating call. Similarly, *L. bicolor* is most abundant amongst the thick *Eleocharis* sedges of paperbark swamps, but is also a characteristic inhabitant of *Pandanus* fronds in the lawn and whitegum habitats. Both species frequently feed on Odonata nymphs which are emerging from the water to metamorphose into adult dragonflies and damselflies. The larger species, *L. rothii*, also feeds on juvenile *L. bicolor* (Brady 1982). *L. tornieri, L. inermis, L. nasuta* and *L. convexiusculus* are usually associated with the edges of such swamps and shallow flooded lawn areas, while *U. inundata* is most commonly heard calling from inundated soils (as the specific name suggests) in whitegum areas.

Reptiles, on the other hand, are better adapted for existence in drier areas, and consequently the abundance of many species declines on the margins during the wet months, depending on their particular life histories, and degree of aquatic adaptation. Reptilian species diversity declines in all margin habitats except whitegum on higher ground, where flooding is usually only temporary. This paucity is particularly marked in the paperbark areas, where extensive flooding occurs.

The ground-dwelling scincid lizards (e.g. *Carlia gracilis, Ctenotus essingtonii*) become less common and most of the *Carlia* specimens observed are juveniles produced from the recent breeding season in the late dry (October-November). However snakes such as the Water python *B. fuscus*, Slaty-grey *S. cucullatus*, King brown *P. australis* and Carpet python *M. spilota* become more active and evident. Other species which prefer an aquatic habitat such as the Northern snake-necked turtle *C. rugosa*, the Freshwater snake *S. mairii* and Mitchell's water monitor *V. mitchelli* also may be observed at this time in the flooded paperbarks. These latter three species are inconspicuous in the dry season, although the large number of empty shells of the turtle amongst paperbarks suggests that mortality from predation or by other means (e.g. desiccation) is high.

Many of the species that occur on the margins and

sedgelands also occur in the rarely inundated habitats, and their habits and adaptations are further discussed in the following section.

HERPETOFAUNA OF RARELY INUNDATED HABITATS

These habitats are the domain of the reptiles but frogs are also common. Regionally, the number of species recorded is as shown in Table 22.

During the dry season active frogs can only be reliably found near permanent water, such as around the small perennial streams flowing from the escarpment. *Litoria coplandi, Litoria personata, Litoria meiriana* and *Megistolotis lignarius* are sandstone rock-dwelling specialists which withdraw into crevices in those places with an ephemeral water supply. Some species which are found in the lowlands are also in the escarpment; for example *Litoria wotjulumensis*, a common inhabitant of rock pools.

Although 95% of the rain falls between November and April, rain does occasionally fall during the dry season, particularly the late dry, and on the escarpment may be sufficient to induce breeding (Tyler et al. 1983). On such occasions frogs appear as if from thin air but only a few species are involved. *L. ornatus* readily breaks dry season aestivation under the soil, and *L. convexiusculus* and *U. inundata* appear from their refuges in damp soil. *L. bicolor* emerge from the base of *Pandanus* leaves. The large Green tree frog *L. caerulea* (Photograph 7) begins calling from its arboreal hollow with its deep resonant croaking. *R. bilingua* will also call during the dry and apparently attempt to breed at any time. Although *L. rothii* are active for much of the year in damp human habitation (e.g. bathrooms), or in paperbark swamps, they restrict their breeding to the wet season.

With the arrival of the wet, frogs seem ubiquitous, and the cacophony of calls of breeding frogs from billabongs can be deafening. It is during actual rain that frogs are most active in these more elevated habitats. Small frogs like *Litoria dorsalis, Litoria rubella* and *L. bicolor*, can be seen climbing amongst thick vegetation, both on the ground and in trees. On the ground *Cyclorana australis* sits statue-like in the open waiting for its varied insect prey. *L. tornieri* is a more active hunter and can be seen chasing small grasshoppers and other insects amongst the ground vegetation. Most amphibians are unspecialized feeders but a local exception is *U. inundata* which consistently takes 90% termites and 10% ants (Brady 1982).

Most frogs have irregular patterns and colouration which allow them to blend with their background. By changing the shape of dark cells in the skin they are able to change colour, further facilitating camouflage. For example one of the most abundant species, *U. inundata*, is rarely seen but large numbers reveal their presence by being caught in pitfall traps. While most species are cryptic some species advertise their presence. For example, the ground-dwelling *Notaden melanoscaphus* (Plate 15) has a colourful appearance which presumably warns would-be predators of its toxicity. It exudes copious amounts of sticky yellow secretion from dorsal glands when handled, and the more brightly coloured juveniles are conspicuously active in daylight hours, demonstrating the effectiveness of their warning colouration.

Although breeding of frogs is almost exclusively during the wet season, there is considerable variation in its duration and intensity during this period by the different species; and there is considerable variation from year to year (Tyler et al. 1983). Some species appear to breed early in the wet season in November-December (e.g. *C. australis, C. longipes, L. meiriana, M. lignarius*) whereas others may breed in January-February (e.g. *N. melanoscaphus, U. inundata*). However, many species can breed early in one year but late in another. For example, *L. rubella* near Jabiru bred from November to February in 1978-79 but only in March during the 1979-1980 wet season (Tyler et al. 1983). Presumably these differences are related to rainfall-habitat conditions.

As the wet season proceeds, tadpoles are a common sight. Many are in temporary pools and will not survive to metamorphose to adulthood. Predation by birds, snakes and frogs is heavy, and even the normally herbivorous tadpoles of some species turn to cannibalism to maximize developmental rates as food abundance or quality declines (e.g. *L. ornatus*). During the day puddles in the sun may reach high temperatures and many dry up before development is complete. Even so, tadpoles have been recorded locally in water at 42.7°C (Tyler et al. 1983). Clearly there is a wide range of environmental conditions prevalent in different water bodies, but we know little of the preferences of different frog species for breeding in such water bodies.

Tyler et al. (1983) have shown that some species only take about a month to develop (e.g. *L. ornatus, L. nasuta, L. rubella*), whereas other species have large egg capsules (two or three times larger) and take twice as long to develop (e.g. *L. inermis, L. rothii*). The

Table 22. Number of species of reptiles and amphibians in rarely inundated habitats.

	Lowland monsoon forest	Woodland	Open forest	Escarpment habitats
Reptiles	27	48	46	46
Frogs	15	15	17	18

number of eggs laid in a group usually ranges from 100-300 but *U. inundata* normally lays only between one and five, while a *C. australis* egg clump of 7000 has been recorded (Tyler et al. 1983). It is unclear to what extent individuals of different species lay many clutches in different places in quick succession.

The development of *Sphenophryne robusta*, an inhabitant of the moist monsoon forest pockets in the escarpment, is likely to be similar to other members of its genus in which eggs are laid in a moist place on land and the tadpole develops to adult form inside its egg capsule (see Tyler 1976, Barker & Grigg 1977). Tadpoles of some species have deep bodies and are strong swimming forms adapted to feeding on suspended microscopic algae (phytoplankton). A flatter body, as with *L. ornatus*, allows more effective feeding under rocks and other bottom substrate feeding. However, most species have a more generalized body form and feed on detritus and encrusting algal films.

The lowland monsoon forest does not have a distinct amphibian and reptile fauna as occurs in rainforest habitats in north Queensland. The same species of broad tolerance that occur in open forest and woodland are found in monsoon forest. The lack of a distinct monsoon forest herpetofauna may be due to the high light penetration in this vegetation which is structurally simple relative to similar vegetation types in north Queensland (Kikkawa & Monteith 1980). However, this habitat was locally uncommon in the past (see Nix & Kalma 1972), and is now probably less than 1000 km² in area throughout the 'Top End'. The small extent and scattered distribution of lowland monsoon forest may have precluded either the persistence or evolution of a distinct herpetofauna.

Whereas frogs are most active at night, reptiles tend to split up the twenty-four hours according to various specializations. Geckoes like *G. australis* have large eyes and foot pads which are used for chasing insects in tree trunks of the open forests and woodlands at night. The most abundant gecko, *H. binoei*, lacks the well developed foot pads and does its nocturnal hunting on the ground. Numerous skinks are active during the day. Various *Carlia* species (usually *C. amax* in open forest and escarpment habitats, *C. foliorum* and *C. gracilis* in woodland and monsoon forest, and *C. triacantha* (Plate 16) in woodland) are obvious on the ground. *C. plagiocephalus* are commonly seen on tree trunks sunning themselves in all habitats. As the temperature increases and the humidity drops during the morning, the *Carlia* species become less active and retreat to shelter under litter and logs. Similarly the *Cryptoblepharus* shift onto the shady side of the tree or into cracks. At this time the *Ctenotus* skinks, such as *C. essingtonii* become more active, and move across the ground searching for active grasshoppers, but if disturbed dash for the safety of a nearby small hole in the

ground. The agamids or dragons, such as *Diporiphora bilineata* are commonly seen at this time, but they tend to sit cryptically on a shrub or log and wait for their prey (frequently ants) to come to them. In the middle of the day, the reptiles are generally inactive but an occasional *V. gouldii* may be seen quietly searching an area using routes of maximum shade.

In the late afternoon the reverse of the morning activity prevails. The larger more heat-tolerant species are replaced by the small skinks as sunset approaches. Some fossorial skinks, such as *Sphenomorphus isolepis* and *Lerista karlschmidti* are often active on the surface at night and shortly after sunrise. The latter species is extreme in its adaptation to a burrowing way of life; it has an extremely reduced external ear opening and a smooth elongated body with no front limbs, while the back limbs are reduced to a single toe on each foot.

Snakes are also most active at night because, like lizards, they must avoid the high temperatures in the middle of the day. The refuges from the sun that monsoon forest provides are ideal for keeping body temperatures low during the dry season, and consequently snakes tend to be more common in this habitat, particularly near Banyan fig (*Ficus virens*) trees with their complex recesses. Snakes are also more commonly seen close to and in the escarpment country.

During the cooler early dry season mornings, larger reptiles can be seen sunning themselves; for example, the small goanna *V. timorensis* is frequently seen high up on tree trunks during the dry season. However, the Frilled lizard *Chlamydosaurus kingii* (Plate 17) disappears completely during the early dry and only comes out of its hollow log or branch after rain during the late dry season (from August). Juveniles first appear during the late wet season.

On the other hand, some species are most active during the wet season. This is particularly noticeable with intermittent feeders like snakes which can go for weeks without feeding. They are more commonly sighted at night hunting the abundant frogs during the wet season. There are at least thirty snake species in the region but sightings are not frequent, particularly in drier areas and those more remote from the escarpment. The most commonly observed species in the rarely inundated habitats are *Demansia olivaceus, S. cucullatus* and *P. australis*.

In the dry season the daily pattern is quite predictable but in the wet, the weather is much more variable. On overcast days it is cooler, and with the great abundance of ground vegetation during the wet season, activity of many species can continue throughout the day. The protection from predation due to this ground vegetation, greater abundance of insect food, long periods within the thermal tolerance of many species available for hunting, and ready availability of water, mean it is an ideal time to raise young. Consequently, for species such as those of the genus *Carlia*, mating usually takes place towards the end of the dry season when males exhibit their vivid breeding colouration, and juveniles are seen throughout the wet. Other wet season breeders are the three *Sphenomorphus* species, *D. bilineata* and *Lophognathus gilberti* (James 1983). All the *Ctenotus* spp., the *Morethia* spp., *H. binoei* and *L. karlschmidti* breed during the dry season. *C. plagiocephalus* appears to breed all year round (James 1983) and at least some of the snakes do (e.g. Shine 1981).

The number of eggs produced at one time (clutch size) is relatively constant at two eggs with *C. plagiocephalus* and the *Carlia* spp., but it is likely that individual females produce two clutches during a breeding season (James 1983). On the other hand, *S. isolepis* and *S. douglasi* produced between three and eight eggs in a clutch depending on the body size of the female. Similarly *C. essingtonii* produces between one and four eggs depending on female size. Mean clutch size is also positively related to mean species size within the genus *Ctenotus*. However, although *D. bilineata* has a clutch size between four and eight, this is not correlated with body length. With both *C. plagiocephalus* and *S. isolepis*, mean egg volume positively relates to female size (James 1983). Obviously a variety of life history strategies is being used. Some species will be found to be short-lived with high juvenile survival while other species will prove to be

longer-lived and will apportion their reproductive effort differently.

As with frogs, some reptiles have strong preferences for certain vegetation formations while others are more general in their requirements. For example, *Diplodactylus occultus* (Plate 18) has only been recorded in woodland in the West Alligator River area (King et al. 1982); *C. plagiocephalus* is common in all habitat types with trees; *S. crassicaudus* is known in all of the habitats on the 'margins' and in open forest; and *Morethia storri* is found in open forest and woodland. In general, the greatest habitat fidelity is found among escarpment species including *Gehyra pamela, Pseudothecadactylus lindneri, Ctenotus coggeri, C. saxatilis, Morethia ruficauda* and *Morelia oenpelliensis*. The last mentioned species is a large python which was only recently recognised and scientifically described (Gow 1977); an indication that there is still much to learn about the fauna of the region.

Among both frogs and reptiles, some species have very small known geographic distributions (e.g. *M. lignarius*), while others are northern tropical species shared with north Queensland (e.g. *Sphenophryne robustus*), and still others are very widespread and are generally thought of as desert animals (e.g. *L. rubella, Diplodactylus ciliarus*).

The influence of human activities in forested habitats is seen with the establishment of the Asian house gecko (*Hemidactylus frenatus*) in human dwellings. In

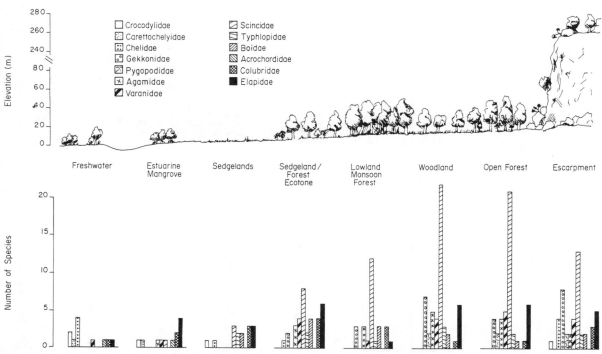

Figure 46. Number of species in reptile families in eight habitat types in the Alligator Rivers region.

the Alligator Rivers area it has now been recorded at Cooinda and Jabiru. The native House gecko (*Gehyra australis*), although found in open forest and woodland also occurs in human dwellings, but from some observations, appears to be reduced in numbers where *H. frenatus* is abundant.

SPECIES RICHNESS OF THE HABITAT TYPES

In Figures 46 and 47 respectively, the records of reptile and frog species have been tallied according to families in the eight habitat types. The area covered is the Alligator Rivers area (Stages 1 and 2 of Kakadu National Park). Such a tally is obviously incomplete and is drawn from sources which result in overall geographic and habitat type biases in the coverage of the region. Nonetheless it is a useful comparison which summarizes some of the observations in the text.

Clearly the drier end of the moisture gradient is richer in both reptile species and families. Specifically, the sedgeland/forest ecotone is much richer than the sedgelands, and the woodland and open forest are considerably richer than the lowland monsoon forest. The escarpment information is a little misleading; the tally is high because a number of different types of vegetation comprise this rather heterogeneous habitat type. In general, the sandstone monsoon forest is very

depauperate of herpetofauna whereas the sandstone spinifex and sandstone woodland have quite a variety of species.

The aquatic families of Crocodylidae, Carettochelyidae, Chelidae and Acrochordidae are concentrated at the wet end of the gradient, as expected. Geckoes are most diverse in the woodland and escarpment, whilst pygopodids (legless lizards) appear only to occur in woodland, open forest and escarpment; and dragons and skinks are most rich in woodland and open forest. Goannas are most diverse in open forest and escarpment whereas the pythons are richest in the margin. The solid-toothed snakes are best represented by species in the mangroves, sedgelands, margin, monsoon forest and escarpment but front-fanged snakes have most species in the margin, woodland, open forest and escarpment.

With the frogs, freshwater, estuarine mangroves and sedgelands are poor in species compared with the other habitat types. The sedgeland/forest ecotone and the escarpment appear to be slightly richer in species than the other upland habitat types, perhaps because water may be retained in rock crevices, buffalo wallows or other depressions in these habitats until well into the dry season. In addition to being closely associated with water bodies, these two habitat types are also highly heterogeneous with steep, small scale moisture gradients. Both the myobatrachids and hylids

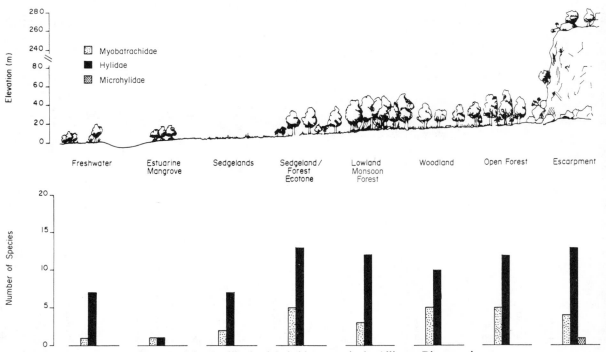

Figure 47. Number of species in amphibian families in eight habitat types in the Alligator Rivers region.

Plate 15. The Northern Spadefoot Toad (*Notaden melanoscaphus*) is a burrowing frog which is conspicuous when it is above ground during the wet weather (Photograph: J.C. Wombey).

Plate 16. The Skink (*Carlia triacantha*) feeds on small insects and spiders in the leaf litter and on open ground in woodland (Photograph: J.C. Wombey).

Plate 17. The Frilled Lizard (*Chlamydosaurus kingii*) aestivates during the early dry season and becomes active again during the first rains of the wet season (Photograph: G.J.W. Webb).

Plate 18. *Diplodactylus occultus* is a newly discovered ground-dwelling gecko from Kakadu National Park (Photograph: J.C. Wombey).

Photograph 6. The Northern Death Adder (*Acanthopis praelongus*) is commonly seen at the beginning of the wet season when it is driven from cracks in the soil of the wetlands (Photograph: J.C.Wombey).

Photograph 7. The Green Tree Frog (*Litoria caerulea*) lives in tree hollows and is common through a wide variety of habitats in northern Australia (Photograph: J.C.Wombey).

show the same habitat trends with the hylids being most diverse, particularly in the freshwater, mangrove and sedgeland habitats.

SOME COMPARISONS WITH OTHER PARTS OF AUSTRALIA

The number of herpetofaunal species occurring in equal-sized areas of several thousand kilometres irrespective of vegetation (species density), generally is highest in the tropics and progressively decreases towards the poles (see Pianka 1966). This is true for Australian snakes, varanids and turtles; but for geckoes and agamids the peak is in the middle latitudes of Australia. Skinks tend to increase from north to south and then to decrease precipitously in far southern Australia (Braithwaite et al. 1977, Cogger & Heatwole 1981, Pianka & Schall 1981). The high densities of lizards in the middle latitudes are due to the great success of this group in the Australian deserts (see Pianka & Schall 1981). Frogs are bimodal with peaks in the north and at the latitude of northern New South Wales; and the distribution of these high values relates well to the distribution of high values of Fitzpatrick & Nix's (1970) average moisture index values for the driest 16-week period of the year. In other words, amphibian species density is highest in areas of Aus-

tralia with the shortest and least intense dry season (Braithwaite et al. 1977). Thus the diversity of the herpetofauna of the Kakadu region would be expected to differ from that of southern Australia, but differences would lie in the importance of the various families. Previous regional comparisons of Australian herpetofauna have not considered differences in habitat use.

To this end, the numbers of species in the various families recorded in the Alligator Rivers region are presented in Figures 46 and 47 in a form comparable with Figure 48 (from Jenkins & Bartell 1980: 28-29) for southern Australia. Figure 49 has been compiled for the amphibians of the same area of southern Australia from the records of one of us (JCW). The comparison is not a well balanced one as the Australian high country is about four to five times the area of the Alligator Rivers region. Nevertheless there are some interesting contrasts.

The number of reptile families is considerably greater in the Kakadu region (13) than in the southern high country (8). Crocodiles and carettochelyid turtles are absent in the south, and the number of chelid turtles is greater in Kakadu. Comparable habitat types at Kakadu are richer in gecko, varanid and typhlopid species. The numbers of dragons appear similar in the two areas. The snake group comprising the pythons, file snakes and colubrids is absent from the high coun-

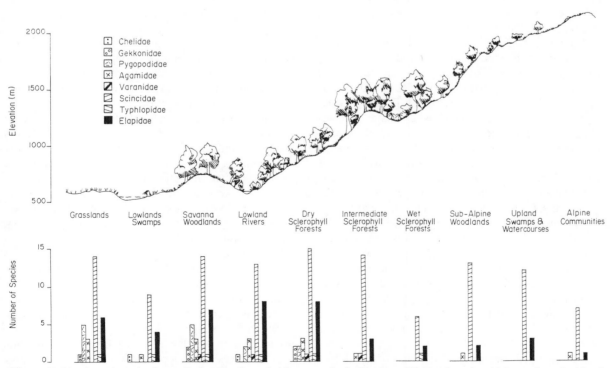

Figure 48. Number of species in reptile families in ten habitat types in the Southern Highlands region (redrawn from Jenkins & Bartell 1980).

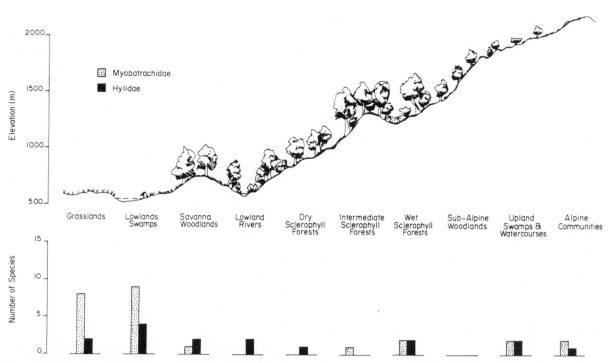

Figure 49. Number of species in amphibian families in ten habitat types in the Southern Highlands region.

try, but species of legless lizards and front-fanged snakes are more numerous in the high country. Similarly skink species are more numerous in the wetter habitats of the high country than those of the Kakadu region, but for the drier habitats the reverse is true.

The success of reptiles in southern cold situations may be limited by their mode of reproduction in many cases. As Greer (1976) points out, although adult lizards are able to maintain their temperatures by behavioural means, their eggs cannot. Thus viviparous reptiles may have a considerable advantage in colder climates, as reflected by the success (high species richness) of the scincids and elapids in southern Australia. However, although viviparity evolved in the colder climates, many such groups have successfully radiated into the tropics (see Shine & Berry 1978). Concomitantly, the oviparous groups, the chelids, agamids, gekkonids, varanids, boids and typhlopids (Rawlinson 1971) are most successful in northern Australia. The oviparous pygopodids, however, are anomalous with a strong southern Australian representation of species. Storr (1964) has suggested that pygopodids appear to be a declining group, and consequently they might be viewed as poorly adapted in competition with the colubrid snakes and burrowing skinks.

The poor representation of the large-sized skinks, such as those in the genera *Tiliqua* and *Egernia*, both

in number of species and number of individuals, is marked in the north. The reason for this is not known.

The smallest skinks, such as those in the genera *Cryptoblepharus, Morethia, Menetia* and *Proablepharus*, have a clear spectacle permanently fused over the eye. Greer (1976) has suggested that this is a means of further limiting water loss compared with the moveable eyelid of many skinks and thus is an adaptation to arid conditions. These are certainly a successful group in the Top End with its seasonal aridity.

In comparing the amphibian faunas of the Kakadu region with the high country of south-eastern Australia (Figs 47 and 49) some differences are immediately apparent. In the Top End, species belonging to the family Hylidae are well represented, particularly in the forested lowlands and the escarpment. The family Myobatrachidae is also well represented in the northern forested lowlands. In the southern high country, representation of both these families is poor; except in the lowland swamps and grasslands where species richness within the Myobatrachidae is high. In the south, the Myobatrachidae is the dominant amphibian family, whereas in the north the Hylidae is dominant. Lower temperatures and a lack of free-standing freshwater are presumably prime factors limiting the amphibian fauna of the southern highlands, although some species have succeeded in invading even alpine communities (e.g. *Philoria frosti*). As Porter (1972) has generalized, reptiles are restricted latitudinally and altitudinally primarily by cold while amphibians as a group are less cold-sensitive but tend to be restricted by dry conditions.

The Kakadu region is further distinguished from the southern high country at the amphibian family level by the existence of a representative of the Microhylidae, *Sphenophryne robusta*, which occurs in escarpment monsoon forest pockets. Members of this family are almost exclusively tropical in distribution and reach great abundance in New Guinea with 13 genera and 102 species currently recognized (Tyler et al. 1981). In Australia, all but one of the seven known species are confined to the north of Australia, particularly north-eastern Queensland (Zweifel & Parker 1977, Tyler et al. 1981). *Sphenophryne robusta* also is the only species in the Top End region which lays its eggs on land rather than in water, but several species and genera with such a life history exist in the more mesic areas of eastern and southern Australia. The paucity of such species in the Top End is probably because of the lack of reliable moist conditions under which direct development is possible (see Tyler et al. 1983).

Another way of considering the fauna of a region is to examine the similarities of species composition with other parts of Australia; areas which share a lot of species are said to be similar. In their study of the Cobourg Peninsula, Cogger & Lindner (1974) found that the following percentages of non-marine reptile and frog species are shared with these other regions: 84% with north-western Australia; 53% with southern New Guinea; 77% with north-eastern Queensland; 39% with south-eastern Australia; 35% with central Australia and 26% with south-western Australia. Clearly the greatest affinities are with the Kimberley region of Western Australia; an area which shares the extreme wet-dry climate with the 'Top End', but has a longer dry season and a drier wet season.

Cogger & Heatwole (1981) have recently examined the affinities of the various regional reptile faunas at the family level. The geckoes of the Top End have strong desert affinities and are similar to those from the Kimberley, but are very distinct from the north Queensland fauna. This is also the case for the elapid snakes. The agamids of the Kimberley and Top End form a strongly distinct grouping, while the skinks from these two regions show more similarities with southern Australia than with north-eastern Queensland. The goannas again show the Top End and Kimberley similarity but there is a stronger similarity with north Queensland this time. Only the legless lizards show a completely balanced similarity between the three northern regions.

Similarly, recent zoogeographical studies on the amphibians (Tyler et al. 1981) have shown that there are strong faunal affinities between the Top End and the Kimberley, but that these areas have fewer affinities with northern Queensland, the latter being more similar to New Guinea. These differences are due mainly to the several taxa endemic to the Kimberley-Top End region (e.g. *L. coplandi, L. meiriana, L. personata,* and *M. lignarius*) as distinct from those restricted to north Queensland, but shared with New Guinea (e.g. *Litoria eucnemis, Litoria infrafrenata, Nyctimystes* spp., *Cophixalus* spp. and *Rana daemeli*). A few further taxa (e.g. the *L. dorsalis* group and *Sphenophryne* spp.) are shared between these regions, but are endemic to the north Australian tropics and New Guinea.

It is thus apparent that the herpetofauna of the wetter tropics of north Queensland shows little similarity with the herpetofauna of the Top End and Kimberley. As the species compositions are so dissimilar, it is reasonable to infer that the problems of adaptation for frogs and lizards in the wet-dry tropics are very different from those of the wet tropics. A further point (as raised by Tyler et al. 1981 for amphibia) is the realization of the northern tropics of Australia as a region of not only high herpetofaunal diversity, but also of a high level of endemism; and the Top End area is no exception. These trends have only recently become apparent in the light of better documentation of the herpetofauna following recent, more intensified survey work in the Australian tropics.

ACKNOWLEDGEMENTS

We thank Drs Grahame Webb and Rick Shine for their helpful comments on a draft of the chapter. RWB and JCW gratefully acknowledge the financial support of the Australian National Parks and Wildlife Service through the Kakadu Fauna Survey consultancy agreements. We also thank the many people who assisted with field work.

REFERENCES

Barker, J. & G.Grigg 1977. *A field guide to Australian frogs.* Rigby, Adelaide.
Brady, P.J. 1982. Frog guts. Unpublished project report. Darwin Community College, Darwin.
Braithwaite, R.W., G.D.Sanson & S.R.Morton 1977. Patterns of species density in Australian vertebrates. Unpublished manuscript.
Cann, J. 1978. *Tortoises of Australia.* Angus & Robertson, London.
Cogger, H.G. 1983. *Reptiles and amphibians of Australia* (3rd ed.). Reed, Sydney.
Cogger, H.G. & D.A.Lindner 1974. Frogs and reptiles. In H.J.Frith & J.H.Calaby (eds.), *Fauna survey of the Port Essington District, Cobourg Peninsula, Northern Territory of Australia.* CSIRO Div. Wildl. Res. Tech. Paper 28: 63-107.

Cogger, H.G. & H.Heatwole 1981. The Australian reptiles: origins, biogeography, distribution patterns and island evolution. In A.Keast (ed.), *Ecological biogeography of Australia*: Junk, The Hague: 1333-1373.

Fitzpatrick, E.A. & H.A.Nix 1970. The climatic factor in Australian grassland ecology. In R.M.Moore (ed.), *Australian grasslands*: Australian National University Press, Canberra: 3-26.

Friend, G.R. 1981. A note on cannibalism in the frog *Litoria dahlii. Northern Territory Naturalist* 1: 23.

Friend, G.R. & K.M.Cellier 1990. Wetland herpetofauna of Kakadu National Park, Australia: Seasonal richness trends, habitat preferences and the effects of feral ungulates. *J. Trop. Ecol.* 6 (in press).

Gow, G.F. 1976. *Snakes of Australia*. Angus & Robertson, London.

Gow, G.F. 1977. A new species of python from Arnhem Land. *Aust. Zool.* 19: 133-141.

Greer, A.E. 1976. A most successful invasion: the diversity of Australian skinks. *Aust. Nat. Hist.* 18: 428-433.

Hegerl, E.J., D.J.F.Davie, G.F.Claridge & A.G.Elliot 1979. *The Kakadu National Park mangrove forests and tidal marshes, Vol. 1: A review of the literature and results of a field reconnaissance*. Australian National Parks and Wildlife Service Consultancy Report.

Hill, R. & G.J.W.Webb 1982. Floating grass mats of the Northern Territory floodplains – an endangered habitat? *Wetlands* 2: 45-50.

James, C.D. 1983. Reproduction in lizards from the wet-dry tropics of Australia. Unpublished B.Sc. (Hons) thesis, University of Sydney.

James, C.D., S.R.Morton, R.W.Braithwaite & J.C.Wombey 1984. *Dietary pathways through lizards of the Alligator Rivers Region, Northern Territory*. Supervising Scientist for the Alligator Rivers Region Technical Memorandum 6: 1-11.

Jenkins, R.W.G. & R.J.Bartell 1980. *A field guide to reptiles of the Australian High Country*. Inkata, Melbourne.

Kikkawa, J. & G.Monteith 1980. *Animal ecology of monsoon forests of the Kakadu Region, Northern Territory*. Australian National Parks and Wildlife Service Consultancy Report.

King, M., R.W.Braithwaite & J.C.Wombey 1982. A new species of *Diplodactylus* (Reptilia: Gekkonidae) from the Alligator Rivers Region, Northern Territory. *Trans. R. Soc. S.Aust.* 106: 5-18.

Legler, J.M. 1980. *Taxonomy, distribution, and ecology of freshwater turtles in the Alligator Rivers Region, Northern Territory*. Report to Office of the Supervising Scientist, Sydney.

Nix, H.A. & J.D.Kalma 1972. Climate as a dominant control in the biogeography of northern Australia and New Guinea. In D.Walker (ed.), *Bridge and barrier: the natural and cultural history of Torres Strait*. Australian National University Press, Canberra: 61-91.

Pianka, E.R. 1966. Latitudinal gradients in species diversity: a review of concepts. *Am. Nat.* 100: 33-46.

Pianka, E.R. & J.J.Schall 1981. Species densities of Australian vertebrates. In A.Keast (ed.), *Ecological biogeography of Australia*. Junk, The Hague: 1677-1694.

Porter, K.R. 1972. *Herpetology*. Saunders, Philadelphia.

Rawlinson, P.A. 1971. Reptiles. Amphibians and Reptiles of Victoria, In V.H.R.Arnold (ed.), *Victorian Year Book 85*. Govt. Printer, Melbourne.

Shine, R. 1981. Ecology of Australian elapid snakes of the genera *Furina* and *Glyphodon. J. Herpet.* 15: 219-224.

Shine, R. & J.F.Berry 1978. Climatic correlates of live-bearing in squamate reptiles. *Oecologia (Berl.)* 33: 261-268.

Storr, G. 1964. Some aspects of the geography of Australian reptiles. *Senck. Biol.* 45: 577-589.

Swanson, S. 1976. *Lizards of Australia*. Angus & Robertson, London.

Tyler, M.J. 1976. *Frogs*. Collins, Sydney.

Tyler, M.J., G.A.Crook & M.Davies 1983. Reproductive biology of the frogs of the Magela Creek System, Northern Territory. *Rec. S.Aust. Mus.* 18: 415-446.

Tyler, M.J., G.F.Watson & A.A.Martin 1981. The amphibia: diversity and distribution. In A.Keast (ed.), *Ecological biogeography of Australia*. Junk, The Hague: 1277-1301.

Zweifel, R.G. & F.Parker 1977. A new species of frog from Australia (*Microhylidae; Cophixalus*). *Am. Mus. Novit.* 2614: 1-10.

CHAPTER 8

The influence of season on Australian crocodiles

GRAHAME J.W. WEBB

INTRODUCTION

Two species of crocodilians occur in Australia, both true crocodiles and distinct from the broad-snouted alligators and caimans (China, North and South America) and the extremely narrow-snouted gharials and *Tomistoma* (India, Malaysia and Indonesia). One of the Australian species, the freshwater or Johnston's crocodile *Crocodylus johnstoni* (Photograph 8), is endemic to Australia, whereas the other, the estuarine or saltwater crocodile *C. porosus* (Photographs 9 and 10), is found from India across through South-east Asia to the Philippines, and through Malaysia, Indonesia and Papua New Guinea to Australia. Australia represents the southernmost extent of the range of *C. porosus*.

C. johnstoni and *C. porosus* are both restricted to the northern coastal belt within Australia (Cogger 1979), and are thus in a climate dominated by the monsoonal weather regime. The distinct 'wet' and 'dry' seasons influence both the extent and productivity of the wetlands, and this influence is ultimately expressed in most aspects of crocodilian biology. 'Seasonal' influences, however, are not restricted to precipitation or the lack thereof. Like all reptiles, crocodiles are ectothermic, and are equally responsive to the annual cycle of ambient temperatures. Because temperature and rainfall cycles are not in phase (Table 23), it is often difficult to separate the extent to which crocodiles are responding to either 'wet-dry', 'hot-cold', or an interaction between the two conditions. For the present discussion there is utility in defining 'seasons' – ones that appear meaningful to the crocodiles – in terms of both temperature and rainfall. Three can be recognised (Table 23):

The wet season (November-December to April). Generally warm, beginning to cool in April; rainfall is common and often intense; wetlands contain abundant water; most tidal rivers contain predominantly fresh water once the wet season rains become regular, but in November-December, maximum salinities may occur if there is a slow start to the wet season.

The early dry season (May to July-August). The col-

dest period of the year, beginning to warm in August; rainfall rarely occurs; wetlands usually have abundant water in May but become dry throughout this period; in tidal rivers the salt wedge moves upstream, increasing salinities.

The late dry season (August-November). Temperatures steadily escalate to give the hottest period of the year in October-November; wetlands continue to dry out and water availability reaches its lowest levels; salinities increase in tidal rivers, although depending on the amount of rain in November, maximum levels may not be reached; occasional localized rains from convectional storms.

The manner in which these 'seasons' appear to influence the biology of Australian crocodiles is discussed below.

CROCODILE HABITATS

C. johnstoni and *C. porosus* tend to be allopatric in their distribution, although they occur together in parts of the freshwater and saline wetlands (Webb et al. 1983d). The respective common names of 'freshwater' crocodile and 'estuarine' (or 'saltwater') crocodile are thus rather misleading, because neither species appears to be restricted by salinity in the manner that their common names imply. This is especially the case with *C. porosus*. Although a common inhabitant of tidal rivers, the species is equally common in some coastal freshwater wetlands well separated from both salinity and tides (Webb et al. 1983e).

To assess the effects of season on Australian crocodiles, it is important to have at least a rudimentary understanding of the range of habitats occupied by each species. This can be achieved by considering 'typical' wetland systems, from the sea to the upstream headwaters (Fig. 50), and identifying where each species is found. Both species of crocodile tend to be most abundant where there is permanent water (saline or fresh), although they may utilise annual water bodies.

Table 23. Meteorological data from Middle Point, 50 km south-southeast from Darwin. Data accumulated over at least 15 years by the Australian Bureau of Meteorology.

Season	Month	Temperature (°C) Daily Max.	Daily Min.	9 a.m.	Relative humidity (%)	Rainfall Mean (mm)	Median (mm)	Number of rain days
Wet	January	32.8	23.9	27.6	84	323	289	20
	February	31.7	23.9	26.8	87	281	262	20
	March	31.9	23.7	27.0	85	262	229	18
	April	33.0	22.1	26.8	77	87	61	7
Early dry	May	32.0	19.4	24.8	71	24	3	3
	June	31.1	17.0	22.8	67	4	0	0
	July	30.9	15.1	21.5	61	0	0	0
Late dry	August	33.0	17.7	24.3	67	1	0	0
	September	34.7	20.3	26.9	65	10	6	2
	October	35.5	22.9	28.5	68	54	34	6
Wet	November	35.5	23.8	29.1	70	116	113	12
	December	33.7	23.9	28.2	79	245	224	16

The coast is occupied by *C. porosus*, but it is unclear as to whether coastal habitats are occupied permanently, or only by individuals in transit between non-coastal sites of occupation. Billabongs adjacent to the coast (Fig. 50, A) may be regularly visited by *C. porosus* from the sea (Messel et al. 1979), and the occasional females nesting in such billabongs may well be permanent residents. These sites tend to contain fresh water emanating from coastal springs, but may be inundated with saline water on the spring tides.

The majority of *C. porosus* are in tidal rivers (Fig. 50, B, C, D); in floodplain creeks (E), billabongs (F) and swamps (G); and in silted river channels (H) which were once meandering tidal rivers (see Chapter 1). In waterways under tidal influence (B, C, D), the banks are muddy and lined with mangroves and/or floodplain sedges and grasses. Where tidal rivers abut elevated land (B), or are isolated branches of the mainstream (E, F), continual evaporation throughout the dry season can result in salinities exceeding 70‰ (sea water is usually about 32‰). In contrast, long meandering mainstream channels (Fig. 50, C; Plate 19) typically have decreasing salinity with increasing distance upstream, because the upstream areas receive an input of fresh water for at least part of the dry season. Nevertheless, a salt wedge moves progressively upstream, increasing salinities as the dry season progresses. Perhaps the most important seasonal changes that affect crocodiles in saline and tidal habitats are: the flushing of saline water once the wet season is well underway; the uniting of channels (upstream E to mainstream C) which have been dry throughout the dry season; and, the flooding of the floodplain river banks.

Where permanent springs are associated with elevated land beside tidal rivers (Fig. 50, G) there are often isolated patches of freshwater swamp (Plate 20). Although small, these patches are particularly important to *C. porosus*, as they are a favoured site for nesting. They are elevated above the general floodplain level, and are thus not inundated by saline water during spring tides. More importantly, they are not subjected to the level of flooding that characterises the floodplain adjacent to meandering tidal rivers (C).

Where a meandering tidal river has silted, there typically remains a short tidal segment (D) and a series of isolated non-tidal billabongs (H) representing the meanders of the original tidal river (C). These contain fresh water throughout the year, although it does not flow during the dry season. Where undisturbed by feral and domestic stock, extensive floating rafts of vegetation extend from the shoreline out over the water (Plate 21); these are important *C. porosus* nesting sites. During the wet season, there is widespread flooding throughout these systems.

The above waterways are occupied mainly by *C. porosus*, whereas those further upstream are occupied mainly by *C. johnstoni*. Upstream waterways typically contain non-tidal fresh water, which may be permanently or seasonally flowing in river channels through either rocky escarpments, plateaux (Fig. 50, I) or upstream floodplains (Fig. 50, J, Plate 22). On the floodplains, creek lines often contract to series of isolated permanent and semi-permanent billabongs during the dry season. Substrates vary but usually contain sand rather than the mud of tidal downstream wetlands. During the wet season, there is widespread flooding throughout the upstream areas, and particularly high water levels occur where rivers pass through rocky gorges.

As indicated on Figure 50, the ranges of *C. johnstoni* and *C. porosus* overlap in many river systems, giving a recognisable zone of sympatry. In addition, indi-

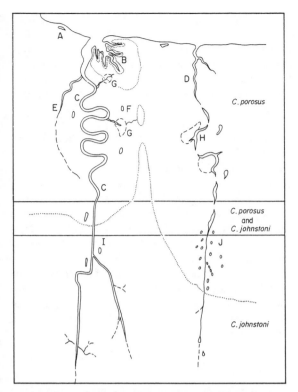

Figure 50. A schematic representation of the types of Northern Territory waterways occupied by crocodiles: (A) freshwater billabongs behind beachlines on the coast; (B) a many-branched tidal river penetrating elevated land; (C) a meandering tidal river transecting a floodplain; (D) the remnant of a meandering tidal floodplain river which has become silted; (E) a tidal floodplain creek with no freshwater input during the dry season; (F) an isolated floodplain depression containing fresh or saline water during the dry season; (G) a spring-fed freshwater swamp adjacent to a tidal river but elevated above the high tide level; (H) isolated sections of a silted meandering tidal river which are now non-tidal, contain freshwater and often floating rafts of vegetation; (I) the non-tidal upper reaches of a tidal river which drain rocky escarpment and contain freshwater; (J) a seasonally flowing mainstream channel through an upstream floodplain which has numerous isolated freshwater billabongs associated with drainage lines into the mainstream. Dotted lines indicate elevated land.

viduals of one species are sometimes found in areas dominated by the other species, either well upstream or downstream of the sympatric zone. Recent evidence (Webb et al. 1983d) suggests that *C. johnstoni* may be gradually being forced from downstream areas as numbers of the larger *C. porosus* increase as a result of legal protection (1971).

NESTING BIOLOGY

With the exception of a single species, the American crocodile *Crocodylus acutus*, the nesting strategies of extant world crocodilians are of two kinds (Greer 1970). They either construct a mound of live or dead vegetation into which they lay their eggs, or they excavate a hole in friable substrates (usually sand) and lay their eggs there (*C. acutus* appears to be essentially a hole nester which in some regions constructs a mound similar to that of megapode birds). Mound nesting is usually associated with 'wet' season nesting, whereas 'hole' nesting typically takes place in 'dry' seasons. The nesting strategies of Australian crocodiles have been reasonably well studied (see Magnusson 1979a, b, c, 1980a, b, 1981, 1982, Magnusson & Taylor 1980, Magnusson et al. 1978, 1980, Webb et al. 1977, 1983c, e).

C. johnstoni is a hole nester. Females seek out sand banks or similar areas of friable substrate next to permanent water, excavate a hole, and deposit a clutch of about 13 eggs. In many areas nest sites are selected next to each other, and sometimes one female will dig into the nest of another (Plate 23). Nesting occurs in a short three weeks during the late dry season, in August-September, and incubation typically takes 70-90 days. During this period, ambient temperatures steadily increase, and nest temperatures increase from about 30°C (laying) to 34-35°C (hatching; metabolic heat from the embryos may contribute to this increase). The synchrony between nesting and increasing temperatures appears to be important. High temperatures early in embryonic development cause developmental abnormalities, yet later in incubation they may enhance yolk utilisation and growth. In addition, ambient temperatures between one third and one half way through development have a direct sex determining function; below about 31°C and above 32°C females are produced, whereas in the 31-32°C bracket males are produced (Webb & Smith 1984).

Although occasional storms during the late dry season can cause inundation of the hole nests of *C. johnstoni*, and drown the developing embryos, the synchrony of hatching with the start of wet season rains (November) probably has a number of advantages. Insects are a major food item in the hatchling diet, and they are more abundant in the wet season than in the late dry season. The wet season also creates a large number of temporary water bodies into which hatchlings and older animals can disperse, although it is not known whether or not this influences hatchling survival.

The hole nesting strategy and the need for optimal temperatures for development and sex determination both appear to limit the seasons in which *C. johnstoni* can nest. Substrates are permanently wet or flooded during the wet season, and nesting earlier in the dry season would result in lower temperatures throughout incubation.

In contrast to *C. johnstoni, C. porosus* is a mound nesting crocodile which nests over the six month extent of the wet season. Females select nest sites close to permanent water, usually among floodplain grasses, in patches of freshwater swamp (Plate 20), or on floating rafts of vegetation (Plate 21). Females appear to be somewhat intolerant of each other, and nests are accordingly well spaced and rarely within 50 m of each other. About a week before egg-laying, females construct a mound about 1.8×1.6×0.5 m high, from the vegetation and often muddy substrate in the site chosen. Into this each deposits an average of 50 eggs.

Incubation takes from 75-95 days, and the sex of embryos is determined by the incubation temperature and perhaps also by other aspects of the nest environment. Flooding is a major cause of mortality, but frequent rain may also play an important cooling role. If conditions during incubation become too dry, some evidence indicates nest temperatures may exceed levels that are lethal to the embryos.

It can be seen that the two nesting strategies of Australian crocodiles are markedly different, and are adapted to, or possibly constrained by, the different seasons within the year.

FEEDING AND GROWTH

Feeding and growth in *C. johnstoni* and *C. porosus* are both influenced by season (Taylor 1979, Webb et al. 1978, 1982, 1983a), but not in identical ways.

Where they have been studied in the McKinlay River area, *C. johnstoni* congregate in isolated pools (Fig. 50, J) during the dry season, and eat little. Growth is negligible, and in fact most animals lose weight between successive wet seasons. With the onset of the wet season, feeding increases dramatically and most annual growth takes place. *C. johnstoni* between 56 and 116 cm snout-vent length contain an average of 6 g of food in their stomachs during the dry season (N = 20), and 35 g during the wet season (N = 45). The major food items are insects and small fish.

This seasonal variation could in part reflect the availability of prey, but other variables appear to be involved. In captivity, where food is permanently available, feeding tails off dramatically during the early dry season when conditions are cool (Table 23). It picks up steadily with increasing ambient temperatures during the late dry season. But higher temperatures and increased food availability in the field do not necessarily result in feeding during the dry season. In November 1982, which was a particularly dry year, *C. johnstoni* were in contracted pools with what appeared to be an abundance of some known prey items. Yet in one such pool observed, all except the largest crocodiles were essentially inactive, and when all crocodiles were caught, it became clear that none had been feeding. One dead crocodile whose head had been crushed by a larger crocodile was found in the pool. Thus although food was available, feeding activity within the limited space may have been avoided in case it elicited aggressive responses from the larger dominant animals.

Juvenile *C. porosus* in tidal rivers (Fig. 50, C) eat mainly crabs, prawns and insects, but there are no seasonal differences in the amounts of food eaten (Taylor 1979). Seasonal differences in growth rates, however, have been demonstrated (Webb et al. 1978). As in *C. johnstoni*, growth is more rapid during the wet season, but there is also a temperature-body-size effect during the dry season. The growth of larger crocodiles is constrained during the dry season, to a greater degree than that of smaller ones. Seasonal variation in the amounts of food eaten by large crocodiles (which were not examined by Taylor) may be more pronounced than in small ones. In captivity, *C. porosus* of all sizes greatly reduce food intake during the cool early dry season.

MOVEMENT AND DISPERSAL

Some data on movement and dispersal are available for both species of crocodiles (Magnusson 1979a, Messel et al. 1981, Webb & Messel 1978, Webb et al. 1983a, b), but much remains to be discovered.

Crocodiles appear to prefer locations in which little energy has to be expended to remain in the same position. They tend to avoid strong currents, and the most commonly adopted postures are those in which the feet and ventral surfaces are in contact with the substrate: i.e. in shallow water on the edge of a bank; exposed on a bank; exposed on a floating mat. In tidal areas, where the position of the water level relative to the bank is continually changing, most crocodiles rise and fall with the tide, remaining at the water's edge. Spotlight counts in tidal rivers are usually carried out at low tide when there is an exposed muddy bank below the mangrove fringe. At high tide, they can still be seen at the water's edge among the mangroves, but many have their eyes shielded from the light by the vegetation.

Wet season flooding is similar to a prolonged high spring tide. Juvenile crocodiles in particular tend to follow the water's edge away from the deeper areas, and they are usually found on the edge of the nearest emergent land. Some observations suggest that the larger crocodiles are more likely to remain in the fringe of flooded vegetation surrounding the deep and permanent water, but this needs to be investigated further. Nonetheless, movement and dispersal appear to be generally enhanced during wet seasons and restricted during dry seasons.

Photograph 8. Freshwater crocodile *Crocodylus johnstoni* (Photograph: G.J.W.Webb).

Photograph 9. Estuarine crocodile *Crocodylus porosus*, about 3 m long. Note the broader more powerful snout than the freshwater species (Photograph: G. Miles).

Photograph 10. A large estuarine crocodile *C. porosus* shot in 1916 on the Mary River (Photograph: Ryko, by courtesy of Institute of Aboriginal Studies Canberra).

Plate 19. Meandering tidal river: the Liverpool River (Photograph: G.J.W.Webb).

Plate 20. Nest of estuarine crocodile on an isolated freshwater swamp which drains into a tidal river: Melacca Swamp (Photograph: G.J.W.Webb).

Plate 21. Nest of estuarine crocodile on a floating raft of vegetation: Finniss River (Photograph: G.J.W.Webb).

Plate 22. Isolated upstream freshwater billabongs inhabited by the freshwater crocodile: McKinlay River (Photograph: G.J.W.Webb).

Plate 23. Nest of a freshwater crocodile, the eggs of which have been brought to the surface by another mother excavating her nest (Photograph: G.J.W.Webb).

A relatively minor flood can enhance long distance movements by connecting seasonal channels which for the majority of the dry season have been dry (E to upstream C on Fig. 50). In these cases, a relatively minor movement by a crocodile (for example in upstream C on Fig. 50) can place it in a channel which, if followed as it eventually dries, will take the crocodile well away from its original location (Webb & Messel 1978).

Both *C. porosus* and *C. johnstoni* have well developed homing abilities, but the extent to which animals of different ages, sexes and sizes utilize these abilities is largely unknown. Females of both species tend to nest in the same areas year after year, so they either restrict movement or 'home' to their area of nesting activity.

With little data available on territory sizes, for either species, caution needs to be exercised when interpreting the movement data that are available. For example, 83% of *C. johnstoni* which were marked in one year and recaught the next, were within 1 km of their original capture site, but not necessarily in the same pool (Webb et al. 1983a). However during the wet season, chains of pools are united, so that crocodiles may simply have been found in a different part of the same pool. The fact that one pool in the wet season subdivides into three isolated pools in the dry, may seem significant to us, but be irrelevant to crocodiles.

Movement of both species of crocodiles appears to be restricted during the early dry season, when temperatures are low and a considerable portion of each day is spent basking. Movements may increase during the late dry season when permanent water becomes scarce in some habitats, although *C. porosus* in some freshwater swamps excavate burrows beneath paperbark roots at this time. In contrast, rains at the start of the wet season, as distinct from flooding during the wet season, appear to stimulate both species to move, but in different directions. *C. johnstoni* disperse from dry season refuges when the first heavy rains occur, whereas adult *C. porosus* appear to congregate in tidal rivers and similar open water channels.

THERMOREGULATION

Limited data suggest both *C. johnstoni* and *C. porosus* have preferred body temperatures of about 30-33°C (Johnson et al. 1976), but no detailed studies have been undertaken. Nevertheless, thermoregulatory basking is obvious in the field, and shows distinct seasonal patterns.

During the wet season, neither species appears to spend appreciable portions of the day basking, and body temperatures are usually close to water temperature, often 30°C or higher. This situation changes with the onset of cold weather during the late wet season (April/May) and early dry season (Table 23). Both species spend large proportions of the day on banks or in shallow water, fully exposed to the sun. In areas where *C. porosus* females nest on floating mats of vegetation, a high proportion of eggs are drowned at this time of year, because females often climb up on the nest to bask, and in so doing sink it (Webb et al. 1983e).

During the late dry season, when ambient temperatures peak (Table 23), crocodiles usually avoid direct exposure to the sun, and either remain in the water or seek banks that are shaded (as do some feral mammals, see Ridpath, Chapter 11).

Seasonality in thermoregulatory behaviour is well exemplified by the behaviour of *C. porosus* in tidal rivers. During May, June and July crocodiles climb onto the muddy banks exposed at low tide, and lie in the sun; the most common track is 'U' shaped (from the water to the exposed bank and back to the water). Around August-September, crocodiles still bask on exposed mudbanks, but many either remain in the water or among the mangroves. If crocodiles climb from the water at low tide, they cross the exposed mud banks and seek the shade of the mangroves; the tracks are straight rather than 'U' shaped. At this time of year more crocodiles are also found buried in mud. The significance of this behaviour is unknown, although it could be thermoregulatory. As peak temperatures are reached at the end of the dry season, fewer and fewer crocodiles are seen basking, and more and more are shaded or buried. These behaviours extend into the wet season, and it is not until April-May that basking becomes commonplace again.

SALT AND WATER BALANCE

C. johnstoni and *C. porosus* both inhabit waters that may become excessively saline, particularly in the late dry season and early wet season. Both species are equipped with glands in the tongue that can secrete a concentrated salt solution (Taplin & Grigg 1981, Taplin et al. 1982), but this mechanism alone may not be sufficient in highly saline conditions. Additional freshwater *per se* is often unavailable (Grigg et al. 1980), and crocodiles may need the water obtained by feeding on hypotonic prey, i.e. prey with a lower salt content than themselves.

Some evidence suggests that female *C. porosus* select territories in habitats that do not become excessively saline during the dry season. Thus, although nesting occurs during the wet season when fresh water is abundant, nests are most abundant in areas that are permanently fresh or brackish during the dry season (Webb et al. 1977, 1983e).

DENSITIES

The extent of wetlands available for occupation by both species of crocodile is highly seasonal, and thus has a seasonal effect on crocodile density. *C. johnstoni* densities during the dry season can reach remarkable levels, for instance, one hundred individuals in a pool 75 m long by 25 m wide. Intraspecific fighting is greater when the animals are concentrated, judging by the frequency of fresh injuries (Webb & Manolis 1983), and such interactions can be fatal. It is likely that a significant mortality is both density and seasonally dependent.

SUMMARY

With both species of Australian crocodile, the cool conditions of the early dry season are associated with an increased incidence of basking, a movement back into permanent water bodies, an increase in densities, a reduction in feeding and growth, and a general reduction in mobility. As conditions warm, thermoregulatory behaviours switch to heat avoidance, densities increase as available water decreases further, salinity increases in tidal areas, and *C. johnstoni* nest. The start of the wet season is associated with increased mobility, a marked increase in feeding and growth, an increase in the extent of wetlands available for occupation, and a reduction in salinities and nesting by *C. porosus*.

REFERENCES

Cogger, H.G. 1979. *Reptiles and Amphibians of Australia* Sydney: Reed.

Greer, A.E. 1970. Evolutionary and systematic significance of crocodilian nesting habits. *Nature* 227: 523-524.

Grigg, G.C., L.E.Taplin, P.Harlow & J.Wright 1980. Survival and growth of hatchling *Crocodylus porosus* in saltwater without access to fresh drinking water. *Oecologia (Berl.)* 47: 264-266.

Johnson, C.R., G.J.W.Webb & C.Tanner 1976. Thermoregulation in crocodilians. II. A telemetric study of body temperature in the Australian crocodiles, *Crocodylus johnstoni* and *Crocodylus porosus*. *Comp. Biochem. Physiol.* 53A: 143-146.

Magnusson, W.E. 1979a. Dispersal of hatching crocodiles (*Crocodylus porosus*) (Reptilia, Crocodilidae). *J. Herpetol.* 13: 227-231.

Magnusson, W.E. 1979b. Incubation period of *Crocodylus porosus*. *J. Herpetol.* 13: 362-363.

Magnusson, W.E. 1979c. Maintenance of temperature of crocodile nests (Reptilia, Crocodilidae). *J. Herpetol.* 13: 439-443.

Magnusson, W.E. 1980a. Habitat required for nesting by *Crocodylus porosus* (Reptilia: Crocodilidae) in northern Australia. *Aust. Wildl. Res.* 7: 149-156.

Magnusson, W.E. 1980b. Hatching and creche formation by *Crocodylus porosus*. *Copeia 1980*: 359-362.

Magnusson, W.E. 1981. Suitability of two habitats in northern Australia for the release of hatchling *Crocodylus porosus* from artificial nests. *Aust. Wildl. Res.* 8: 199-202.

Magnusson, W.E. 1982. Mortality of eggs of the crocodile *Crocodylus porosus* in northern Australia. *J. Herpetol.* 16: 121-130.

Magnusson, W.E., G.C.Grigg & J.A.Taylor 1978. An aerial survey of potential nesting areas of the saltwater crocodile *Crocodylus porosus* Schneider, on the north coast of Arnhem Land, northern Australia. *Aust. Wildl. Res.* 5: 401-415.

Magnusson, W.E., G.C.Grigg & J.A.Taylor 1980. An aerial survey of potential nesting areas of *Crocodylus porosus* on the west coast of Cape York Peninsula. *Aust. Wildl. Res.* 7: 465-478.

Magnusson, W.E. & J.A.Taylor 1980. A description of developmental stages in *Crocodylus porosus*, for use in aging eggs in the field. *Aust. Wildl. Res.* 7: 479-486.

Messel, H., G.C.Vorlicek, A.G.Wells & W.J.Green 1981. *Surveys of tidal river systems in the Northern Territory of Australia and their crocodile populations, Monogr. 1: The Blyth-Cadell River systems study and the status of Crocodylus porosus in tidal waterways of northern Australia. Methods for analysis, and dynamics of a population of C. porosus.* Sydney: Pergamon.

Messel, H., A.G.Wells & W.J.Green 1979. *Surveys of tidal river systems in the Northern Territory of Australia and their crocodile populations, Monogr. 6: Some river and creek systems on Melville and Grant Islands.* Sydney: Pergamon.

Taplin, L.E. & G.C.Grigg 1981. Salt glands in the tongue of the estuarine crocodile, *Crocodylus porosus*. *Science* 212: 1045-1047.

Taplin, L.E., G.C.Grigg, P.Harlow, T.M.Ellis & W.A.Dunson 1982. Lingual salt glands in *Crocodylus acutus* and *C. johnstoni* and their absence from *Alligator mississippiensis* and *Caiman crocodilus*. *J. Comp. Physiol.* 149: 43-47.

Taylor, J.A. 1979. The foods and feeding habits of subadult *Crocodylus porosus* Schneider in northern Australia. *Aust. Wildl. Res.* 6: 347-359.

Webb, G.J.W., R.Buckworth & S.C.Manolis 1983a. *Crocodylus johnstoni* in the McKinlay River area, NT. III. Growth, movement and the population age structure. *Aust. Wildl. Res.* 10: 383-401.

Webb, G.J.W., R.Buckworth & S.C.Manolis 1983b. *Crocodylus johnstoni* in the McKinlay River area, NT. IV. A demonstration of homing. *Aust. Wildl. Res.* 10: 403-406.

Webb, G.J.W., R.Buckworth & S.C.Manolis 1983c. *Crocodylus johnstoni* in the McKinlay River area, NT. VI. Nesting biology. *Aust. Wildl. Res.* 10: 607-637.

Webb, G.J.W. & S.C.Manolis 1983. *Crocodylus johnstoni* in the McKinlay River area, NT. V. Abnormalities and injuries. *Aust. Wildl. Res.* 10: 407-420.

Webb, G.J.W., S.C.Manolis & R.Buckworth 1982. *Crocodylus johnstoni* in the McKinlay River area, NT. I. Variation in the diet and a new method of assessing the relative importance of prey. *Aust. J. Zool.* 30: 877-899.

Webb, G.J.W., S.C.Manolis & G.C.Sack 1983d. *Crocodylus johnstoni* and *Crocodylus porosus* coexisting in a tidal river. *Aust. Wildl. Res.* 10: 639-650.

Webb, G.J.W. & H.Messel 1978. Movement and dispersal patterns of *Crocodylus porosus* in some rivers of Arnhem Land, northern Australia. *Aust. Wildl. Res.* 5: 263-283.

Webb, G.J.W., H.Messel & W.E.Magnusson 1977. The nesting biology of *Crocodylus porosus* in Arnhem Land, northern Australia. *Copeia* 1977: 238-249.

Webb, G.J.W., H.Messel, J.Crawford & M.J.Yerbury 1978. Growth rates of *Crocodylus porosus* (Reptilia: Crocodilia) from Arnhem Land, northern Australia. *Aust. Wildl. Res.* 5: 385-399.

Webb, G.J.W., G.C.Sack, R.Buckworth & S.C.Manolis 1983e. An examination of *Crocodylus porosus* nests in two northern Australian freshwater swamps, with an analysis of embryo mortality. *Aust. Wildl. Res.* 10: 571-607.

Webb, G.J.W. & A.M.A.Smith 1984. Sex ratio and survivorship in the Australian freshwater crocodile, *Crocodylus johnstoni.* In M.W.J.Ferguson (ed.), *The Structure, Development and Evolution of Reptiles:* 318-355. London: Academic Press.

CHAPTER 9

Birds

S. R. MORTON & K. G. BRENNAN

INTRODUCTION

In 1953, a promising new agricultural industry began on the subcoastal floodplains of the Northern Territory. Experimental rice-farms were laid out near Humpty Doo, just to the south-east of Darwin, and optimism about the success of the new venture was widespread. However, agriculturalists had not reckoned with the Magpie Goose, a large herbivorous bird which descended in substantial numbers onto the new crops and posed an immediate threat to the industry. In 1955, therefore, the Wildlife Survey Section of CSIRO was asked to undertake ecological studies of Magpie Geese, principally to assess potential control measures. The work carried out under this brief (Frith & Davies 1961a, b, Davies 1962, 1963) marked the beginning of ecological ornithology in monsoonal northern Australia.

The avifauna of the Top End had not been ignored before that time, of course, but previous ornithological investigations had been almost entirely of a distributional or taxonomic nature. The first important collections of birds were made in 1840 by John Gilbert (see Calaby 1974), a collector employed by the eminent John Gould, and he was followed by numerous observers over the next 100 years (notably W.H.D.Le Souef, J.T.Tunney and W.R.Maclennan) who gradually established the character of the avifauna. Basic taxonomic listing of northern birds was completed with the description of the White-lined Honeyeater (White 1917), although the study of intraspecific variation and of speciation continues today (e.g. Ford 1978, Schodde et al. 1979) and will need to proceed into the future. Distributional data continued to be made available by scientific expeditions such as the American-Australian Scientific Expedition to Arnhem Land of 1948 (Deignan 1964) and the fifth phase of the Harold Hall Australian Expedition (Freeman 1971). By the time of the latter expedition, however, ecological work on northern birds had begun, and it accelerated as CSIRO's Division of Wildlife Research expanded its efforts (Frith & Hitchcock 1974, Schodde 1973), as uranium mining led to the creation of the

Alligator Rivers Region Research Institute within the Office of the Supervising Scientist, and as the Conservation Commission of the Northern Territory rapidly developed expertise in research and management.

Today we are in the midst of unprecedented ecological work on birds of the Top End, and it is an appropriate time to review what is known of biogeography, of habitat relationships, of the trophic structure of communities, of the effects of seasonality, and of the place of birds in ecosystems of the Top End. Our review will by no means answer all questions, but we hope it will help focus attention on current knowledge and outstanding problems. Common names for birds are used throughout our text, with scientific names being listed in Appendix 1. A convenient entry to the literature of the Northern Territory's birds may be found in Estbergs (1980).

BIOGEOGRAPHY OF NORTHERN BIRDS

The Northern Australian tropics are recognized as one of the three main biogeographic provinces of Australia, the Torresian zone. This province is a belt of forests and woodlands dominated by the genus *Eucalyptus* which includes the Kimberley, the Top End of the Northern Territory, a belt of woodlands interspersed along the Gulf of Carpentaria, and the woodlands and forests of eastern Queensland. Over most of its southern boundary, the Torresian province is bounded by the shrublands and grasslands of the arid Eyrean province. Within the Torresian province, Schodde & Calaby (1972) have recently defined two additional, distinctive avifaunal elements that are found in moist forests. The Tumbunan element comprises birds of temperate, montane rainforests confined to the ranges of eastern Australia and the central highlands of New Guinea. The Irian element consists of birds of lowland rainforests, which are well developed in north-eastern Queensland and New Guinea, but present only as small, disjunct pockets in suitably moist locations across the Top End and the Kimberley (see Chapter 4, Photograph 1, Plate 5). To understand the present-day composition of the northern Australian avifauna, and particularly its relationships with the Irian and Tumbunan elements, it is necessary to consider briefly the origins of the Australian avifauna and the palaeohistory of climate and habitats within which it developed.

When the Australian continental plate split from the ancient Gondwanan land mass about 90 million years ago, a much wetter and more uniformly temperate climate prevailed. As Australia drifted northwards towards the Asian plate, mild conditions persisted until the Pliocene epoch, and the continent was covered with temperate rainforest. From the Pliocene onwards, cycles of aridity began to occur which resulted from global glacial activity (see also Chapter 1). During dry cycles, much of the former rainforest was eliminated and replaced by vegetation tolerant of drier conditions. In addition, the continent's northward drift gradually brought it into the seasonal tropics. Thus, northern Australia developed a eucalypt-based flora adapted to a summer wet season, southern Australia a eucalypt-based flora experiencing rainfall in winter, and the inland of the continent a xeric flora adjusted to low and erratic rainfall. These three broad regions constitute the basic biogeographic provinces. Some patches of the primal rainforests persisted in scattered refuges along the Great Dividing Range of eastern Australia and in the Highlands of New Guinea, but where the Australian plate penetrated the humid tropical latitudes, lowland rainforests developed.

By the time Australia came into close contact with New Guinea and South-east Asia, birds must have left the ancient temperate rainforests and evolved into forms that could successfully occupy the eucalypt woodlands of the north. Thus, the structure of the avifauna at this time probably resembled its present form (Keast 1981, Schodde 1982). The climatic fluctuations of the Pleistocene continued to alter the distributions of habitat types, however, and had considerable effects on birds. Birds endemic to the temperate rainforests were confined to remnants along the east coast, and now form the Tumbunan element. As the land bridge with New Guinea was established, birds of lowland rainforests were able to enter Australia where they now persist as the Irian element. During dry cycles of the Pleistocene, the rainforests contracted into pockets and eucalypt forests also shrank coastward, while in moister cycles the reverse occurred. For birds, therefore, it was a time of environmental flux during which their ranges were contracting and expanding, often isolating populations of many species in pockets of suitable habitat and thereby stimulating speciation. Since the most recent phase of aridity some 17,000 years ago (Bowler 1982), the climate has moderated to that of today. In association, the sea-level has risen, cutting off Cape York from New Guinea.

In broad terms, then, the Torresian avifauna contains those species that occupied the eucalypt woodlands and forests throughout the Miocene and Pliocene, and survived the exigencies of the glacial period. The Tumbunan element comprises species with their origins in the cool, wet rainforests of eastern Australia and New Guinea, now isolated in Australia in patches along the eastern coast. Finally, the Irian element contains species with origins in warm, moist forests, many of which survive today in patches of suitable habitat scattered across northern Australia.

The role of climatic change in promoting habitat selection and speciation is best understood by considering specific examples. The Red-winged Parrot of the genus *Aprosmictus*, which is a widespread Torresian species, can be readily linked to the polytelitine parrots, which include the genera *Alisterus* (the Australian King-Parrot) and *Polytelis* (the Regent, Superb and Alexandra's Parrots). Schodde (1982) suggests that the ancestral stocks of this group inhabited the Tumbunan rainforests, and are represented today by the Australian King-Parrot of eastern Australia. At the drier boundaries of the rainforests, these stocks appear to have differentiated into the modern *Aprosmictus* and *Polytelis* as rainforests contracted and eucalypt forests expanded. Today, *Aprosmictus* occupies *Eucalyptus* and *Melaleuca* forests of northern Australia, while the three species of *Polytelis* have occupied drier woodland and desert in the south-west, south-east and centre of the continent.

Reconstruction of the evolutionary history of the Red-winged Parrot shows us how ancient events continue to reverberate through time to affect the structure of present-day communities. These events culminate in the striking effects on speciation of the expansion and contraction of habitats in northern Australia. For example, the Helmeted Friarbird occupies moist forests throughout northern Australia. During the Pleistocene glaciations, rainforests or monsoon forests in the Top End were fragmented by arid cycles, apparently confining populations of Helmeted Friarbirds to refuges in Cape York and southern New Guinea, in protected gorges along the Arnhem Land escarpment, and in patches of forest in the Kimberley of north-western Australia. Severe aridity in the Kimberley resulted in the gradual disappearance of monsoon forests, but Helmeted Friarbirds successfully adapted to the change, evolving in the process into a different species, the Silver-crowned Friarbird (Schodde 1982). Evidently in subsequent interglacial periods, when the climate again became wetter, Silver-crowned Friarbirds spread eastward and occupied the drier forests and woodlands across the Top End and Cape York. Residual populations of Helmeted Friarbirds in Arnhem Land have never regained contact with their conspecifics on the east coast (the subspecies *Phile-*

mon buceroides yorki), and in fact continuing climatic change isolated two major populations within the Top End (Schodde et al. 1979). Dry glacial periods, it seems, resulted in the retreat of some Helmeted Friarbirds to sandstone gorges, while others were able to occupy coastal monsoon forests and mangroves. Because sea levels dropped during dry glacial periods, the populations inhabiting coastal forests and mangroves almost certainly shifted northwards with the coastline away from the populations living in their refuges within the Arnhem Land escarpment. Sea levels have risen within the past 15,000 years, thus bringing the two populations geographically close once more, but coastal birds (*P. b. gordoni*) remain recognizably different from escarpment birds (*P. b. ammitophila*).

These examples show some of the processes by which evolutionary history is linked directly with distribution and habitat preferences today. In particular, the example of the friarbirds highlights the important point that many Torresian and Irian birds have discontinuous distributions throughout northern Australia because of the disruptive effects of climatic change, particularly throughout the Pleistocene. Two barriers characterized by relatively arid habitats, the hinterlands of the Gulf of Carpentaria and the Joseph Bonaparte Gulf, split the avifauna into three subregions, Cape York, the Top End and the Kimberley (see Ford 1978, Keast 1981). The effectiveness of these barriers in isolating components of the avifauna can be gauged by tabulating the numbers of species that fail to range across them (Table 24). This analysis shows that Cape York has a much larger number of endemic species than either of the other two subregions, reflecting both its greater diversity of habitats and the substantial Tumbunan and Irian elements in its rainforests. Only ten species with discontinuous ranges are shared with the Top End, and only three species are confined to the Top End (the Banded Fruit-Dove, the White-throated Grasswren and the Chestnut-quilled Rock-Pigeon). The Top End shares with the Kimberley thirteen species with discontinuous ranges; examples are Green-backed Gerygones and Purple-crowned Fairy-wrens. The Kimberley, however, contains only one endemic species, the Black Grasswren, itself a vicariant of the White-throated Grasswren endemic to Arnhem Land. Table 24 shows, in summary, that refuges favouring speciation in northern Australian birds of Irian and Tumbunan elements have been heavily concentrated in Cape York and New Guinea, and that such refuges and the consequent degree of endemic speciation decreased dramatically westward from Cape York with increasing aridity and environmental uniformity.

Species with discontinuous ranges comprise only a small proportion of the avifauna of the Top End, and a more complete picture of the content of the fauna can be gained by examining the overall proportions of species shared between Cape York, the Top End and the Kimberley (Table 25). Species inhabiting sandy, muddy or rocky coastline are excluded from Table 25, but mangrove inhabitants are included. We have included a few of the Northern Hemisphere waders which we judge to be regular visitors (e.g. Asian Dowitcher) but ignored others that are probably vagrants (e.g. Little Ringed Plover). Table 25 shows that the Top End and the Kimberley have avifaunas of comparable richness, but Cape York has a substantially richer fauna. The Top End's birds are generally widely distributed in northern Australia, for more than 90% of the species occur either in Cape York or the Kimberley in addition to the Top End. The table emphasizes the point made at the beginning of this discussion of biogeography, which is that the Top End's avifauna is characterized by widely distributed species of Torresian origin.

BIRDS AND HABITATS OF THE TOP END

In considering ecological relationships within a diverse avifauna of more than 250 species in the Top End, our immediate question is: in what habitats do all these birds fit? A preliminary answer is summarized in Figure 51, which shows the distribution of the species

Table 24. Numbers of species of birds having discontinuous ranges in northern Australia tabulated according to occurrence in the three mesic Torresian subregions.

Habitats	Subregions Cape York only	Cape York and the Top End	Top End only	Top End and the Kimberley	Kimberley only
Mangroves	2	1	0	1	0
Floodplains	3	2	0	3	0
Riparian forests	12	3	0	2	0
Rainforests or monsoon forests	57	1	1	1	0
Woodlands and open forests	2	3	0	2	0
Sandstone escarpments	0	0	2	4	1
Total	76	10	3	13	1

Table 25. Numbers of species of birds found in the three major subregions of northern Australia, and the percentages of the Top End's avifauna also occurring in the other two subregions.

	Subregion Top End	Kimberley	Cape York
Number of species	251	241	323
Percentage of the Top End's avifauna also occurring in other subregions	–	93	90

among seven basic habitats that seem to be recognized by birds in the area. Each species was allocated to what we consider to be its primary habitat, but many species occur so widely that a category of 'widespread' was included. Such widespread species constitute nearly a quarter of the total, and include abundant birds like the Brown Honeyeater, the Black Kite and the Peaceful Dove. These widespread birds are broadly distributed elsewhere throughout Australia and do not have such strict habitat requirements as do other birds.

The birds occurring primarily in open forest (see Chapter 4, Plate 7) and woodlands also tend to be widely distributed throughout Australia because the eucalypt woodlands they occupy are themselves spread over large areas of the continent; consequently, only 24% of species occurring in this habitat are confined to the Torresian province. Examples of birds that are widespread outside the Torresian province are Red-tailed Black-Cockatoos and White-winged Trillers. As a corollary, very few species are confined to northern Australia, examples being Partridge Pigeons and Long-tailed Finches, and even these have vicariants in the eastern Australian woodlands. Open

forests and woodlands account for more than 95% of the area of the Top End. Despite their similarity in structure to the eucalypt forests and woodlands in the Bassian and Eyrean provinces, they are floristically distinctive, particularly in the dominance and diversity of understorey grasses (Clifford & Simon 1981, Pryor & Johnson 1981, Schodde 1982). Thus, one unique aspect of the northern woodland avifauna is a relatively rich community of granivorous birds, particularly finches.

Other habitats shown in Figure 51 are smaller in area but are more characteristic in the species they contain of northern Australia in general and of the Top End in particular. Riparian forests are variable in floristic composition, often containing eucalypts or species more characteristic of monsoon forests, but they are dominated by paperbark trees, *Melaleuca* spp., especially in forests around freshwater wetlands and floodplains. This habitat is primary habitat for a small but distinctive group of birds. The rich, loud calls of the White-gaped Honeyeater, for example, can frequently be heard among dense vegetation lining watercourses. Some of the species find their prey in waterholes scattered along creeks lined with riparian forest, e.g. the Black Bittern and the Azure Kingfisher. About half of the riparian birds are endemic to northern Australia, contrasting with the low proportion of purely Torresian species in open forests and woodlands.

Many species found in riparian forests also occur commonly in monsoon forests, and our allocation of several species to one or other of the two habitats is arbitrary. Nevertheless, the analysis suggests that similar numbers of species choose the two forests as primary habitat (Fig. 51). Monsoon forests are limited to pockets and occur patchily throughout the Top End, but the structural and floristic distinctiveness of their

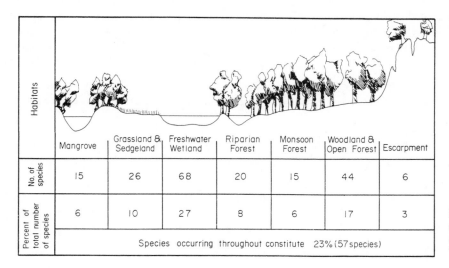

Habitats	Mangrove	Grassland & Sedgeland	Freshwater Wetland	Riparian Forest	Monsoon Forest	Woodland & Open Forest	Escarpment
No. of species	15	26	68	20	15	44	6
Percent of total number of species	6	10	27	8	6	17	3

Species occurring throughout constitute 23% (57 species)

Figure 51. Numbers of species occupying as their primary habitat each of seven major habitats in the Top End.

vegetation result in occupation by a discrete suite of birds. As noted above, many of the birds belong to the Irian avifaunal element. Nearly all of the species are confined in their Australian distribution to the north, e.g. Green-backed Gerygone, Rainbow Pitta and Banded Fruit-Dove, but a few range widely throughout the eastern Australian forests, namely the Rose-crowned Fruit-Dove and Emerald Dove.

The sandstone escarpment of Arnhem Land, in the centre of the Top End, is one of the most unusual parts of the region. The steep cliffs and tumbled boulders at the margins of the plateau form spectacular rock faces and gorges, and, together with the dissected tableland behind the escarpment, form a habitat complex for a small but unique set of birds. The two principal elements of the complex are the open woodland and hummock grassland formations on the plateau and the dense forests of the gorges. Six species of birds occupy the complex as primary habitat: Chestnut-quilled Rock-Pigeon, Sandstone Shrike-thrush, White-throated Grasswren, White-lined Honeyeater, a subspecies of the Variegated Fairy-wren called the Lavender-flanked Fairy-wren, and the sandstone-dwelling subspecies of the Helmeted Friarbird. All of these taxa are confined to northern Australia, although the latter two are merely distinctive subspecies of widely-distributed species. Together with the Kimberley (see Table 24), the Arnhem Land escarpment therefore supports a small but uniquely characteristic group of birds.

Subcoastal floodplains are scattered across the Top End, particularly in its western half. No other habitat undergoes such marked seasonal changes, from verdant, lily-covered waters in the wet season to brown, desiccated grasslands and sedgelands in the dry. Figure 51 shows that nearly a quarter of the species occurring in the Top End occupy inundated floodplains as their primary habitat, thereby indicating the importance of this type of ecosystem in allowing the development of a regionally rich avifauna. However, almost all of the species are widespread throughout Australia or Asia, and few are endemic (Table 24). Dominant species are Magpie Geese (Plate 24), Wandering Whistling-Ducks (Plate 25), Intermediate Egrets and Glossy Ibis. The contribution that birds of the inundated floodplains make to the Top End's avifauna is inflated by migratory waders of the families Charadriidae and Scolopacidae, which constitute 18 of the species considered in Figure 51. These birds arrive from the Northern Hemisphere in the late dry season on their way to eastern and southern Australia, and few of them are resident in the Top End. The most common species in freshwater wetlands are Sharp-tailed Sandpipers, Curlew Sandpipers and Red-necked Stints.

The grasslands and sedgelands of the drying floodplains are occupied by a substantial number of birds, some of them resident and others migrants in the dry season. Where vegetation is dense, small passerines like the Golden-headed Cisticola and the Chestnut-breasted Mannikin may be resident. Less dense grassy areas harbour migratory species from the inland such as the Australian Pratincole (Plate 26) and the Australian Bustard.

The thin strip of mangroves lining tidal channels of the north coast is primary habitat for about 15 species (Fig. 51). A few are confined to the north coast in Australia (see Table 24), e.g. the Red-headed Honeyeater and the Broad-billed Flycatcher, but most are distributed along a considerable extent of the mangrove fringe elsewhere around the continent, e.g. the Collared Kingfisher and the Striated Heron. We will not discuss the mangrove avifauna in any detail in this chapter; readers are referred to Ford (1982) and Schodde et al. (1982) for further information.

Consideration of habitat selection has so far been focussed on primary habitats, but as most observers are aware, nearly all species can be found in more than one habitat. In order to examine the distribution of species among habitats in a comprehensive way, we scored each species in every habitat in which it regularly occurs and simultaneously divided the species into nine trophic groups or feeding guilds (Table 26). The proportions of the total avifauna occurring in the major habitats are greatest in open forests and woodlands and in riparian forests; this result again emphasizes the importance to the avifauna of eucalypt forests and the structurally similar paperbark forests. Each of the other five habitats is suitable for about a third of the regional avifauna. Thus, the differences described above in the relative importance of primary habitat are largely removed because of flexibility in habitat use by many birds. Some of the flexibility actually results from seasonal changes in habitat use, and we will return to this important topic below. It is also worth noting that species diversity is frequently higher in ecotones than in pure habitats because of the admixture of birds occurring there (Taylor et al. 1984); the richest avian communities, therefore, are frequently found at the edges of the primary habitats.

Distribution of the numbers of species among trophic groups shows several important relationships. In all habitats except freshwater wetlands, insectivores dominate the avian communities. The insectivores comprise a wide variety of morphological types, and consequently utilize a broad range of foraging zones and techniques. It is the partitioning of the diverse and abundant insect fauna through the use of different foraging zones that results in insectivorous birds reaching greater species richness than other trophic groups of terrestrial birds, and morphology and foraging technique are intimately involved in this resource partitioning. Wherever there are trees, six broad forag-

Table 26. Percentages of the avifauna within each habitat divided among nine trophic groups, together with an estimate of the percentage of the total avifauna occurring in each habitat.

Habitats	Foods Vegetation	Aquatic invertebrates	Fish	Seeds	Fruit	Nectar*	Insects*	Vertebrates	Wide variety**	Percentage of total avifauna
Escarpment	–	–	–	10	–	17	54	16	3	29
Woodland and open forest	–	–	–	20	1	11	48	12	8	53
Monsoon forest	–	–	–	10	4	14	55	11	6	33
Riparian forest	–	5	2	10	2	11	49	15	6	50
Freshwater wetlands	8	53	12	–	–	–	8	7	12	30
Sedgelands and grasslands	–	–	–	16	–	–	46	22	16	30
Mangroves	–	7	2	–	–	18	59	10	4	28

* Honeyeater species that eat both nectar and fruit appear under both headings.
** Principally a mixture of seeds, fruit and insects.

ing zones are available to insectivorous birds (Schodde 1981): ground zone; tussock and shrub zone; trunk and branch zone; foliage and twig zone; sub-canopy free space; and above-canopy free space. Where the forest is relatively dense, the trunk and branch zone and the foliage and twig zone may be further subdivided by birds (Crome 1978). After choosing to forage in a particular zone, birds then further partition the insect fauna by adopting certain types of foraging techniques (Crome 1978, Schodde 1981). In probing, birds poke their slender, down-curved bills into bark or bunches of leaves to extract prey. Search-gleaners move deliberately through vegetation or along the ground, picking insects off the substrate as they are encountered. Perch-sallying birds stay on a perch until prey is sighted, fly down to seize it, and fly back to a perch to eat it. Hover-salliers dive from the air onto the substrate to catch their prey. Many species hawk for insects, usually perching on a branch and then swooping out to catch insects in flight. Finally, terrestrial birds may scratch in the litter to expose the insects sheltering therein. It is easy to imagine that use of one or two of these six foraging techniques by a species would focus on a particular segment of the insect fauna; thus, partitioning of the food resource would result. No detailed foraging studies of the insectivorous avifauna of the Top End have been published, and we cannot confidently classify all species by foraging zone and technique. Nevertheless, in Table 27 are representative examples of birds that use each zone and technique. In addition to such partitioning of resources, birds of different sizes almost certainly eat insects of different sizes and select insects that are particularly palatable or behave in certain ways (Hespenheide 1975), thereby partitioning the resource further.

The dominance of insectivores in dry land habitats is clear, but to some extent it is exaggerated in Table 26 because most of the honeyeaters included as insectivores satisfy much of their energy requirements from nectar (Pyke 1980). Nectarivores, including Varied Lorikeets and Red-collared Lorikeets as well as honeyeaters, are very abundant birds despite the fact that they are not as rich in species as are the insectivores. A relative lack of richness may reflect the reduced number of ways in which the nectar resource can be partitioned, compared to the multitude of methods available for partitioning insects. Nevertheless, whenever trees or shrubs are flowering, as many as six species of nectarivores can be seen squabbling over the flowers. The most abundant and widespread of these birds are the Varied Lorikeet, the Brown Honeyeater, the White-throated Honeyeater, the Little Friarbird, and, particularly in southern parts of the Top End, the Banded and Yellow-tinted Honeyeaters.

To a large extent, the factor underlying the structure of avian communities in woodlands and forests is the supply of insects and nectar. Birds in these two trophic groups constitute about 60% of the total numbers of species in such habitats, and probably comprise close to 90% of the avian biomass. Only one other trophic group, the granivores, is important in terms of biomass. In all habitats except mangroves and freshwater wetlands, between 10 and 20% of species are granivores, ranging in size from the tiny Double-barred Finch (about 10 g) up to the Red-tailed Black-Cockatoo (about 650 g). The highly seasonal monsoonal environment favours annual grasses and herbs that persist through the dry season as seeds; prominent among these is annual sorghum, *Sorghum intrans*. This seasonally regular production of seeds supports an array of small granivores, the finches, pigeons and small parrots. The cockatoos include in their diets a greater proportion of the seeds of trees, often gathering the seeds directly from the branches. Red-tailed Black-Cockatoos, in particular, appear to cause considerable damage to eucalypts by ripping apart hundreds of seed-capsules from individual trees.

Frugivory is a way of life for few species in woodlands and forests, although many of the species listed

Table 27. Examples of birds that utilize specific foraging zones and foraging techniques in woodlands and forests of the Top End.

Foraging zones	Foraging techniques					
	Probe	Search-glean	Perch-sally	Hover-sally	Hawk	Scratch
Ground	–	Bush Thick-knee	Pallid Cuckoo	Restless Fly-catcher	–	Rainbow Pitta
		Sandstone Shrike-thrush	White-browed Robin	Shining Fly-catcher		Orange-footed Scrubfowl
Tussocks and shrubs	–	Red-backed Fairy-wren White-throated Grasswren	–	–	–	–
Trunks and branches	Black-tailed Treecreeper Varied Sitella	Grey Shrike-thrush Rufous Whistler	–	Black-faced Cuckoo-shrike	–	–
Foliage and twigs	–	White-throated Gerygone White-throated Honeyeater	–	–	–	–
Sub-canopy free space	–	–	–	–	Lemon-bellied Flycatcher Northern Fantail	–
Above-canopy free space	–	–	–	–	Little Wood-swallow Tree Martin	–

as omnivores in Table 26 do consume fruits to some extent. Specialized frugivores are concentrated in monsoon forests, a habitat in which trees with fruit adapted to consumption by birds are relatively common. In this regard, monsoon forests of the Top End are similar to the rainforests of north Queensland, although the richness of fruit-doves and pigeons in the latter region is much greater (Crome 1975).

All dry land habitats in the Top End contain a considerable number of carnivores: hawks, falcons, eagles, harriers and owls. Most species are uncommon, and indeed several are rare (Red Goshawk, Letter-winged Kite, Grey Falcon, Black Falcon, Peregrine Falcon and Eastern Grass Owl), but two scavenging species, Black Kites and Whistling Kites, are particularly common. It is worth noting that the large number of carnivorous species recorded in sedgelands and grasslands is partly due to the occurrence of several basically inland species that range into the Top End along dry floodplains; examples are Spotted Harriers and Letter-winged Kites. Omnivorous birds are also comparatively rich in sedgelands and grasslands (Table 26), but in this case the richness is caused by the presence of species in rank grasses on the fringes of floodplains (e.g. Brown Quail, Red-backed Button-quail and Red-chested Button-quail). Many species selecting rank grasses can be found equally plentifully in thick grass within woodlands. Thus, the apparent concentration of omnivores in sedgelands and grasslands does not reflect a greater biomass relative to other habitats. One omnivore, the Australian Bustard, migrates from inland Australia onto sedgelands and grasslands during the dry season, but it is uncommon in the Top End.

Our discussion so far has not considered the freshwater wetlands because they support a unique assemblage of birds, most species of which do not occur in any other habitat. The principal differences from the trophic structures of avifaunas in other habitats are threefold: a switch from insectivory to consumption of aquatic invertebrates; a concentration of fish-eating species; and the presence of a guild of herbivorous birds. As a corollary, no birds are specialized frugivores, nectarivores or granivores (although all of the herbivorous birds eat seeds as well as vegetation). That such distinctive differences should emerge is not surprising, for freshwater wetlands present an array of food resources to birds that differs dramatically from that of dry land habitats. Over thousands of years, nutrients leached from terrestrial habitats have been deposited by floodwaters in freshwater wetlands; thus, floodplain sediments are capable of supporting a large, annual biomass of vegetation. The macrophytic vegetation is frequently soft and succulent when actively growing, and is therefore readily consumed. Further, its resources persist through the dry season as seed (e.g. wild rice, *Oryza sativa*) or as underground bulbs (e.g. spike-rushes, *Eleocharis dulcis* and *E. sphacelata*). When aquatic, the vegetation includes as well tiny, floating plants and algae that can be eaten by birds or support aquatic invertebrates. With such an array of food available to them, the herbivorous Magpie Geese and Wandering Whistling-Ducks are usually the most common birds in wetlands.

Aquatic vegetation supports invertebrate animals that are distinctively different from those in terrestrial environments. Because there is little complexity in the emergent macrophytic vegetation, there is little scope for partitioning of their insect resources by birds. Thus, there are few specialized insectivorous species, and the insects of emergent vegetation are most often eaten by waterbirds that also eat other aquatic invertebrates (e.g. Whiskered Terns, Comb-crested Jacanas). In contrast, the invertebrates inhabiting the water column and the submerged substrate present an array suitable for partitioning by predators. Table 28 details representative foraging strategies used by different species. The table shows that waterbirds have developed a similar range of strategies for utilizing aquatic invertebrates as have land birds for capturing insects. Waterbirds probably also partition prey by size, but the study of resource specialization in the Top End's wetlands is only just beginning and patterns are not yet clear.

Many birds that eat aquatic invertebrates will capture small fish as well, but several species subsist almost entirely on fish. Prominent in this group are the Little Black Cormorant, Darter (see Introduction to Part 3, Plates 11a, b), Australian Pelican, Black-necked Stork, and Great Egret. Almost all piscivores hunt in open water, where hundreds of individuals may congregate together in schools. Piscivores also occur to a small extent in riparian forests (e.g. Azure Kingfisher) and mangroves (e.g. Great-billed Heron).

To what extent do the Top End's birds differ in trophic structure from avifaunas elsewhere on the continent? With the exception of the freshwater wetlands, particularly with their unique aggregation of species eating aquatic invertebrates or vegetation, the avifauna is very similar to other forested regions in Australia.

The importance of insectivores and nectarivores, for instance, is universal among Australian birds.

Five points serve to summarize our discussion of habitat selection and trophic structure of avian assemblages in the Top End. First, freshwater wetlands present a distinctive set of resources, and support communities dominated in richness by species consuming aquatic invertebrates and in biomass by herbivores. Second, all other habitats are dominated in richness and biomass by insectivores and nectarivores. Third, riparian forests have the most diverse range of trophic groups because they support some species that utilize the associated waterbodies for food. Fourth, sedgelands and grasslands have reduced structural complexity and consequently support the smallest range of trophic groups. Finally, the avifauna is fundamentally similar to those in other forested parts of Australia, the major exception being that its aquatic elements are uncommonly diverse and well developed in the freshwater wetlands of the Top End.

SEASONALITY: MOVEMENTS AND BREEDING

Our discussions of biogeography, habitat relationships, and trophic structure of communities have hinted at the importance of the strongly seasonal climate in governing the lives of birds. Seasonality affects birds most obviously in their patterns of movements and of reproduction, and we will now examine these phenomena in turn. There are two major components to seasonal movements by birds: migration into and out of the monsoonal tropics, either from the north or from southern Australia; and local movements by resident birds within the region. We will discuss migration first.

Migratory birds account for 24% (61 species) of the total number of species occurring in the Top End. By far the most significant group within this number is the

Table 28. Examples of birds that utilize specific foraging zones and foraging techniques while searching for invertebrates and small fish in freshwater wetlands of the Top End.

| Foraging zones | Foraging techniques | | | | | |
	Probe	Peck or chase	Sweep	Filter	Dive from the air	Dive into the water column
Mud	Black-fronted Plover	Masked Lapwing	–	–	–	–
Shallow water	Glossy Ibis	Little Egret Intermediate Egret	Royal Spoonbill	Radjah Shelduck	Whiskered Tern	–
Deep water with dense vegetation	–	Intermediate Egret	–	–	–	–
Deep water with sparse vegetation	–	–	Red-necked Avocet	Pink-eared Duck	Whiskered Tern	Australasian Grebe

waders. From late August and early September, post-breeding adult and juvenile waders from the Northern Hemisphere undertake a general migration, not only to Australia but to all southern continents, for the duration of the northern winter. The movement to Australia, which regularly involves about 35 species, can take place to many parts of the continent, but several species such as the Little Curlew, Oriental Plover, Large Sand Plover and Mongolian Plover are more common in the north than the south. From our unpublished aerial estimates of waterbirds on saline and freshwater floodplains of the Alligator Rivers region, it appears that populations of waders reach a peak both immediately prior to the onset of the wettest months of the wet season (January-March) and immediately following the wettest months. The initial increase in numbers, principally of Sharp-tailed Sandpipers, Curlew Sandpipers and Red-necked Stints, indicates that the Top End is used by several species as a resting area before the migration continues to southern Australia. The proximate stimulus for their rapid, mass departure does not seem to be the flooding of the plains, because this takes place after most of the birds have left. The peak in populations seen in April or May marks the return passage of birds which briefly stage in the Top End before continuing north to their breeding grounds. Not all non-breeding migrants from the north are waders. Species such as the Yellow Wagtail and the Barn Swallow visit wetlands and associated grassy areas; they are restricted to northern areas of Australia and are usually uncommon.

As floodplain waters recede during the dry season, a number of non-breeding waterfowl from southern Australia can be found in the Top End's wetlands. Prominent are Grey Teal, Pink-eared Duck, Hardhead, and Purple Swamphen. Strictly speaking, most of these species are not migratory because they do not undertake wholesale movements from southern Australia. Rather, they are nomadic species adapted to seeking out any available shallow, freshwater swamps throughout the continent (Frith 1962). Similarly, the Spotted Harrier is nomadic from the northern part of its inland range and is infrequently seen during the dry season on grassy floodplains.

Migratory birds in other habitat types can be grouped into two main categories. Open forests and woodlands are occupied during the dry season by a suite of non-breeding migrants from southern Australia, but during the wet season, the riparian and monsoon forests receive predominantly breeding migrants from New Guinea, Indonesia and Asia.

All of the dry season migrants to woodlands and forests have extensive distributions within the interior of Australia, but unfortunately their migration to the Top End is not well described. It is reasonably well established that Rainbow Bee-eaters and Pallid Cuckoos, for example, migrate from south-eastern Australia northward along the eastern coast to Cape York, New Guinea and eastern Indonesia. The situation in the Top End, however, is unclear. In the Alligator Rivers region, the Rainbow Bee-eater is largely a resident, whereas the Pallid Cuckoo is uncommon in the dry season and apparently is a seasonal nomad. A diverse assemblage of other dry season migrants does have a more clear-cut and regular presence in the Top End, most of them probably arriving from arid inland Australia. Examples are White-winged Trillers, Diamond Doves, Cockatiels, Tree Martins and White-browed Woodswallows. Further data and ideas concerning serial, partial and total migrations are given by Crawford (1972) and Nix (1976).

Five species arrive on migration during the wet season from New Guinea, Indonesia and Asia to occupy the riparian and monsoon forests of the Top End. The Torresian Imperial-Pigeon, Channel-billed Cuckoo and Common Koel are predominantly frugivorous, but the Oriental Cuckoo and Dollarbird feed on large insects. These foods are probably most abundant during the wet season, thereby favouring breeding migrants; the Oriental Cuckoo is the only non-breeding member of this group.

Migration is in many ways the most noticeable form of movement because observers immediately notice the arrival of unusual birds, such as the Common Koel with its loud, ringing calls. However, local movements within the monsoonal lowlands are equally important ecologically in determining the structure of avian communities. These movements are best documented for waterfowl of the wetlands, which exhibit considerable within-habitat movement throughout the year, and for some populations of honeyeaters which show marked seasonal patterns of movement within forests and woodlands. We will examine the honeyeaters first. Most of the following discussion is based upon our unpublished studies in the Alligator Rivers region. From regular censuses, the seasonal patterns of occupation of different areas in the region become obvious, but without extensive studies involving banding and telemetry it must be recognized that the movement of individuals can only be inferred.

Most of the eighteen species of honeyeaters in the Top End are to some extent nectarivorous, and consequently respond to the flowering of nectar-producing trees in different habitats. The Brown Honeyeater, undoubtedly the most common honeyeater, is a good example of a 'floral nomad'. At the end of the wet season in March or April, most individuals are located in open woodlands where eucalypts such as *Eucalyptus porrecta* are flowering. A regular succession of eucalypts and other trees such as *Grevillea pteridifolia* and *Xanthostemon paradoxus* continues to produce nectar in woodlands until late August. In September, a flowering sequence begins in riparian forests, beginning with *Melaleuca argentea*, and in response there is a virtually complete movement of Brown Honeyeaters from woodlands and open forests into the riparian forests. Densities of Brown Honeyeaters, indeed of all honeyeaters at this time, become very high as birds converge on this spatially restricted habitat. Perhaps because they are crammed in at unusually high densities, the Brown Honeyeaters begin to move away as soon as the *M. argentea* cease to flower. They search out other flowering aggregations, either in nearby riparian forests where *M. viridiflora* or *M. leucadendron* begin flowering, or in more distant woodlands of the south where eucalypts that are locally uncommon in the Alligator Rivers region form more substantial tracts and flower throughout the wet season. At the end of the wet season, Brown Honeyeaters begin returning to the woodlands and open forests once more. From year to year, Brown Honeyeaters appear to respond not only to the presence of nectar-bearing flowers, but also to the relative abundance of flowers in different areas. Because there is great variation in floral abundance between years and between sites, the density of Brown Honeyeaters at any particular place is unpredictable even though their seasonal presence may be predictable.

Although seasonal, interhabitat movements of Brown Honeyeaters are evident in our local studies in the Alligator Rivers region, other species such as the Banded Honeyeater and Rufous-throated Honeyeater only appear in significant numbers in woodlands during peaks in the flowering of eucalypts. These two Honeyeaters seem to be more dependent upon eucalypts, and they rarely cross into other habitats, such as riparian forests, even though trees other than eucalypts may be flowering there. Thus, the Banded and Rufous-throated Honeyeaters probably move greater distances than the Brown Honeyeater in their search for nectar because they choose to stay within one broad habitat type. At the other end of the scale of movements, the Rufous-banded Honeyeater and the Bar-breasted Honeyeater are generally confined to wetter, non-eucalypt forests, particularly riparian and monsoon forests and mangroves. Floral cycles of the paperbarks, *Melaleuca* spp., probably play a large part in determining their movements.

Widespread seasonal movements of other groups of birds in the monsoonal tropics, particularly of granivorous species of finches and parrots, have so far not been described, but they could prove to be as extensive as those shown by nectarivorous birds. Of course all the movements described above, which are characterized by the need to undertake long forays in search of a specialized food in a strongly seasonal environment, form only a subset of the foraging strategies used by birds. Identification of the seasonal changes in forag-

ing locations exhibited by the many sedentary species awaits more detailed study.

Seasonal movements are equally striking among waterbirds, and are probably more widely appreciated because even casual observers notice the huge aggregations which gather on the few remaining waterbodies late in the dry season. But the movements of waterbirds are not simply caused by the availability of water, for each species has to integrate a variety of requirements in evaluating the suitability of a particular wetland at each phase of the seasonal cycle. As an example, we will consider the best known and most abundant of the Top End's waterbirds, the Magpie Goose. Our account is based mainly upon Frith & Davies (1961a).

At the end of the dry season, Magpie Geese move to the swamps and billabongs that still contain water, and await the arrival of the rains. They feed principally by grazing on the blades of terrestrial and semi-aquatic grasses that grow on the fringes of the swamps (Fig. 52). Grass is an important food source until the floodplains fill with water, and especially until aquatic plants produce seeds. The early rains in November and December trigger enlargement of gonads in both males and females, and in January the development of sperm and oocytes is virtually completed. By this time, the swamps have usually filled following monsoonal rain, and Magpie Geese begin to congregate where there is a moderately dense growth of spike-rushes and wild rice in water of about 0.8 m depth. As soon as the spike-rushes grow, the Magpie Geese begin building platforms by bending the clumps over and weaving them into mats. Shortly before the birds are ready to lay, the platform-building becomes more elaborate, with spike-rushes being dragged up into the platform and formed into walls around a deep cup (Plate 27).

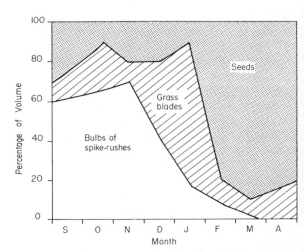

Figure 52. Major components of the diet of Magpie Geese throughout the year (modified from Frith & Davies 1961a).

The precise timing of laying varies considerably between years and localities. The Magpie Geese appear to defer laying eggs until the water depth is appropriate and the growth of spike-rush suits their needs, but because there is variation between years in the patterns of rainfall within the wet season, laying may begin as early as January or as late as May. The bulk of breeding takes place in the late wet season, however, when water levels are relatively stable and vegetation in the swamps is at a peak. The eggs (Plate 28) take about two weeks to hatch, and the goslings fledge in ten weeks. During this time, aquatic grasses such as wild rice, *Echinochloa* spp., *Panicum* spp., and *Paspalum* spp., begin to produce abundant seeds, and both adult and juvenile birds are thereby able to obtain food without having to move great distances across the swamps. Seeds constitute more than 80% of the diet of the birds at this stage of the year (Fig. 52).

As the dry season progresses, and as the goslings fledge, the Magpie Geese begin to leave their breeding areas and seek out deeper swamps where grasses may still be seeding. They also begin to aggregate around the drying edges of swamps where they can dig up and consume the bulbs of spike-rushes. Bulbs dominate the diet from the middle of the dry season onwards, and in their search for this food the Magpie Geese gather in huge mobs on extensive monospecific stands of spike-rushes known locally as 'goose camps'. By the end of the dry season, Magpie Geese have virtually exhausted the supply of bulbs, for the swamps where spike-rushes grow have dried and the mud has baked hard, thus preventing the birds digging into the sediment. The Magpie Geese move to the perennial swamps and billabongs, begin subsisting largely on grass, and await the arrival of the rains once more.

Our description shows that the Magpie Geese regularly move between swamps to satisfy their nutritional and reproductive requirements. In broad terms, the birds move in the wet season to swamps that satisfy their nesting requirements, but in the dry season they distribute themselves according to the availability of food and water. The changes in numbers that result can be dramatic, as indicated by our unpublished studies in the Alligator Rivers region (Fig. 53). These data show that Magpie Geese invade the Nourlangie Floodplain in large numbers in the middle of the dry season, probably from many parts of the Top End. After the bulbs of the spike-rushes have been eaten and the swamps of the goose camps have dried, the birds leave the Nourlangie Floodplain until the following July. The persistent swamps of the Magela Floodplain provide an important refuge for many Magpie Geese in the late dry season, as evidenced by the consistent increase in numbers until November. This floodplain does not provide large areas of habitat suitable for nesting, however, and almost all Magpie Geese leave it

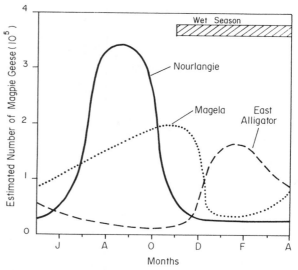

Figure 53. Seasonal changes in the numbers of Magpie Geese occurring on three floodplains in the Alligator Rivers region. The estimates are based on aerial surveys during 1981-1984 (S.R.Morton, K.G.Brennan & M.D.Armstrong, unpubl. data).

when other floodplains begin to fill. Many of the departing birds probably fly to the nearby East Alligator Floodplain, where larger areas are available for nesting. The East Alligator Floodplain is relatively shallow, though, and as soon as the goslings fledge most Magpie Geese proceed to other floodplains in the vicinity. This example confirms that seasonal movements are a way of life for Magpie Geese. No single locality is occupied by the birds throughout the year; rather, the different resources available at different localities are used in a seasonal progression. Almost all species of waterbirds have adopted a way of life similar to that of Magpie Geese, although the precise pattern of movements shown by each species reflects its requirements for specific resources.

Discussion of the movements of birds in response to a strongly seasonal environment almost always leads to consideration of breeding patterns, and the example of the Magpie Goose shows how intimately movement and breeding are intertwined. It is convenient to describe reproduction separately, however, and we will now consider in general terms this second major response to seasonality. Detailed studies of reproduction in birds of the Top End are very limited; among those studied are Magpie Geese (Frith & Davies 1961a, D.G.Tulloch, unpubl. data), Whistling Kites (J.A.Estbergs, unpubl. data) and White-bellied Sea-Eagles (A.L.Hertog, unpubl. data) (Plate 29). The summaries of breeding provided by Frith & Davies (1961b) and by Nix (1976) are most easily discussed under the trophic groupings that we used earlier (Fig.

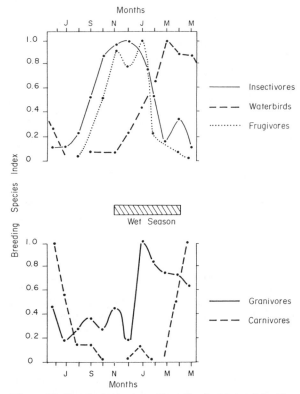

Figure 54. Seasonal patterns of breeding in birds of the Top End (re-drawn from Nix 1976 for data based on Darwin from Crawford 1972). The breeding species index was derived by normalizing the number of species within each trophic group that was found breeding in any one month.

54). Insectivorous, nectarivorous, frugivorous and omnivorous birds breed throughout the wet season. Although phenological studies of insects are lacking, it is certain that arthropods greatly increase in abundance early in the wet season, and so a concentration of breeding in this period by insectivores – or at least the rearing of their young then – is not surprising. Similarly, many nectarivores and omnivores feed their young on insects, and therefore have a similar reproductive pattern. Frugivores appear to breed later in the wet season, but otherwise have similar reproductive cycles to those of insectivores. Granivorous birds have more extended breeding seasons than the groups discussed above, but the proportions of species breeding reach a maximum in the late wet season. This pattern presumably ensures that breeding birds can find immature seeds to feed to their young, as well as ensuring that breeding takes place when seeds are either more abundant or more easily located.

Waterbirds, like granivores, breed mostly in the late wet season. Greatest breeding effort occurs when water levels have stabilized and when aquatic vegeta-

tion, invertebrates and fish have reached maximum abundance. Within this general trend, however, there are significant variations. Cattle Egrets, for instance, breed in the early wet season, presumably because they obtain most of their food in the wet grasslands that reach maximum development in that period. In contrast, Great Egrets breed in the late wet season, when water levels begin to recede and fish are easier to locate.

Carnivores have a breeding cycle distinctly different from other trophic groups because reproduction is almost entirely confined to the dry season. The ultimate causes of this are unclear, but four can be suggested. First, raptors appear to breed in the cooler winter months in most parts of Australia (Nix 1976); thus, explanations of the breeding pattern are unlikely to be based entirely on characteristics of the monsoonal environment. Second, prey may be easier to find in the dry season, or generally during periods of reduced vegetation growth, because the ground cover is sparse; this pattern may lead ultimately to adoption of winter breeding. Third, Nix (1976) suggested that during the dry season numbers of immature prey are at a maximum and concentration of prey species around permanent waters is increasing, thereby providing a more accessible food supply. Finally, the critical period for breeding carnivores may be when the juvenile birds are learning to hunt for themselves. If breeding begins in the dry season, therefore, young birds may be able to begin learning to hunt in the wet season when prey are generally more abundant. It is not yet possible to state which combination of these possibilities provides the best explanation.

In summary, it is difficult to over-emphasize the importance of the extreme monsoonal seasonality to avian assemblages in the Top End. The seasonality causes movements, both localized and migratory; it causes habitat shifts in birds of disparate ways of life such as Magpie Geese and Brown Honeyeaters; and it inevitably plays a critical role in setting the stage for reproductive strategies of a variety of types.

THE PLACE OF BIRDS IN ECOSYSTEMS OF THE TOP END

What effects do birds have upon the ecosystems in which they live, and what role do they play in the biological cycling of nutrients and energy? Such fundamental, functional questions are a great deal easier to ask than to answer, unfortunately, and it will be many years before any understanding develops. Nevertheless, it is worth mentioning several possible relationships, some of which may be unique to the Top End and others probably common to avian communities throughout Australia.

Plate 24. Magpie Geese *Anseranas semipalmata* and Whistling-Ducks, alighting near the East Alligator River, February 1984 (Photograph: G.J.W.Webb).

Plate 25. The Wandering Whistling-Duck *Dendrocygna ar-cuata* (Photograph: G.J.W.Webb).

Plate 26. Australian Pratincole *Stiltia isabella*, a dry season visitor to the monsoonal wetlands from inland Australia (Photograph: G. Miles).

Plate 27. Magpie Goose on nest at Kapalga, April 1981 (Photograph: A. L. Hertog).

Plate 28. Nest of Magpie Goose, with eggs being taken by the Water Python *Bothrochilus fuscus* (Photograph: D.G. Tulloch).

Plate 29. The White-bellied Sea-Eagle *Haliaeetus leucogaster* a common predator of fish and other aquatic vertebrates (Photograph: A.L. Hertog).

Much recent work has been conducted upon the effects of insectivorous birds upon herbivorous insects grazing in forests. In south-eastern Australia, grazing by insects on *Eucalyptus* trees may be more intense than it is on trees of forests in North America and Europe (Morrow 1977, Springett 1978, Fox & Morrow 1983), and although there are no quantitative data on this aspect of eucalypts in monsoonal Australia, trees in the Top End undoubtedly do suffer similar chronically high levels of attack. We have observed, for example, outbreaks of caterpillars on *E. miniata*, and several other trees such as *E. confertiflora* are consistently damaged by herbivorous insects. Otvos (1979) suggested that insectivorous birds have a significant influence on the population dynamics of many forest insects, especially by suppressing and delaying the build-up of populations to epidemic levels; recent experimental studies show that this is indeed the case (Holmes et al. 1979, Loyn et al. 1983). What, then, is the effect of the extensive assemblage of insectivorous birds in the Top End on populations of herbivorous insects? There is insufficient information to allow assessment of this question, but we can point out that the impact of insectivorous birds on herbivorous insects will largely depend on the birds' foraging strategies and their preferences for certain types of trees. For instance, in parts of the Top End the Weebill forages preferentially on *Terminalia ferdinandiana* when it is in leaf, thereby potentially exerting strong predatory pressure on the insect fauna of that species of tree. We do not yet know whether insectivorous species in the monsoonal tropics play a significant role in reducing damage to trees by grazing insects, as they appear to do in some temperate regions, but the results of work on this question would be of considerable interest.

The effects of frugivorous birds in the Top End are similarly poorly known. Considerable effort has been expended in tropical Central America in understanding the significance to trees and shrubs of dispersal of their seeds by birds. There is a complete lack of information on this topic in the Top End, although Crome (1975, 1976) conducted some related work in tropical Queensland. Again, studies of frugivory and dispersal of seeds by birds would be of much interest.

Many workers have suggested that nectarivorous birds are important as pollinators. Honeyeaters are frequent visitors to many plants in the Top End, and as we described above, their numbers at flowering trees can be very large. Thus, birds are almost certainly major pollinators of many of the plants in the region, although the relative importance of insects is unknown. Apart from our own unpublished studies of the general relationships between abundance of flowers and nectarivorous birds, there has been no quantitative work on birds as pollinators in the Top End.

The greatest concentration of avian biomass in the Top End occurs on the freshwater wetlands. Because of the concentration of nutrients in floodplain sediments, it is not surprising that avian abundance is unparalleled elsewhere in northern Australia. The very quantity of the birds, however, results in a substantial impact on the environment, as evidenced by the rice industry's fears of Magpie Geese in the 1950s. Magpie Geese almost certainly have a marked effect on the patterning of some vegetational communities through their consumption of certain plants. For example, D.A. Lindner (pers. comm.) showed us a backswamp of the South Alligator River in which local storms had produced substantial sprouting by spike-rush bulbs late in one dry season. The partial filling of the swamp at such an early stage of the seasonal cycle attracted large mobs of Magpie Geese, which were able to locate the bulbs with ease due to the presence of the sprouts. Thus, the birds virtually exterminated spike-rushes from the swamp, and the rushes are still only slowly regenerating. The interplay between herbivore and vegetation in the monsoonal wetlands would obviously repay investigation.

Much less is known of relationships between waterbirds that consume fish or aquatic invertebrates and the populations of their prey, but as with herbivorous birds, the abundance of the birds is such that they almost certainly do have a significant impact on population dynamics of aquatic animals. For example, L. Baker, D. Clay, B. Bailey and D. Walden (pers. comm.) have suggested that piscivorous birds contribute to the decline of fishes in billabongs of the Alligator Rivers region during the late dry season (see also Bishop & Forbes, Chapter 6). Of course, the fish may die in an extended dry season even if no birds were present, so without experimentation it is difficult to know precisely what effects birds might have on fish species with different strategies for persisting through the dry season.

Our discussion of the functional role of birds in Top End ecosystems has, of necessity, been inconclusive. Until a larger body of information builds up, answers to such questions will remain difficult to obtain.

PROBLEMS IN AVIAN CONSERVATION

None of the Top End's habitats have been subjected to extensive clearing, either for agriculture or for other purposes, and consequently the most basic problem in conservation – outright habitat destruction – has largely been avoided. Many areas are set aside for conservation in the region, notably Kakadu and Katherine Gorge National Parks, but several other rather subtle problems remain. Much concern about the potential effects of uranium mining was voiced during the

Ranger Uranium Environmental Inquiry (Fox et al. 1977), with the results that stringent controls have been placed on mining, and continuing research and monitoring covering the whole environment (including birds) have been underway in the vicinity of Kakadu National Park for several years. Three further conservation problems may be more widespread in their effects: buffalo, fire and weeds.

The effects of buffalo on birds are the subject of work by CSIRO's Division of Wildlife and Ecology. Friend & Taylor (1984) showed that some species of land birds appeared to have been disadvantaged by environmental changes caused by buffalo, but they argued that other species have benefited from the same changes. In particular, birds that prefer open ground may have expanded their local ranges, whereas those needing dense cover may have declined in abundance. It seems certain that buffalo have had substantial effects on some types of land birds, but it is difficult to predict the long-term effects once buffalo have been removed.

Similarly, identification of the impact of buffalo on waterbirds is difficult. Buffalo have almost certainly affected waterbirds indirectly by altering patterns in the vegetation upon which birds depend. Species that nest on the ground may have been severely disadvantaged by the trampling of buffalo which is severe on the edges of floodplains. Other feral animals, especially pigs, have probably compounded the problems caused by buffalo. The difficulties of attributing causative effects to buffalo are demonstrated by the long-term changes in abundance of Magpie Geese described by Tulloch & McKean (1983). They showed that populations declined ten-fold from the early 1950s to the late 1960s, and that numbers subsequently increased again until by 1980 they were at the level of the 1950s. Tulloch & McKean (1983) discussed several possible causes of the trend, but were unable to reach a firm conclusion. It seems likely that buffalo were involved in some way in causing the decline, because Magpie Geese began to increase in numbers at about the time that shooting and catching of buffalo became more intensive in the 1960s (see Ridpath, Chapter 11). However, the overall effects of buffalo on waterbirds remain unclear.

Fire is another widespread phenomenon in the Top End but as in the case of buffalo, its detailed effects on birds have not yet been unequivocally identified. There seems little doubt that altered patterns of burning have affected birds (Ridpath 1974, Crawford 1979). But what precisely are the long-term effects of various fire regimes? If current burning practices differ from the Aboriginal pattern by increasing the proportion of widespread, hot fires in the late dry season, then woodland and open forests may now have reduced middle-storeys of shrubs and perennial grasses (Haynes, Chapter 5). Thus, birds occurring predominantly in shrubby layers may have declined considerably in richness and diversity. Little is known about this important topic, although it has been conjectured that Northern Scrub-robins have been eliminated (Bennett 1983) and that Gouldian Finches are in danger of declining because of changes similar to those mentioned previously. Research into the problems of fire management needs to be given priority.

Finally, the spread of weeds throughout the Top End is cause for concern. Compared with southern Australia, the Top End has been remarkably free of weeds until recent times. Now, three are worth mentioning: *Mimosa pigra*, *Salvinia molesta* and *Hyptis suaveolens*. *Mimosa* forms dense thickets on floodplains, and is spreading rapidly in many parts of the Top End. It threatens to make large areas unsuitable for waterbirds, which cannot feed among such dense growth. *Salvinia* has become established in some freshwater creeks and floodplains, but there is a good chance that its effects will be limited because of control by an introduced weevil, *Cyrtobagous salviniae*. *Hyptis* is a densely-growing terrestrial weed; its effects on birds of the shrub and ground layers are unknown. As in the case of fire, the problems of weeds and their control need to be given immediate attention.

CONCLUSION

The birds of the Top End represent a unique conjugation of Torresian and Irian species. The resultant avifauna is rich in species, and far outstrips that of any other vertebrate class in the Top End. One ecological factor above all others seems to affect the lives of birds in this region: strong seasonality caused by the monsoonal climate. In consequence, detailed ecological investigation of a species or group of species is frequently an inquiry into the strategies used by the birds to make best use of the seasonal fluctuations of resources and those strategies allowing persistence during each seasonal decline. Full understanding of such strategies will require more ecological effort, and although information on the avifauna is expanding at an encouraging rate, problems of research and management often demand more intimate knowledge than ecologists can currently muster. In short, there is still much to learn about the intriguing birds of the Top End.

ACKNOWLEDGEMENTS

Some of the work discussed in this chapter formed part of our research programme at the Alligator Rivers

Region Research Institute, Jabiru. We are grateful to Martin Armstrong and Peter Dostine for assistance with this programme. Other colleagues in the Institute, and in the Division of Wildlife and Ecology, CSIRO, Darwin, gave freely of their advice. Dr R. Schodde commented constructively on a draft of the chapter.

REFERENCES

Bennett, S. 1983. The Northern Scrub-robin *Drymodes superciliaris* in the Northern Territory. *Emu* 83: 105-107.

Bowler, J. M. 1982. Aridity in the late Tertiary and Quaternary of Australia. In W. R. Barker & P. J. M. Greenslade (eds), *Evolution of the flora and fauna of arid Australia*: 35-45. Frewville: Peacock.

Calaby, J. H. 1974. Historical background. In H. J. Frith & J. H. Calaby (eds), *Fauna survey of the Port Essington district, Cobourg Peninsula, Northern Territory of Australia*: 7-19. CSIRO Division of Wildlife Research Technical Paper 28.

Clifford, H. T. & B. K. Simon 1981. The biogeography of Australian grasses. In A. Keast (ed.), *Ecological biogeography of Australia*: 537-554. The Hague: Junk.

Crawford, D. N. 1972. Birds of the Darwin area, with some records from other parts of Northern Territory. *Emu* 72: 131-148.

Crawford, D. N. 1979. Effects of grass and fires on birds in the Darwin area, Northern Territory. *Emu* 79: 150-152.

Crome, F. H. J. 1975. The ecology of fruit pigeons in tropical northern Queensland. *Australian Wildlife Research* 2: 155-185.

Crome, F. H. J. 1976. Some observations on the biology of the Cassowary in northern Queensland. *Emu* 76: 8-14.

Crome, F. H. J. 1978. Foraging ecology of an assemblage of birds in lowland rainforest in northern Queensland. *Australian Journal of Ecology* 3: 195-212.

Davies, S. J. J. F. 1962. The nest building behaviour of the Magpie Goose, *Anseranas semipalmata*. *Ibis* 104: 147-157.

Davies, S. J. J. F. 1963. Aspects of the behaviour of the Magpie Goose, *Anseranas semipalmata*. *Ibis* 105: 76-98.

Deignan, H. G. 1964. Birds of the Arnhem Land Expedition. In R. L. Specht (ed.), *Records of the American-Australian Scientific Expedition to Arnhem Land*, Vol. 4: Zoology: 345-425. Melbourne: Melbourne University Press.

Estbergs, E. 1980. Bibliography on Northern Territory birds. *Northern Territory Naturalist* 1 (3): 20-36.

Ford, J. 1978. Geographical isolation and morphological and habitat differentiation between birds of the Kimberley and the Northern Territory. *Emu* 78: 25-35.

Ford, J. 1982. Origin, evolution and speciation of birds specialized to mangroves in Australia. *Emu* 82: 12-23.

Fox, R. W., G. C. Kelleher & C. B. Kerr 1977. *Ranger Uranium Environmental Inquiry*, Second Report. Canberra: Australian Government Publishing Service.

Fox, L. R. & P. A. Morrow 1983. Estimates of damage by herbivorous insects on *Eucalyptus* trees. *Australian Journal of Ecology* 8: 139-147.

Freeman, D. J. 1971. The nest of the White-lined Honeyeater. *Emu* 71: 86-87.

Friend, G. R. & J. A. Taylor 1984. Ground surface features attributable to feral buffalo, *Bubalus bubalis* II. Their relationship to the abundance of small animal species. *Australian Wildlife Research* 11: 311-323.

Frith, H. J. 1962. Movements of the Grey Teal, *Anas gibberifrons* Müller (Anatidae). *CSIRO Wildlife Research* 7: 50-70.

Frith, H. J. & S. J. J. F. Davies 1961a. Ecology of the Magpie Goose, *Anseranas semipalmata* Latham (Anatidae). *CSIRO Wildlife Research* 6: 91-141.

Frith, H. J. & S. J. J. F. Davies 1961b. Breeding seasons of birds in subcoastal Northern Territory. *Emu* 61: 97-111.

Frith, H. J. & W. B. Hitchcock 1974. Birds. In H. J. Frith & J. H. Calaby (eds), *Fauna survey of the Port Essington district, Cobourg Peninsula, Northern Territory of Australia*: 109-178. CSIRO Division of Wildlife Research Technical Paper 28.

Hespenheide, H. A. 1975. Prey characteristics and predator niche width. In M. L. Cody & J. M. Diamond (eds), *Ecology and evolution of communities*: 158-180. Cambridge, Massachusetts: Belknap.

Holmes, R. T., J. C. Schultz & P. Nothnagle 1979. Bird predation on forest insects: an exclosure experiment. *Science* 206: 462-463.

Keast, A. 1981. The evolutionary biogeography of Australian birds. In A. Keast (ed.), *Ecological biogeography of Australia*: 1587-1635. The Hague: Junk.

Loyn, R. H., R. G. Runnalls, G. Y. Forward & J. Tyers 1983. Territorial Bell Miners and other birds affecting populations of insect prey. *Science* 221: 1411-1413.

Morrow, P. A. 1977. The significance of phytophagous insects in the *Eucalyptus* forests of Australia. In W. J. Mattson (ed.), *The role of arthropods in forest ecosystems*: 19-29. New York: Springer.

Nix, H. A. 1976. Environmental control of breeding, post-breeding dispersal and migration of birds in the Australian region. *Proceedings of the 16th International Ornithological Congress*: 272-305. Canberra: Australian Academy of Science.

Otvos, I. S. 1979. The effects of insectivorous bird activities in forest ecosystems: an evaluation. In J. G. Dickson, R. N. Conner, R. R. Fleet, J. C. Kroll & J. A. Jackson (eds), *The role of insectivorous birds in forest ecosystems*: 341-374. New York: Academic Press.

Pryor, L. D. & L. A. S. Johnson 1981. *Eucalyptus*, the universal Australian. In A. Keast (ed.), *Ecological biogeography of Australia*: 499-536. The Hague: Junk.

Pyke, G. H. 1980. The foraging behaviour of Australian honeyeaters: a review and some comparisons with hummingbirds. *Australian Journal of Ecology* 5: 343-369.

Ridpath, M. G. 1974. The ecological consequences of fire for animal communities. In R. E. Fox (ed.), *Report on the use of fire in national parks and reserves*: 48-53. Darwin: Department of the Northern Territory.

Royal Australasian Ornithologists Union 1978. Recommended English names for Australian birds. *Emu* 77: 245-307.

Schodde, R. 1973. Birds. *Alligator Rivers Region fact-finding study*, Section 5. Canberra: Division of Wildlife Research, CSIRO.

Schodde, R. 1981. Bird communities of the Australian mallee: composition, derivation, distribution, structure and seasonal cycles. In F. di Castri, D. W. Goodall & R. L. Specht (eds), *Mediterranean-type shrublands*: 387-415. Amsterdam: Elsevier.

Schodde, R. 1982. Origin, adaptation and evolution of birds

in arid Australia. In W. R. Baker & P. J. M. Greenslade (eds), *Evolution of the flora and fauna of arid Australia*: 191-224. Frewville: Peacock.

Schodde, R. & J. H. Calaby 1972. The biogeography of the Australo-Papuan bird and mammal faunas in relation to Torres Strait. In D. Walker (ed.), *Bridge and barrier: the natural and cultural history of Torres Strait*: 257-300. Canberra: Research School of Pacific Studies, Australian National University.

Schodde, R., I. J. Mason & H. B. Gill 1982. The avifauna of the Australian mangroves – a brief review of composition, structure and origin. In B. F. Clough (ed.), *Mangrove ecosystems in Australia*: 141-150. Canberra: Australian Institute of Marine Science and Australian National University Press.

Schodde, R., I. J. Mason & J. L. McKean 1979. A new subspecies of *Philemon buceroides* from Arnhem Land. *Emu* 79: 24-30.

Springett, B. P. 1978. On the ecological role of insects in Australian eucalypt forests. *Australian Journal of Ecology* 3: 129-139.

Taylor, J. A., G. R. Friend & M. L. Dudzinski 1984. Influence of sampling strategy on the relationships between fauna and vegetation structure, plant lifeform and floristics. *Australian Journal of Ecology* 9: 281-287.

Tulloch, D. G. & J. L. McKean 1983. Magpie Goose populations on the coastal plains of the Northern Territory (1958-1980). *Corella* 7: 32-36.

White, H. L. 1917. Description of new honey-eater of the genus *Ptilotis*, from north Australia. *Emu* 16: 165.

APPENDIX 1: SCIENTIFIC NAMES FOR ALL SPECIES MENTIONED BY COMMON NAME IN THE TEXT, LISTED BY FAMILY (Nomenclature follows Royal Australasian Ornithologists Union 1978)

Podicipedidae
Australasian Grebe — *Tachybaptus novaehollandiae*

Pelecanidae
Australian Pelican — *Pelecanus conspicillatus*

Anhingidae
Darter — *Anhinga melanogaster*

Phalacrocoracidae
Little Black Cormorant — *Phalacrocorax sulcirostris*
Little Pied Cormorant — *Phalacrocorax melanoleucos*

Ardeidae
Great-billed Heron — *Ardea sumatrana*
Cattle Egret — *Ardeola ibis*
Great Egret — *Egretta alba*
Little Egret — *Egretta garzetta*
Intermediate Egret — *Egretta intermedia*
Striated Heron — *Butorides striatus*
Black Bittern — *Dupetor flavicollis*

Ciconiidae
Black-necked Stork — *Xenorhynchus asiaticus*

Plataleidae
Glossy Ibis — *Plegadis falcinellus*
Royal Spoonbill — *Platalea regia*

Anatidae
Magpie Goose — *Anseranas semipalmata*
Wandering Whistling-Duck — *Dendrocygna arcuata*
Radjah Shelduck — *Tadorna radjah*
Grey Teal — *Anas gibberifrons*
Pink-eared Duck — *Malacorhynchus membranaceus*
Hardhead — *Aythya australis*

Accipitridae
Letter-winged Kite — *Elanus scriptus*
Black Kite — *Milvus migrans*
Whistling Kite — *Haliastur sphenurus*
Red Goshawk — *Erythrotriorchis radiatus*
White-bellied Sea-Eagle — *Haliaeetus leucogaster*
Spotted Harrier — *Circus assimilis*

Falconidae
Black Falcon — *Falco subniger*
Peregrine Falcon — *Falco peregrinus*
Grey Falcon — *Falco hypoleucos*

Megapodiidae
Orange-footed Scrubfowl — *Megapodius reinwardt*

Phasianidae
Brown Quail — *Coturnix australis*

Turnicidae
Red-backed Button-quail — *Turnix maculosa*
Red-chested Button-quail — *Turnix pyrrhothorax*

Rallidae
Purple Swamphen — *Porphyrio porphyrio*

Otididae
Australian Bustard — *Ardeotis australis*

Jacanidae
Comb-crested Jacana — *Irediparra gallinacea*

Burhinidae
Bush Thick-knee — *Burhinus magnirostris*

Charadriidae
Masked Lapwing — *Vanellus miles*
Little Ringed Plover — *Charadrius dubius*
Mongolian Plover — *Charadrius mongolus*
Large Sand Plover — *Charadrius leschenaultii*
Oriental Plover — *Charadrius veredus*
Black-fronted Plover — *Charadrius melanops*

Recurvirostridae
Red-necked Avocet — *Recurvirostra novaehollandiae*

Scolopacidae
Little Curlew — *Numenius minutus*
Asian Dowitcher — *Limnodromus semipalmatus*
Sharp-tailed Sandpiper — *Calidris acuminata*
Red-necked Stint — *Calidris ruficollis*
Curlew Sandpiper — *Calidris ferruginea*

Glareolidae
Australian Pratincole — *Stiltia isabella*

Laridae
Whiskered Tern — *Chlidonias hybrida*

Columbidae
Banded Fruit-Dove — *Ptilinopus cinctus*
Rose-crowned Fruit-Dove — *Ptilinopus regina*
Torresian Imperial-Pigeon — *Ducula spilorrhoa*
Peaceful Dove — *Geopelia placida*
Diamond Dove — *Geopelia cuneata*
Emerald Dove — *Chalcophaps indica*
Partridge Pigeon — *Petrophassa smithii*
Chestnut-quilled Rock-Pigeon — *Petrophassa rufipennis*

Cacatuidae
Red-tailed Black Cockatoo — *Calyptorhynchus magnificus*

Loriidae
Red-collared Lorikeet — *Trichoglossus rubritorquis*
Varied Lorikeet — *Psitteuteles versicolor*

Polytelitidae
Australian King-Parrot — *Alisterus scapularis*
Red-winged Parrot — *Aprosmictus erythropterus*
Superb Parrot — *Polytelis swainsonii*
Regent Parrot — *Polytelis anthopeplus*
Alexandra's Parrot — *Polytelis alexandrae*
Cockatiel — *Nymphicus hollandicus*

Cuculidae
Oriental Cuckoo — *Cuculus saturatus*
Pallid Cuckoo — *Cuculus pallidus*
Common Koel — *Eudynamis scolopacea*
Channel-billed Cuckoo — *Scythrops novaehollandiae*

Tytonidae
Eastern Grass Owl — *Tyto longimembris*

Alcedinidae
Azure Kingfisher — *Ceyx azurea*
Collared Kingfisher — *Halcyon chloris*

Meropidae
Rainbow Bee-eater — *Merops ornatus*

Coraciidae
Dollarbird — *Eurystomus orientalis*

Pittidae
Rainbow Pitta — *Pitta iris*

Hirundinidae
Barn Swallow — *Hirundo rustica*
Tree Martin — *Cecropis nigricans*

Motacillidae
Yellow Wagtail — *Motacilla flava*

Campephagidae
Black-faced Cuckoo-shrike — *Coracina novaehollandiae*
White-winged Triller — *Lalage sueurii*

Muscicapidae
Northern Scrub-robin — *Drymodes superciliaris*
Lemon-bellied Flycatcher — *Microeca flavigaster*
White-browed Robin — *Poecilodryas superciliosa*
Rufous Whistler — *Pachycephala rufiventris*
Sandstone Shrike-thrush — *Colluricincla woodwardi*
Grey Shrike-thrush — *Colluricincla harmonica*
Broad-billed Flycatcher — *Myiagra ruficollis*
Shining Flycatcher — *Myiagra alecto*
Restless Flycatcher — *Myiagra inquieta*
Northern Fantail — *Rhipidura rufiventris*

Sylviidae
Golden-headed Cisticola — *Cisticola exilis*

Maluridae
Purple-crowned Fairy-wren — *Malurus coronatus*
Lavender-flanked Fairy-wren — *Malurus lamberti dulcis*
Red-backed Fairy-wren — *Malurus melanocephalus*
Black Grasswren — *Amytornis housei*
White-throated Grasswren — *Amytornis woodwardi*

Acanthizidae
Weebill — *Smicrornis brevirostris*
Green-backed Gerygone — *Gerygone chloronota*
White-throated Gerygone — *Gerygone olivacea*

Neosittidae
Varied Sitella — *Daphoenositta chrysoptera*

Climacteridae
Black-tailed Treecreeper — *Climacteris melanura*

Meliphagidae
Helmeted Friarbird — *Philemon buceroides*
Silver-crowned Friarbird — *Philemon argenticeps*
Little Friarbird — *Philemon citreogularis*
White-lined Honeyeater — *Meliphaga albilineata*
White-gaped Honeyeater — *Lichenostomus unicolor*
Yellow-tinted Honeyeater — *Lichenostomus flavescens*
White-throated Honeyeater — *Melithreptus albogularis*
Brown Honeyeater — *Lichmera indistincta*
Bar-breasted Honeyeater — *Ramsayornis fasciatus*
Rufous-banded Honeyeater — *Conopophila albogularis*
Rufous-throated Honeyeater — *Conopophila rufogularis*
Banded Honeyeater — *Certhionyx pectoralis*
Red-headed Honeyeater — *Myzomela erythrocephala*

Ploceidae
Double-barred Finch — *Poephila bichenovii*
Long-tailed Finch — *Poephila acuticauda*
Chestnut-breasted Mannikin — *Lonchura castaneothorax*
Gouldian Finch — *Erythrura gouldiae*

Artamidae
White-browed Woodswallow — *Artamus superciliosa*
Little Woodswallow — *Artamus minor*

CHAPTER 10

Adaptation in native mammals

C. K. WILLIAMS & A. E. NEWSOME

INTRODUCTION

Documentation of the mammalian fauna of the monsoonal lowlands in northern Australia began during the first substantial European settlement at Port Essington between 1838 and 1849 (Chisholm 1955, Calaby 1974). It was another 120 years before ecological studies began and then not of the native species but of the Asiatic swamp buffalo *Bubalus bubalis* (Linnaeus) released from that settlement (Tulloch 1969, 1970, 1978).

There is a current upsurge in scientific research in this region, but at the time of writing comprehensive ecological studies existed for only four native mammals in the lowlands*, the wallaby *Macropus agilis* (Gould) (Bolton 1975), two rodents, *Rattus colletti* (Thomas) (Redhead 1979, Williams, unpubl., 1987 and Friend et al. 1988) and *Melomys burtoni* (Ramsay) (Taylor & Horner 1970a, Begg et al. 1983, Williams, unpubl.), and the minute dasyurid marsupial *Planigale maculata* (Gould) (*sinualis* form) (Taylor et al. 1982, Williams, unpubl.). Only *Rattus colletti* is restricted to the central northern monsoonal lowlands. The others also inhabit monsoonal north-western Australia, and the eastern coastal lowlands of Queensland where the climate, though monsoonal, is influenced also by the south-east trade winds which reduce the severity of the dry season. *Macropus agilis* also occurs in the coastal lowlands of New Guinea where the rainfall pattern is similar to that of north-eastern Australia. The studies on these four species exemplify a range of problems and adaptations in several aspects of biology of mammals living in the monsoonal lowlands.

The monsoonal environment sets northern Australia apart from all other regions in Australia. The only other strictly seasonal climate in Australia is in the south-west and central south of the continent where

most of the rain falls in the winter semester. The antithesis of this Mediterranean climate can be seen in northern Australia in the generally high temperatures, the summer monsoonal rain and floods, and the dry winter. The wet season floods and the dry season aridity are predictably cyclical, though the length of each phase and the intensity of the flooding and aridity vary from year to year.

In this chapter we discuss the solutions used by the four mammalian species to overcome these annual ecological extremes. It will be shown that some aspects resemble those of opportunistic desert species and that the habitats of the mammals on the subcoastal riverine plains which are flooded in the wet season differ from those on the wooded lowlands which are not flooded.

Climate

The annual cycle of water balance in the monsoonal lowlands is illustrated in Figure 55 using the monthly hydric efficiency ratio, $r/e^{0.7}$, where r = rainfall, e = pan evaporation in the same units, and the exponent 0.7 converts the latter to a standardised soil surface (Gentilli 1971). Also shown are the equivalent values for Australian environments ranging from 'per-humid' to 'per-arid' (Gentilli 1971). The wet season easily qualifies as per-humid and the dry season as arid and per-arid for a long period. Thus with an annual cycle, the average climate of the monsoonal lowlands varies from that associated with wet rainforests to one characteristic of hot dry deserts. The change between these extremes is very rapid.

The seasonal variation is compared from north to south in Australia, in Table 29, especially by the coefficient of variation of the pasture growth index per week, based on Nix (1976). Seasonal variation of the monsoonal climate is considerably greater than that of the Mediterranean climate but the same as that of arid regions.

The extremes of this cyclical monsoonal climate are enhanced by uncertainty in the times of onset and duration of the wet and dry seasons. Late arrival of the

---

*Subsequent comprehensive studies include the following on a native rodent *Mesembriomys gouldii* (Friend 1987), a bandicoot *Isoodon macrounus* (Friend, in press), a dasyur *Antechinus bellus* (Friend 1985) and a possum (Kerle 1983).

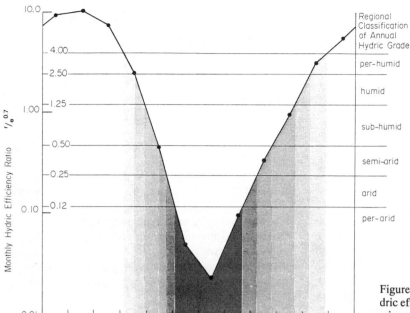

Figure 55. Long-term average of monthly hydric efficiency ratio (logarithmic scale) for Darwin and climate classification (see text).

wet season prolongs the arid phase, and the rains may cease early or late. The pattern of rainfall in the wet season is also important and, with cyclones, can cause severe and sudden flooding of the landscape. McAlpine (1969) presents a model of this variability between years in the form of probabilities of rainfall sufficient to promote pasture growth during the months of transition to and from the wet season.

Phenology

The monsoonal lowlands have a very low relief, com-

prising treeless riverine plains and wooded interfluvial rises. Consequently ecotones tend to be subtle or broad and extensive with vegetation types grading from one form to another. Exceptions are areas where water remains for considerable time, such as permanent seepage areas which may support rainforest, and the edges of the plains which are flooded for six months or longer annually.

The effects of the seasonal cycle on the vegetation and landscape are indeed remarkable (see Preface, Plates 1-3). At the end of the dry season the plains are dry and deeply cracked and covered by dry matted sedges. Fires are frequent on the wooded lowlands so that the lateritic soil is often bare. During the November-December storms which precede the wet season, the shallow soils and deep lateritic hardpan in the woodlands absorb little of the torrential rains which rapidly drain away in sheet flooding and streams to the alluvial clay plains. As the wet season approaches the rain increases in frequency and distribution. On the wooded lowlands the annual grasses and herbs germinate and commence growth. On the plains the dry matted sedges and grasses are broken down biochemically and tend to disintegrate in the early rains. The clay soil swells as it gradually becomes moister and the cracks eventually disappear and the sedges and grasses resprout. After the clay soil becomes saturated, the plains flood, often appearing as extensive open water until the sedges grow well above it, and through the wet season the various sedge and grass species flower and seed in variable sequences.

Table 29. Seasonal variation of regional climates.

Locality	Latitude °S	Description of climate	Mean annual rainfall (mm)	Index of variability of weekly climate*
Northern Australia	12-14	Monsoonal, torrid; high summer rainfall; distinct dry seasons	1500	0.75-1.0
Central Australia	22-24	Arid; rainfall low and uncertain	250	0.75-1.0
Southern Australia	34	Mediterranean; winter rainfall; distinctly dry summer	500	0.5-0.75
Tasmania	40	Temperate cool; mostly winter rainfall	1000	0.5-0.75

*Coefficient of variation of weekly growth indices throughout a year (Nix 1976, Newsome 1983).

Photograph 11. The dusky rat *Rattus colletti* (Photograph: I. Morris).

In the woodlands the annual grasses grow to about two metres in height. Towards the latter part of the wet season the grasses again rapidly increase in height as flower and seed heads protrude and open. At the beginning of the dry season, the tall grasses cast their seed, dry out and fall over. This forms a low tangle of straw which may burn at any time through the dry season, either early in creeping, cool and patchy fires, or more commonly later in hotter fires driven in broad fronts by dry continental winds.

At the onset of the dry season those parts of the plains which are slightly elevated drain and dry rapidly, and the deeper water generally remains for

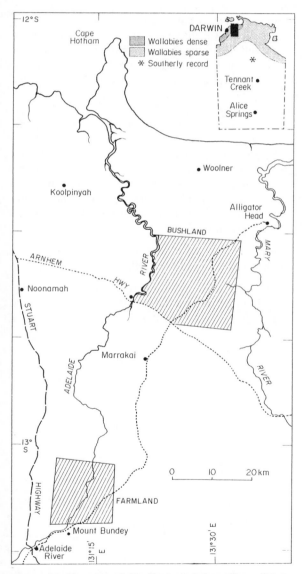

Figure 56. The general distribution of the agile wallaby, a southern record in the N.T., and sites of ecological study there.

between three and eight months. Much of the drying plains become heavily trampled and grazed by feral swamp buffalo (see Ridpath, Chapter 11), possibly increasing the rate of drying and extent of cracking of the soil over the dry season. On the exposed plains the sedges and grasses themselves dry and may fall and form a mat of straw litter. The drying soil shrinks and cracks ubiquitously in polygonal shapes from the surface downwards, exposing the damp clay beneath. During the dry season these cracks continue to enlarge to widths of 10 cm and more and depths of 50 cm and more. Sometimes the dry vegetation on the plains burns during the dry season, but in present times much less frequently than on the rises. As described by Haynes (Chapter 5), the frequency, pattern and timing of burning on the plains and lowland rises depends much on the influences of Aborigines and recent European settlers (see also Chapter 12).

In the remainder of this chapter we shall examine known aspects of life history, behaviour, physiology and population dynamics of the four species mentioned, in relation to the environmental conditions of the monsoonal lowlands.

THE AGILE WALLABY *MACROPUS AGILIS*

The agile wallaby (weight, male, 30 kg; female, 16 kg) (see Introduction to Part 3, Plates 12a, b) is the most common tropical macropodid and one of the most widely distributed (Fig. 56). The wallaby is most abundant on the sub-coastal lowlands of the Northern Territory and on other floodplains of the coastal hinterland around northern Australia. However, they extend further inland, and there is a record from a road-kill at Newcastle Waters, N.T., 500 km inland and well into the semi-arid zone (Fig. 56) (Newsome 1983).

Elsewhere in Australia, the most common wallaby or kangaroo troubles farmers and graziers at times. The agile wallaby is a problem in the Kimberley and on the few farms in the Northern Territory. A collaborative project established in 1966, lasted for five years (Bolton 1975, Bolton et al. 1982, Dudzinski et al. 1977, Merchant 1976, Newsome et al. 1977). Results here are mostly drawn from Bolton et al. (1982).

The breeding cycle

Wallabies become sexually mature between one and two years old. In the wild they mature earlier on more nutritious pastures (Table 30). As with all marsupials, gestation is short, only 29 days (Merchant 1976). If the female does not mate, the interval between one oestrus and the next is 33 days. With pregnancy intervening, that interval is shortened by two to three days (Merchant 1976). The female usually mates a few days after giving birth but a normal pregnancy does not result. As

Table 30. Reproduction in the agile wallaby.

	In captivity*	On native pasture**	On farm land**
Mean age of male (yrs)	1.0	1.9	1.6
Sexual maturity (yrs) of female	1.0	2.2	1.7
Seasonality of birth	None	May-June	None
% Breeding	100	94	99
% with pouch-young	–	89	90
% with young-at-foot	–	32	36
Mean no birth/100 females/month	–	18	18
% survival of young to 7 months	–	35	35
Period of marked loss of breeding	–	Wet season floods	None
Period of highest mortality of pouch young	–	Wet season floods	Wet season floods

* Merchant (1976).
** Field study 1967-1970; Bolton et al. (1982).

in other kangaroos and wallabies studied (Tyndale-Biscoe et al. 1974), the embryo develops to the blastocyst stage (about 80-100 cells) and then becomes dormant. It does not implant in the uterus until near the end of pouch life of the current young which is then aged around seven months (Merchant 1976). The blastocyst reactivates and the young is born about a month later. The female continues to suckle the larger young from outside the pouch, probably for another two to three months, longer in captivity.

Breeding success

Again as in other kangaroos and wallabies, the unimplanted dormant blastocyst acts as a reserve, and reactivates should the pouch-young die or be lost early from the pouch. Since females can breed continuously (Merchant 1976) the female may produce three young from the pouch in just under two years in favourable conditions. Fecundity is very high. Almost 90% of mature females in our study carried pouch-young and about one in three suckled young-at-foot as well. At any one time about 120 young are being suckled per 100 adult females, with about 18 born per 100 females per month (Table 30). This very high fecundity is reduced by an unexplained mortality of pouch-young in the wild (Table 30). Only about one in three young born survive to be seven months old, both on more nutritious farmland and on native pastures.

Non-breeding

A few adult females were found to be anoestrous. As well, a few others were old enough to breed but remained immature. Their pouches, teats and mammary glands were small, and the ovaries and uteri inactive. On farmland only 1.3% of females were anoestrous, but there were six times as many on native pastures. A pertinent point is that about one third of those females were suckling pouch-young. They had ceased to be able to produce young, but were able to suckle those which had been produced. One female with a pouch-young was found emerging from anoestrus to oestrus, like others which had returned to breeding condition from anoestrus.

Histology of the reproductive systems of all these females collectively indicated that they had failed to have the customary oestrus post-partum. The only other species known to possess this remarkable flexibility is a desert species, the red kangaroo (*M. rufus*) of central Australia (Newsome 1964a, b, 1965). The environment there is risky, with rainfall uncertain and often scant. The adaptiveness of such a process there is clear. Well into a drought, a female can produce yet one more young after a procession of early deaths in previous young. Later, after it rains, such females can return to breeding despite lactation. That about one third of young surviving droughts were suckled by anoestrous mothers indicates the advantage of the process to the red kangaroo.

The advantage of such a process in a tropical wallaby is more difficult to perceive. By analogy with the red kangaroo, these attributes indicate uncertainties in its environment despite the bountiful rainfall. Uncertainty in the midst of plenty is the central theme of the remainder of this section.

Anoestrus and the wet season floods

Unlike the desert red kangaroo, the agile wallaby rarely ceased breeding during the droughts of the dry season. Rather, they ceased in the wet season, the time

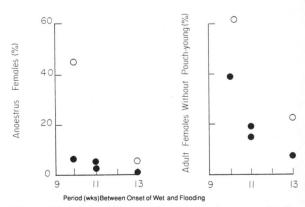

Figure 57. Non-breeding and loss of pouch-young of agile wallaby as functions of the suddenness of flooding in the wet seasons; ●: on farmland, ○: on native pastures.

Table 31. Seasonal weight gains and losses in adult and young agile wallaby.*

Dry season	
Early	Gains
Middle	Gradual losses
Late	Gradual losses
Wet season	
Early rains	Gains
Flooding	Sudden losses**
Water receding	Quick recovery

* More marked on native pastures than on farmland.
** Floods from unseasonally heavy rains in the dry season are most disruptive.

of apparent plenty. The wet season is also the time when large numbers of females lost pouch-young. The actual quantity of rain did not seem to be involved, but rather how quickly the sub-coastal plains were flooded once the wet season began (Fig. 57). With a gradual build-up to the floods, hardly any females became anoestrous and less than 20% lost pouch-young. But 50-60% could be found in either category if flooding quickly followed the onset of the wet season by only nine to ten weeks. The wallabies lost weight quickly as well during the floods (Table 31).

Feeding grounds and pasture nutrition

The grassy sub-coastal plains and grassy fringes along freshwater creeks and billabongs are favourite feeding grounds during the dry season. Both habitats also provide water and shade during the heat of the day. Those feeding grounds are flooded first during the wet season forcing the wallabies into the eucalypt woodlands or to levee banks along the main rivers upstream. Wallabies on flooded land have been seen wading about seeking emerging tips to feed upon. We suspect that it is the sudden disturbance of feeding habits and daily movements that are involved in this stress to reproduction. Wallabies do become stranded by floods and are even caught in them and drowned. Pouch-young of wallabies examined after they swam flooded rivers were drowned.

Table 32. Quality of pastures.*

Season	Native pasture		Introduced and fertilised	
	Late dry	Late wet	Late dry	Late wet
Protein (g%)	1-3	3-6	4-8	8-10
Phosphorus (g%)	0.02	0.04	0.04	0.09-0.14
Digestibility (in vitro %)	32.9	26.8	43.5	50.3

*Data after Wesley-Smith (1972, pers. comm.) and Bolton (1975). Means for 1967-1968.

The nutrition provided by pastures on the Adelaide River farms (Wesley-Smith 1972, Bolton 1975) showed clearly the advantage of introduced, fertilised grasses over native species (Table 32). Nitrogen and phosphorus contents in the former were more than double those of the latter, and introduced and mixed pastures were almost twice as digestible as native species. No wonder the wallabies breed so well on the farmlands. They do well enough as it is on native pastures. Indeed, the fecundity is so high that the species is clearly well adapted.

Other macropodids of the monsoonal lowlands

Two other macropodids occur in the monsoonal lowlands. In the lowlands the gregarious but uncommon antilopine wallaroo *M. antilopinus* lives mainly in woodland and savanna, and the more solitary euro *M. robustus woodwardi* lives in the undulating savanna woodland and rocky hills (Russell & Richardson 1971). Little is known of the reproduction of these lowland species. The ages of a few pouch-young of these two species suggest that births are concentrated near the end of the wet season and that young emerge from the pouch mainly early in the wet season (Russell & Richardson 1971) when herbage is more nutritious (Story 1969). This contrasts with the continuous opportunistic breeding of *M. agilis* on the plains.

Opportunistic breeding of the agile wallaby and lack of it in tropical ruminants

While detailed knowledge of breeding in other tropical kangaroos and wallabies is limited, it is available for the much larger introduced ruminants, cattle (*Bos taurus, B. indicus* and their crosses) and water-buffalo (*Bubalus bubalis*) (Andrews 1976, Tulloch 1974, Tulloch & Grassia 1981). Both cattle and buffalo have seasonal peaks, most calves being dropped early in the wet season with conceptions following late in the wet season. Patterns of conception in the agile wallaby, cattle and buffalo are compared in Figure 58. Note the drop in conception rate for the wallaby (due to flooding) around the time at which it peaks for the other two species. In equatorial East Africa many antelope (also ruminants) breed during the rains. In drier sub-tropical regions, however, some breed all year round like the desert kangaroo (*M. rufus* and *M. robustus*) here (Newsome 1983).

Why, then, is the breeding pattern of the agile wallaby so different from large tropical herbivores and other macropodids in the region? It does not seem to be inability of macropodid marsupials to breed seasonally, because two wallabies in southern Australia are strict seasonal breeders. The tammar (*M. eugenii*)

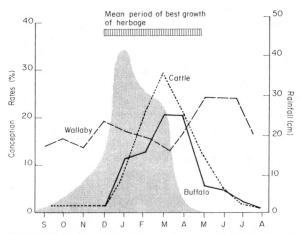

Figure 58. Comparisons of conception rates of the introduced ruminants and the agile wallaby with respect to rainfall (shaded). The best period for pasture growth is also shown (hatched bar).

on Kangaroo Island, and Bennett's wallaby (*M. r. rufogriseus*) in Tasmania are obligatory seasonal breeders (Sadleir & Tyndale-Biscoe 1977, Merchant & Calaby 1981). Moreover, the quokka (*Setonix brachyurus*), a wallaby with the same basic reproductive cycle as the agile wallaby (and the desert red kangaroo and euro too), goes into anoestrus seasonally on Rottnest Island (Sharman 1955, 1959). The western grey kangaroo (*M. f. fuliginosus*) is also a seasonal breeder on Kangaroo Island (Poole 1976). All four species time their breeding so that young leave the pouch after the winter rains when pastures are at their best in that Mediterranean climate (Tyndale-Biscoe 1977).

The question then arises as to what is so special about this seasonally wet tropical environment. Why should this region's most successful wallaby be an opportunistic breeder like the desert kangaroo? The answer appears to lie with neither seasonality nor the quantity of rainfall, but with high variability in the timing and, possibly, the duration of the monsoons. The latter may start at any time over a three month period each year (McAlpine 1969 and Chapters 2, 9 and 11). As well there is the possibility of cyclones whose drenching rains suddenly flood wide stretches of land. Their mean frequency of occurrence is 6.5 per decade (Lourensz 1977). An opportunistic breeding pattern would be advantageous where annual seasonal change is so variable temporally, and frequently changes from extreme drought to deluge.

THE DUSKY RAT *RATTUS COLLETTI*

Rattus colletti (Photograph 11) is limited to the monsoonal sub-coastal lowlands of the Northern Territory

from the Daly River plains in the west to Rose River near the south-western corner of the Gulf of Carpentaria. It is the dominant small mammal inhabiting the sub-coastal plains. Body weight ranges from about 22 to 213 g from independent young to old.

There have been three field studies on the ecology of *R. colletti*. The first was on the sub-coastal plains of the Adelaide River near Humpty Doo from 1972-1976 (Redhead 1979, as *R. sordidus colletti*), the second was at Kapalga on the plains of the South Alligator River, approximately 150 km to the east between 1975 and 1978 (Williams, unpubl.) and the third also at Kapalga between 1979 and 1982 (Friend et al. 1988). In these studies the plains and margins were live-trapped in standard ways at monthly intervals. *Melomys burtoni* and *Planigale maculata sinualis* were also caught. Some physiological aspects were studied also (Williams, unpubl.). Patterns in the biology and ecology of *R. colletti* were similar in both river systems.

Seasonal movements and habitat use

During the dry season *R. colletti* lives only on the black soil plains. They use the maze of irregular cracks in the clay soil for protection from intense tropical insolation, heat, desiccation and occasional cold weather (Table 33). During the early part of the dry season the density of *R. colletti* is usually rather low. On the edges of the plains, by about the middle of the dry season, density declines still further, often to zero. Meanwhile there is sometimes an increase in density in the middle of the plains where shallow gilgai depressions and their heavily cracked soils are common. In the latter half of the dry season the rats seem to favour these depressions more than the surrounding plain and the sexes use them differently but not exclusively at different times. In the late dry season there seems to be a pattern of retreat to these depressions.

As the pre-wet season storms commence there is an upsurge in activity of *R. colletti*. Marked individuals are trapped less frequently and the pattern of residency which developed in the latter half of the dry season breaks down. As the frequency of storms increases in November and December numbers decline in the middle of the plains and increase on the edges.

Table 33. Ambient temperature and relative humidity 20 cm above ground level and 60 cm deep in a large exposed crack in clay soil of the sub-coastal plains of the Adelaide River in mid afternoon, dry season.

	Shaded air temperature	Relative humidity
Above ground	36°C	35%
In soil crack	29°C	75%

Animals can also be trapped on other slightly elevated areas, the levee banks of the rivers and the margins of the wooded lowlands, where none can be caught in most dry seasons. When the soil cracks disappear few *R. colletti* remain in the middle of the plains. Those remaining probably live in the dry, fallen and disintegrating sedge of the previous wet season.

Through the wet season densities are greatest on the levees and shallow flooded edges of the plains and margins of the woodlands, but density declines through the wet season. Immediately the water level recedes, *R. colletti* vacates the higher ground completely and disperses across the plains behind the rapidly receding water. As the plains dry out they shelter in the cracks forming in the clay.

Refuges

Maximum temperatures measured in the cracks (Table 33) remained within or below the thermoneutral range (optimum ambient temperature range at which resting animals have minimal metabolic heat production and 'normal' body temperatures) of equivalent sized Australian *Rattus* species (Collins 1973a, b, Collins & Bradshaw 1973) despite high shaded temperatures above ground. Underground humidities were higher, comparing favourably with those above ground, also reducing evaporative losses. Ambient temperatures are sometimes low at night especially early in the dry season, e.g. 8°C (Redhead 1979), but the underground microclimate remains relatively constant (Table 34). Some adult *R. colletti* caught in traps early in the dry season when minimum ambient temperatures above ground were about 20°C were then stiff, inert and unresponsive with body temperatures as low as 23.9°C. They recovered as body temperatures rose.

During the wet season, on the levees, margins, and cheniers where these occur, *R. colletti* dig burrows, usually at the bases of grass clumps or *Pandanus spiralis*, where the soil is friable and well drained. On the woodland margin where the soil is not friable, exposed root clusters and dense grass clumps may be used as shelter from rain and insolation. On the shallow flooded edges of the plains the rats shelter in dry fallen sedge clumps from the previous wet season and readily move through the warm shallow water. At this time ambient temperatures are generally lower than at other times, and apparently hyperthermia is not a problem as long as insolation is avoided.

Diet

Through the dry season no free water is available on the plains and *R. colletti* eats mainly the underground corms and rhizomes of abundant sedges, more specifically so as the dry season progresses (Redhead 1979, Williams, unpubl.). The gross composition of corms of one sedge species as an example (Table 35) indicates their high pre-formed water content although it declines during the dry season. The proportion of available carbohydrate is very high also. Metabolism of this is economical in water balance as the digestive heat increment (specific dynamic action) is low and there is little metabolic waste. Similarly the low values for protein and ash indicate low expenditure of urinary water in excreting electrolytes and nitrogenous wastes, and the low level of fibre would not incur large losses of water in faeces. Although the protein level is fairly low for the diet of a small mammal, body condition declines only in the latter half of dry seasons and in very wet wet seasons. The diet during the wet season also includes stem bases of monocotyledons, probably new shoots of sedges and grasses.

Body condition and mortality

In dry seasons without rain, especially towards the end of the season, the animals often become thin and infested with the tick *Haemaphysalis ratti*, the fur becomes dry, harsh and very sparse in patches. Growth then slows markedly in some years (Redhead 1979). The loss of body condition is demonstrated by levels of total body fat measured by acetone extraction

Table 34. Pre-dawn and mid-afternoon ambient temperatures above ground level and in soil cracks during the early dry season, Kapalga.

	Above ground	Soil crack
Pre-dawn	19.9°C	24.7°C
Mid-afternoon	35.0°C	24.7°C

Table 35. Proximate analysis of corms of the sedge *Eleocharis dulcis* from the black soil plains, South Alligator River. Water data are % wet weight, and other classes are in % dry weight.

Month	Water (± S.E.)	Protein	Lipid	Crude fibre	Ash	Available carbohydrate
Mid June	59±3	–	–	–	–	–
Mid October	–	12	4	18	2	64
Early December	43±2	–	–	–	2	–

Table 36. Total body fat in *R. colletti* during a dry season and a wet season at Humpty Doo.

	June 1972	July 1972	Oct. 1972	May-June 1973	Aug. 1973	Dec. 1973	May 1974
Total body fat (% of weight ± S.E.)	36 ± 10 n = 4	31 ± 6 n = 10	14 ± 3 n = 8	6 ± 1 n = 4	23 ± 6 n = 5	6 ± 1 n = 4	7 ± 3 n = 6

(Table 36). Measures of rates of water turnover (Williams 1987) suggest low rates of water and food intake during two dry seasons when populations were declining and low. The declining abundance in some dry seasons (Redhead 1979), and declining body condition suggest that there may be fairly high mortality in *R. colletti* during long dry seasons. Both demographic studies found that late in the dry season the population consists entirely of recruits from that year, suggesting that dry season mortality differentially removes the older age class.

Similarly, as the wet season progresses the numbers of *R. colletti* trapped in the refuge areas decline, and there is loss of body condition (Table 36) in very wet wet seasons such as 1973-1974. Following that wet season Redhead (1979) recorded very few *R. colletti* returning to the plains from higher ground compared with previous years. The decline in abundance during the wet season suggests that the elevated habitat progressively becomes unsuitable towards the end of the wet season, and this is confirmed by its immediate and total abandonment as the rains end.

Reproductive biology

No pregnancies were found at the height of the wet season; the rats recolonising the plains at the end of the wet season were all adults. Pregnancies appeared soon after, about May, and in most years breeding was concentrated at the beginning of the dry season. However *R. colletti* may breed throughout the year in captivity (Taylor & Horner 1973b) which indicates the species' ability to breed in the field whenever climatic conditions are suitable, irrespective of the time of year. Taylor & Horner (1973b) recorded large litters of up to twelve young and there are twelve maternal teats. Under suitable conditions females have a post-partum oestrus which enables them to bear a litter while suckling another, resulting in a rapid succession of litters (Taylor & Horner 1973b). Gestation lasts about 21 days and the young may become independent of the mother when aged about three weeks (Watts & Aslin 1981), and females of the litters may breed when aged four to five weeks (Taylor & Horner 1973b). Therefore the high reproductive potential of *R. colletti* is based on high fecundity, a very short generation time, and extension of the variable breeding period when suitable conditions arise and persist.

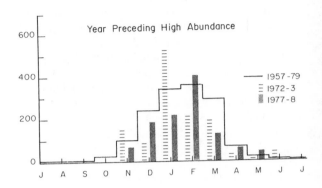

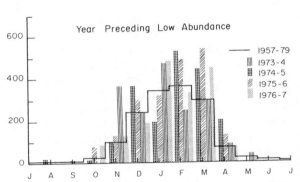

Figure 59. Monthly rainfall (mm) for Oenpelli, N.T., during the years 1972-1978 for years preceding high and low abundance of *Rattus colletti* at Kapalga.

Climatic variation and population fluctuations

As with the agile wallaby, the pattern of reproduction in *R. colletti* may be a sensitive indicator of variation in environmental stresses, and of physiological tolerance. This seems evident from correlations of annual variations in the monsoonal cycle with years of low reproduction and abundance, and years of great proliferation (Fig. 59). The major fluctuations to rarity and great abundance of *R. colletti* between 1971 and 1979 occurred simultaneously on the plains of the Adelaide and South Alligator Rivers, and climatic variations there were similar.

In years with normal dry seasons lasting until late in the year breeding was confined to a short period early in the dry season. Then the species was neither abundant nor rare.

After wet seasons with rainfall well above average,

or when the rain was concentrated at the end of the season, few adults returned to the plains and subsequent reproduction was either insufficient or failed so that the populations declined to rarity. This occurred in 1975 when the populations became virtually undetectable and remained that way for two years at Humpty Doo and Kapalga.

When the wet seasons are mild with rainfall below average, the plains flood less deeply and dry out earlier, resulting in an earlier start to breeding. This occurred in 1973 and 1978 and was followed in both years by rain during the dry season. In 1973 this caused a flush of green growth in June-July, and in 1978 rain in four dry season months maintained some green herbage until near the end of the dry season. On both occasions prolific breeding resulted. With cohorts of the season also contributing offspring, population densities became extremely high. At Kapalga in 1978 the four-night estimates of density increased from 18 animals per hectare in May to 230 animals per hectare in August, with the population doubling every 17 days during the interval.

The demographic studies suggest that reproduction of *R. colletti* is impeded by dry dry seasons and enhanced by wet dry seasons. Low population densities occur after wet wet seasons followed by dry dry seasons, and high densities arise after dry wet seasons followed by wet dry seasons (Fig. 60).

Reproductive strategy

The reproductive pattern in *R. colletti* seems to be influenced by the variable repetition of the two seasonal extremes. Population size is repeatedly reduced during wet season inundation of habitat and during dry

season aridity. The costs of reproduction in metabolism and water balance are usually met only for a short time following the monsoon when the plains again become a suitable open niche. Rapid reproduction before restriction by aridity is therefore highly advantageous. As such suitable conditions occur variably, sometimes early, sometimes late and sometimes briefly preceding the wet season, the ability to breed opportunistically is also advantageous. A consequence of highly opportunistic breeding in *R. colletti* is that there are phases of great abundance and of great rarity. The phases of great abundance tend to favour the persistence of the species for at least two reasons: replenishment of the genetic variation which may have become restricted during phases of rarity (a 'bottleneck' effect), and portions of habitat where local extinctions occurred during unfavourable climatic variations may be repopulated by animals with replenished genetic variation, some of which may be better adapted to subsequent climatic variation.

Other Rattus species

Rattus tunneyi tunneyi (Thomas), the only other native *Rattus* in the region, lives in burrows in sandy soils in woodland savanna habitats (Taylor & Horner 1973a, b, Calaby & Keith 1974). Taylor & Horner (1973b) obtained detailed reproductive information during a short study in the region. They concluded that *R. t. tunneyi* breeds mainly during the early dry season but possibly at other times as well. Ovulation rates and litter sizes are about four but six have been observed and females have ten teats. Similar litter sizes have been observed in captivity (Watts & Aslin 1981). Gestation lasts about 22 days, females have a postpartum oestrus and may breed repeatedly. The young may be weaned at about three weeks and they may become sexually mature at about five weeks (Watts & Aslin 1981). These laboratory findings suggest that *R. t. tunneyi* has a high reproductive potential, though lower than that of *R. colletti*, and field observations suggest that this breeding potential is not generally realised in the monsoonal lowland environment. This may be due to *R. t. tunneyi* in the monsoonal lowlands being relict populations of an essentially arid-adapted species which has recently disappeared from central Australia. Taylor & Horner (1973b) conclude that *R. t. tunneyi* has small litter sizes and a rather long maturation period in the field, contrasting with those of *R. colletti*.

Rattus sordidus (Gould) and *Rattus villosissimus* (Waite), the other two species with *R. colletti* in the *Rattus sordidus* species group (Taylor & Horner 1973a, b), also have very high fecundity and short generation time. *R. sordidus* occurs in the wetter grasslands of the coastal humid tropics of eastern Queens-

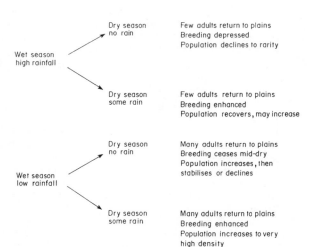

Figure 60. Responses of *Rattus colletti* populations to the rainfall patterns in the sequence of wet and dry seasons.

land (Taylor & Horner 1973b) where summer rainfall is very high and the dry season is much less severe than in the monsoonal tropics (Gentilli 1971: 147-150). *R. villosissimus* inhabits grasslands of the semi-arid and arid inland tropics of northern and central Australia (Taylor & Horner 1973a). The central arid region has low rainfall and the more northern semi-arid part of the range has higher rainfall with most falling in summer (Gentilli 1971: 147-150).

The three species in the *Rattus sordidus* group have similar length oestrous cycles (Breed 1978) and gestation times; the females have twelve teats and a postpartum oestrus and all are capable of breeding whenever conditions are suitable (Taylor & Horner 1973b). Ovulation rates of *R. villosissimus* are greater than those of *R. colletti* and *R. sordidus* (Breed 1978). All three species are precocious breeders and can plague under suitable conditions (McDougall 1946b, Carstairs 1974, 1976, Redhead 1979). They achieve the highest fecundity of all the Australian *Rattus* species, much higher than the species of temperate, mesic and rainforest habitats (Taylor & Horner 1973b).

The grassland environments of the three species in the *R. sordidus* group are all prone to flooding in the wet seasons (McDougall 1946a, Carstairs 1976, Redhead 1979), and all are subject to aridity that is less extreme for *R. sordidus* and more extreme and prolonged for *R. villosissimus* than for *R. colletti* in the monsoonal lowlands. Thus the three species are opportunistic breeders and have great reproductive potential in environments prone to variably frequent and unpredictable catastrophes, with variably ephemeral bountiful periods.

Planigale maculata sinualis

When *Rattus colletti* is rare, the most abundant small mammal on the sub-coastal plains is a minute marsupial carnivore, *P. m. sinualis* (male, 14 g; female, 10 g; independent juveniles as little as 3 g). On the South Alligator River sub-coastal plain in 1976-1978, however, this species was trapped twenty times less frequently than *R. colletti*. Its pattern of occupancy of habitats resembles that of *R. colletti*. Population fluctuations of the two species appear to be out of phase, but *R. colletti* when abundant may pre-occupy traps, giving a false impression.

During the dry season *P. m. sinualis* seems to be only on the sub-coastal plains. It too lives in the soil cracks by day. Numbers trapped decline before the middle of the dry season to apparent rarity until the pre-wet season storms when they also are trapped more frequently on the plains and appear on the margins of the plains and on the levees. During the wet season they can be trapped in moderate abundance in the wooded margins of the plains if *R. colletti* is not in

great abundance. As floodwaters recede at the end of the wet season they disperse immediately across the plains. They seem to advance with the receding water and burrow in the clay soil while it is damp and later they occupy the soil cracks.

Metabolism

Many aspects of the biology of *P. m. sinualis* are influenced by its metabolic rate. Its standard metabolic rate (SMR) measured at rest and in its thermoneutral zone (optimum ambient temperature range) is only 79% of that of many other marsupials when body size is standardised (Morton & Lee 1978, Morton 1980). The SMR of various marsupials tends to be about 30% lower than that of many eutherian mammals (MacMillen & Nelson 1969, Dawson & Hulbert 1970), emphasising the low SMR in *P. m. sinualis*. Two advantages of it in the high thermal range of the monsoonal environment are (a) decreased rates of energy loss so that food requirements would be correspondingly reduced, and (b) reduced rates of evaporative water loss, both advantageous during dry season aridity (see Morton 1980).

In contrast to *P. m. sinualis*, the SMR of *P. m. maculata* was similar to that of other dasyurid marsupials (MacMillen & Nelson 1969), as for *P. ingrami*, *P. tenuirostris* and *P. gilesi* of semi-arid and arid environments (Dawson & Wolfers 1978). These are compared in Table 37.

Body temperature and torpor

The normal resting body temperature of *P. m. sinualis* in the thermoneutral ambient range was found to be about 34°C, similar to other *Planigale* species except

Table 37. Comparison of standard metabolic rate (SMR) and body temperature T_B°C at ambient temperatures within the thermoneutral zone between the two forms of *Planigale maculata* and three other *Planigale* species.

Species and source	SMR (mean ± S.E. ml0$_2$.g^{-1}.hr^{-1})	T_B°C	n	Habitat
P.m.sinualis Morton & Lee (1978)	1.01 ± 0.06	34.2 ± 0.04	9	Coastal monsoonal lowland tropics
P.m.maculata MacMillen & Nelson (1969)	1.26	34.8	2	Coastal humid lowland tropics
P.ingrami and *P.tenuirostris* Dawson & Wolfers (1978)	1.59 ± 0.06	34.5 ± 0.28	4	Inland semi-arid tropics Inland semi-arid and arid subtropics
P.gilesi Dawson & Wolfers (1978)	1.40	35.9	1	Inland temperate arid

for the larger *P. gilesi*. These body temperatures are lower than those of most mammalian and marsupial species and would assist in energy conservation by reducing heat loss.

A possibly greater contribution to energy conservation in *P. m. sinualis* would be from variation or lability in body temperature and metabolism due to frequent torpor. In the laboratory, *P. m. sinualis*, *P. ingrami* and *P. tenuirostris* became torpid daily for several hours at a time, even in the presence of abundant food (Morton & Lee 1978, Dawson & Wolfers 1978). In the torpid state body temperature fell close to ambient temperature, and oxygen consumption dropped to 30-50% of SMR in all of these species. This would conserve both nutrients and body water, reducing requirements for insect prey which are probably the main source of both.

When the results of the different studies are compared, the body temperature and metabolic rates of the other *Planigale* species seem to be less labile than that of *P. m. sinualis*. For *P. m. sinualis* at ambient temperatures of about 32°C, body temperature was maintained at about 34.2°C but this fell to about 30°C at 25-30°C, and for ambient temperatures below 25°C body temperature was usually 2-3°C above the ambient temperature (Morton & Lee 1978). In contrast, *P. ingrami*, *P. tenuirostris* and *P. gilesi* at ambient temperatures of 25°C had metabolic rates about four times SMR, increasing to about six times at 14°C (Dawson & Wolfers 1978).

The metabolic measurements indicate that *P. m. sinualis* has low thermal insulation (high thermal conductance) which suggests that it is adapted to a relatively high and narrow range of ambient temperatures. Ambient temperatures in the soil cracks at Kapalga during 1978 (Table 38) suggest that this is so. In the coldest period of that year the body temperature of *P. m. sinualis* can be expected to have been slightly less than 30 to 34.2°C, except possibly during daily bouts of torpor. Temperatures in the underground environment are probably a rare constraint to normothermic activity, but the minimum ambient temperatures above ground level may deter much surface activity later in the night. This may contribute to the low trap returns during the dry season.

Reproduction

P. m. sinualis is a relatively fecund mammal, females having eight to thirteen teats and litter sizes up to twelve with an average of eight young (Taylor et al. 1982). At Kapalga for the years 1976-78 the average was about seven young per litter (Williams, unpubl.). Breeding can occur in all months with an apparent decline during the height of the wet season when the animals take refuge on higher ground (Taylor et al. 1982). At Kapalga there was a distinct pulse of repro-

Table 38. Pre-dawn ambient temperatures (°C) at 20 cm above ground level and at a depth of 30 cm in an undisturbed natural soil crack in the black soil plains, Kapalga, 1978. Values are means for 3 or 4 consecutive days.

	June	July	Aug.	Early Sept.	Late Sept.
Above ground	20.1	17.5	17.2	16.4	22.6
Below ground	24.8	23.0	24.9	22.8	27.8

duction in the transition periods from wet to dry seasons, similar to *R. colletti* and the agile wallaby in most years. These observations suggest that breeding in *P. m. sinualis* may be opportunistic and strongly influenced by seasonal variations as we have seen in *R. colletti* and *M. agilis*.

Laboratory studies (Aslin 1975) showed that female *P. m. sinualis* (as *Antechinus maculatus*) are polyoestrus and that several litters can be reared each year. The gestation period is 19-20 days, young are independent of the mother 70 days after birth and females may breed from about ten months of age. *P. m. maculata* has a similar gestation period of about 20 days, young are independent of the mother at about 72 days, and this form also appears able to breed throughout the year (Van Dyck 1979). It is possible that *P. m. maculata* may have smaller litter sizes than *P. m. sinualis* as the numbers of teats vary from five to ten (Archer 1976), less than the twelve of *P. m. sinualis*. Reproductive data on arid-adapted *Planigale* are not available, although all apparently have a usual number of twelve teats (Archer 1976, Taylor et al. 1982). It therefore seems that *P. m. sinualis* has a relatively high reproductive output and is probably opportunistic. This is probably true for *P. m. maculata* also.

Other dasyurid species

Calaby & Taylor (1981) and Taylor et al. (1982) contrasted the breeding patterns of dasyurid species from the monsoonal lowlands and the rocky escarpment and suggested that it is distinctive for *P. m. sinualis*. The five woodland and escarpment species are monoestrus and breed during the dry season. This suggests that the distinctive pattern of reproduction in *P. m. sinualis* may be related to its occupancy of the sub-coastal plains rather than the woodlands. Its high fecundity and opportunism in common with *R. colletti* with which it shares the sub-coastal plains, suggest advantages in rapidly repopulating seasonally inundated habitat. Among the dasyurids generally, there seems to be a trend for extended opportunistic breeding and early maturity in arid-adapted species and short breeding seasons in mesic-adapted species with longer times to sexual maturity (see data of Woolley

1973). Therefore in high productivity and opportunistic breeding, *P. m. sinualis, R. colletti* and *M. agilis* resemble arid adapted species suggesting that survival in the sub-coastal plains of the monsoonal lowlands is as precarious as in desert regions.

Melomys burtoni

The *Melomys* species in Australia and New Guinea have been taxonomically unsatisfactory for many years (Keith 1970, Calaby & Keith 1974). Recent studies clarified the relationships substantially (Mahoney 1965, Knox 1976, 1978, Baverstock et al. 1980) and four Australian species are currently recognised, two of them very little known (Baverstock et al. 1980). The *Melomys* in the monsoonal lowlands, *M. burtoni*, is now considered to be a single species extending along the coast from Derby in the west (Mahoney 1965) to northern New South Wales (Keith 1970, as *M. littoralis*). Watts & Aslin (1981) show an apparent break in the known distribution around the Gulf of Carpentaria although it occurs on Groote Eylandt (Baverstock et al. 1977a) and in the Sir Edward Pellew Islands (Keith 1968, as *M. cervinipes*) in the Gulf.

This small rodent (about 50 g) is mainly vegetarian (Watts 1977) and much less abundant on the sub-coastal plains than *P. m. sinualis*. In the monsoonal lowlands it is semi-arboreal (Taylor & Horner 1970a) and also occurs on the treeless plains. It inhabits the wetter types of woodland and forest in the monsoonal lowlands (Begg et al. 1983), occurring in patches of dense monsoon rainforest, paperbark woodlands (*Melaleuca* spp.), woodlands on the edges of the plains, in *Pandanus* woodland on the margins between the plains and wooded lowlands, and J.H.Calaby (pers. comm.) has trapped it in mangroves on the riverine margins of the plains. At Kapalga the trapping grids included only the cyperaceous plains (Cyperus Land System) and the *Pandanus* woodland margin (Kosher Land System) which adjoined a patch of dense rainforest higher on the gentle slope.

Seasonal movements

Behavioural avoidance of seasonal extremes by *M. burtoni* is different from that observed in *R. colletti* and *P. m. sinualis*. At Kapalga *M. burtoni* was not trapped on the black soil plains during the middle and late dry season, but it was on the dry cracked soil near paperbark woodland in early December after the beginning of the early storms. *M. burtoni* appears in the *Pandanus* woodland margin at that time also, probably by dispersal from the patch of monsoon rainforest, and usually increases in abundance there late in the wet season. At the end of the wet season the abundance of *M. burtoni* declines in the *Pandanus* margin and numbers, though low, increase in the middle of the plains near large pools of water in gilgai

depressions. However it disappears from there before the middle of the dry season but it is present then in the rainforest. In 1978 when unseasonal rain fell in four months of the dry season, *M. burtoni* remained in moderate abundance in the *Pandanus* margin whereas it had been absent there in the latter half of the previous dry season. These observations suggest that in this part of the monsoonal lowlands, *M. burtoni* uses the rainforest as a refuge in the dry season. It disperses into the *Pandanus* margins and plains only when there is relaxation of aridity from pre-wet season storms until shortly after the wet season, while parts of the plains remain wet or damp, or during unusual wet dry seasons. Individuals trapped on the plains include a wide range of body sizes and pregnant females, suggesting that the expansion of range is not simply dispersal of population recruits. It may facilitate gene flow between populations or demes occupying the various damp forest or woodland refugia, which generally are limited in extent.

Begg et al. (1983) studied a segment of a population inhabiting forest and woodland near a lagoon on the Cobourg Peninsula. There the more dense forest was the favoured habitat and the population showed little movement. This situation seems similar to the forest refuges of *M. burtoni* bordering the plains at Kapalga, and the population seems to reflect a more stable environment in the forest habitats.

Reproduction

The breeding pattern in *M. burtoni* contrasts in some respects with that of the highly fecund *R. colletti*, although the times of breeding are similar in most years. Observations of Taylor & Horner (1970a), Begg et al. (1983) and Williams (unpubl.) suggest that *M. burtoni* may breed in the monsoonal lowlands at any time of the year with periods of more concentrated breeding. At Kapalga pregnancies were at low incidence but concentrated just before the wet season (December) and mainly in the transition from wet to dry seasons (March to June). No pregnancies were observed there during the dry season from June until the rains were increasing in frequency in December. Therefore the pattern of inhibition of breeding suggests that the monsoonal dry season may impose physiological stresses on this species also. At Cobourg Peninsula in a dense forest habitat, Begg et al. (1983) found a rather similar concentration of pregnancies, but pregnancies continued through the dry season except for the month of August. This pattern seems consistent with the above for a moist refugial habitat.

Litter size in *M. burtoni* is small. Females have only four inguinal teats and the young cling by the mouth to them as she moves about (Taylor & Horner 1970a, Williams, unpubl.). Litter sizes observed in the field vary from two to four (Davies 1960, as *M. cervinipes albiventer*; Keith 1968, Williams, unpubl.). The gestation period is between 30 and 40 days in Queensland *M. burtoni* (Redhead 1973, as *M. littoralis*). Two litters of Northern Territory *M. burtoni* were observed in captivity to cease suckling 27-28 days after birth (Williams, unpubl.). The age at sexual maturity in Queensland *M. burtoni* is about nine to ten weeks (Redhead 1973). Regression of gonads occurs in adults (Taylor & Horner 1970a), presumably during periods of environmental stress such as dry seasons. The low incidence of pregnancy spread over five months of the year and low population densities at Kapalga concur with the opinion of Taylor & Horner (1970a) that *M. burtoni* has a low reproductive output over extended time, and that population density and abundance does not fluctuate greatly. Similarly in the forest habitat where Begg et al. (1983) observed *M. burtoni* breeding throughout the year with little migration, recruitment was low and population density remained fairly stable.

Other Melomys

There is little information available on the other Australian *Melomys* species except for *M. cervinipes* which occupies humid tropical and subtropical rainforest and eucalypt forest in eastern coastal Queensland and northern coastal New South Wales (Taylor & Horner 1970a, Watts & Aslin 1981). Wood (1971) reported breeding in all months in a rainforest population, but a lull occurs during the drier winter months. The breeding records discussed by Taylor & Horner (1970a) show geographic variation but are largely consistent with this seasonal pattern. *M. cervinipes* from rainforest gave birth to an average of 1.8 young per litter and some females bred in the season of their birth (Wood 1971). The litter sizes reported by Taylor & Horner (1970a) are generally similar for *M. cervinipes* and *M. burtoni* of coastal Queensland and for *M. burtoni* of the monsoonal lowlands. Therefore these species of *Melomys* have similar low fecundity and similar patterns of breeding over a variable and extended period of the year.

Other monsoonal woodland rodents

Reproduction has been studied in several other rodents that occur in the monsoonal wooded lowlands, namely *Mesembriomys gouldii* (Crichton 1969), *Conilurus penicillatus* (Taylor & Horner 1971), *Pseudomys nanus* (Taylor & Horner 1972, as *P. gracilicaudatus*), *P. delicatulus* (Taylor & Horner 1970b), and the major aspects of their reproduction are summarised for each species in Watts & Aslin (1981). The general breeding pattern of reproduction in these species is rather similar to that of *M. burtoni*, except for the short gestation period and oestrus cycle in *P. nanus* which resemble those of *Rattus* species (Watts & Aslin 1981). The similar reproductive patterns in these species are also consistent with their common habitat, the monsoonal woodland and savanna, and their respective reproductive potentials appear to be less than those of the small mammals of the sub-coastal plains.

DISCUSSION

These four species of native mammal have behavioural responses to the wet and dry seasons which prolong survival. Their physiological and reproductive responses indicate that the seasonal extremes reduce survival. Variability in the cycle of wet and dry seasons seems to influence reproduction and survival causing population fluctuations which have been assessed in *Rattus colletti*, the most variably abundant native mammal of the monsoonal lowlands.

The wet season annually floods the primary habitat of *R. colletti* and *P. m. sinualis* and, to a lesser extent, that of *M. agilis*. The flooding may be less drastic for *M. burtoni* which similarly retreats from the plains but also expands its distribution from wet forest and woodland refuges into adjacent drier woodland habitat as the rains take effect. Whereas breeding may start in the small mammals after the rains begin, it is depressed at the height of the wet season. Severe flooding causes high losses of pouch-young also in the wallaby *M. agilis*. The environmental stresses are sometimes evident in declining body weight and condition of the individuals. Possible causes may be related to populations being displaced by floods, social crowding at refuges, dietary changes, nutritional problems caused by the superabundance of water or perhaps the growth characteristics of the vegetation, and lowered quality or quantity of shelter.

Behavioural avoidance of aridity during the dry season is evident in the three species of small mammals taking refuge in deep cracks in the soil. The much larger *M. agilis* congregates around the moister cooler areas (although this has not been documented) where they can also drink. As aridity becomes severe *R. colletti* retreats to depressions in the middle of the plains which probably provide a more equable habitat. *P. m. sinualis* seems to be less active on the surface of the plains during the middle and later part of the dry season. Then frequent torpor may aid survival and they may eat the insects also living in the soil cracks. *M. burtoni* may also avoid aridity by retreating to the wetter types of forest and woodland. That the population movements observed in the rodents during the dry seasons are actually aridity-avoidance responses is indicated by use of the wooded margin by low numbers of both *R. colletti* and *M. burtoni* in the milder dry season when it rained.

If there is substantial dry season rainfall *R. colletti* does not lose weight, unlike in severe dry seasons. There is no similar information for *P. m. sinualis* and *M. burtoni*. Physiological torpor in *P. m. sinualis* may help prevent loss of weight except under extremely dry conditions. The large size of *M. agilis* provides advantages of lower requirements for food and water relative to size. Size also gives it the mobility to obtain drinking water, and its ruminant-like digestion allows survival on a wide range of herbage species including some of low quality (Kinnear et al. 1979). Nevertheless, *M. agilis* loses condition during prolonged dry seasons when green herbage dwindles to damp soaks and depressions where they congregate.

Reproductive condition is probably a more sensitive indicator of the onset of environmental stress than loss of weight. In the three small mammals reproduction ceases before loss of condition is evident. During prolonged dry seasons *M. agilis* loses condition but reproduction is barely affected. On two occasions when rain fell during the dry season, reproduction in *R. colletti* was not depressed but actually enhanced. Then *M. burtoni* remaining in the wooded margins of the plains did not continue to breed although they may have done so in the more typical dry season habitats of rainforest and paperbark woodland.

The most favourable periods of the annual cycle are in the transition from one extreme to the other, and mainly from wet season to dry season. The transition from dry season to wet season can be gradual or saltatory, but often it is sudden with torrential rains quickly flooding low-lying land. In the latter case the mammals of the plains are forced rapidly onto higher ground where there may be little herbage due to the dry season aridity and possibly fire. With sudden flooding even the agile wallaby loses weight; pouch-young die and females cease to breed. The transition from wet to dry season is different. Although rains may cease suddenly, the floods retreat slowly, more by evaporation than drainage. The dry parts of the plains then provide dense cover for the small mammals plus virtually the entire annual crop of green herbaceous vegetation of the plains, and the wet season production of prey for *P. m. sinualis*. However reproduction may be prevented or truncated if flooding is prolonged and the cracking of the black soil is delayed, so denying ᵗʰe small mammals their underground refuge when the cool, dry anticyclonic winds from central Australia cause low night temperatures. Successful breeding in the small mammals is probably impossible under such conditions. Then the clear skies, intense insolation and dry winds hasten the arid phase, resulting in little recruitment. Very wet wet seasons or late heavy rains may contribute in this way to the periodic rarity of *R. colletti*.

As a whole, the native mammals of the monsoonal lowlands fall into two natural groupings based on rate of reproduction, and these are respectively associated with the two major landforms, the sub-coastal plains and the wooded lowlands. The annually flooded plains are the major habitat of the two highly fecund species *R. colletti* and *P. m. sinualis*. The plains and margins are also the habitat of *M. agilis* which, for a macropod, appears to have evolved a high reproductive output through virtually continuous reproduction. The fourth native mammal to use the plains, *M. burtoni*, has a modest reproductive output but only short-term seasonal use of the plains and its principal habitat is in wet woodland and forest.

Other native mammals occupy the woodland, all with relatively lower fecundity than those of the plains. *Rattus tunneyi tunneyi*, the only other native *Rattus* of the region, lives in woodland savanna and its reproductive output in the field is modest compared to *R. colletti* (Taylor & Horner 1973b). All the non-*Rattus* species of native rodents living in the monsoonal lowlands occupy the wooded areas (including mangroves). The species are *Mesembriomys gouldi*, *M. macrurus, Conilurus penicillatus, Hydromys chrysogaster, Pseudomys delicatulus, P. nanus, Melomys burtoni* and *Xeromys myoides* (Watts & Aslin 1981). The latter two species have been found in tidal mangroves and little is known of the biology of *X. myoides*. Basic reproductive variables have been determined on most of these species and are summarised in Watts & Aslin (1981). The *Rattus* group has been shown to be more distant genetically from all the other native rodents which include the *Melomys/Uromys* group (Baverstock et al. 1977a, b, 1981). Among all Australian native rodents the highest potential reproductive rates occur among the *Rattus* species, and the other rodent groups including *Melomys* have relatively low reproductive outputs in comparison (Watts & Aslin 1981). Similarly *P. m. sinualis* has a very high reproductive output that is unique among the dasyurids of the region (Taylor et al. 1982). All the other local dasyurids occupy the woodland and rocky escarpment where flooding is not a problem, and have lower reproductive rates (Taylor et al. 1982). Similarly *M. agilis* from the plains is distinctive from the other macropods of the monsoonal woodlands in breeding continuously and opportunistically rather than seasonally as suspected for *M. antilopinus* and *M. r. woodwardi*.

These observations suggest that the species occupying the plains are adapted to rapidly replacing population losses caused by flooding and aridity. Reproductive repopulation of the plains following the wet season must be rapid because of the very short period between flooding and seasonal aridity (Fig. 55), and this can be achieved only by very high fecundity. Equable periods in the desert are usually also ephemeral but less regular than in the monsoonal lowlands. Adapted desert species often have high fecundity, the capability of breeding whenever conditions are suitable and they are often long-lived (Main 1976). On the monsoonal sub-coastal plains longevity does not seem to have been strongly selected, apparently as the monsoonal sequence is a regular annual event. However the species adapted to the plains resemble desert species in rapid maturation, opportunistic breeding and high fecundity.

The generally lower fecundity of the small species in the wooded lowlands is consistent with a more buffered and stable environment than the plains. This is clearly so for flooding but there is no information comparing aridity between plains and woodlands. Possibly the deep-rooted trees provide the woodland animals with greater availability of water in the dry season through frugivores, folivores, sap suckers, and the food web. In contrast soil cracks on the plains provide a humid microclimate for small species, and water from the wet season is stored by sedges in the underground corms and rhizomes, although access to this moisture for mammals and insects declines as the soil hardens during the dry season. From another aspect, the woodlands, with sparse skeletal soils leached of nutrients by the heavy wet season rains and desiccated in the dry season, may be too infertile to support mammal populations which rapidly reproduce to high densities. The plains receive much of these washed-out nutrients and therefore, when the climatic sequences permit, may produce sufficient vegetational resources on which the rapidly expanding animal populations ultimately depend. Nevertheless, on the basis of flooding alone we would expect mammal populations in the woodlands to be less prone to fluctuation than those on the plains, and on the basis of complexity we would expect more native mammal species to occupy the woodlands than the plains.

REFERENCES

Andrews, L.G. 1976. Reproductive performance of beef cattle in the Northern Territory. Unpubl. M. Vet. Sc. thesis, James Cook University of North Queensland.

Archer, M. 1976. Revision of the marsupial genus *Planigale* Troughton (Dasyuridae). *Mem. Qld. Mus.* 17: 341-365.

Aslin, H.J. 1975. Reproduction in *Antechinus maculatus* Gould (*Dasyuridae*). *Aust. Wildl. Res.* 2: 77-80.

Baverstock, P.R., C.H.S.Watts, M.Adams & S.R.Cole 1981. Genetical relationships among Australian rodents (Muridae). *Aust. J. Zool.* 29: 289-303.

Baverstock, P.R., C.H.S.Watts, M.Adams & M.Gelder 1980. Chromosomal and electrophoretic studies of Australian *Melomys* (Rodentia: Muridae). *Aust. J. Zool.* 28: 553-574.

Baverstock, P.R., C.H.S.Watts & J.T.Hogarth 1977a. Chromosome evolution in Australian rodents. I. The Pseudo-myinae, the Hydromyinae and the *Uromys/Melomys* group. *Chromosoma* 61: 95-125.

Baverstock, P.R., C.H.S.Watts, J.T.Hogarth, A.C.Robinson & J.F.Robinson 1977b. Chromosome evolution in Australian rodents. II. The *Rattus* group. *Chromosoma* 61: 227-241.

Begg, R., B.Walsh, F.Woerle & S.King 1983. Ecology of *Melomys burtoni*, the grassland Melomys (Rodentia: Muridae) at Cobourg Peninsula, N.T. *Aust. Wildl. Res.*10: 259-267.

Bolton, B.L. 1975. An ecological study of the agile wallaby on the coastal plains of the Northern Territory with a comparison between improved and unimproved areas. Unpubl. M. Vet. Sc. thesis, University of Queensland.

Bolton, B.L., A.E.Newsome & J.Merchant 1982. Reproduction in the agile wallaby, *Macropus agilis* (Gould), in the tropical lowlands of the Northern Territory: opportunism in a seasonal environment. *Aust. J. Ecol.* 7: 261-277.

Breed, W.G. 1978. Ovulation rates and oestrus cycle lengths in several species of Australian native rats (*Rattus* spp.) from various habitats. *Aust. J. Zool.* 26: 475-480.

Calaby, J.H. 1974. Historical background. In H.J.Frith & J.H.Calaby (eds.), *Fauna Survey of the Port Essington District, Cobourg Peninsula, Northern Territory of Australia*. Div. Wildl. Res. Tech. Pap. 28. CSIRO: Melbourne.

Calaby, J.H. & K.Keith 1974. Mammals. In H.J.Frith & J.H.Calaby (eds.), *Fauna Survey of the Port Essington District, Cobourg Peninsula, Northern Territory of Australia*. Div. Wildl. Res. Tech. Pap. 28. CSIRO: Melbourne.

Calaby, J.H. & J.M.Taylor 1981. Reproduction in two marsupial-mice, *Antechinus bellus* and *A. bilarni* (Dasyuridae), of tropical Australia. *J. Mammal.* 62: 329-341.

Carstairs, J.L. 1974. The distribution of *Rattus villosissimus* (Waite) during plague and non-plague years. *Aust. Wildl. Res.* 1: 95-106.

Carstairs, J.L. 1976. Population dynamics and movements of *Rattus villosissimus* (Waite) during the 1966-69 plague at Brunette Downs, N.T. *Aust. Wildl. Res.* 3: 1-9.

Chisholm, A.H. 1955. *Strange New World*. Sydney: Angus & Robertson.

Collins, B.G. 1973a. Physiological responses to temperature stress by an Australian murid, *Rattus lutreolus*. *J. Mammal.* 54: 356-368.

Collins, B.G. 1973b. The ecological significance of thermoregulatory responses to heat stress shown by two populations of an Australian murid, *Rattus fuscipes*. *Comp. Biochem. Physiol.* 44A: 1129-1140.

Collins, B.G. & S.D.Bradshaw 1973. Studies on the metabolism, thermoregulation, and evaporative water losses of two species of Australian rats, *Rattus villosissimus* and *Rattus rattus*. *Physiol. Zoöl.* 46: 1-21.

Crichton, E.G. 1969. Reproduction in the pseudomyine rodent *Mesembriomys gouldii* (Gray) (Muridae). *Aust. J. Zool.* 17: 785-797.

Davies, S.J.J.F. 1960. A note on two small mammals of the Darwin area. *J. Roy. Soc. West. Aust.* 43: 63-66.

Dawson, T.J. & A.J.Hulbert 1970. Standard metabolism, body temperature, and surface areas of Australian marsupials. *Am. J. Physiol.* 218: 1233-1238.

Dawson, T.J. & J.M.Wolfers 1978. Metabolism, thermoregulation and torpor in shrew sized marsupials of the genus *Planigale*. *Comp. Biochem. Physiol.* 59A: 305-309.

Dudzinski, M.L., A.E.Newsome, J.C.Merchant & B.L.Bolton 1977. Comparing the two usual methods for ageing Macropodidae on tooth-classes in the agile wallaby. *Aust. Wildl. Res.* 4: 219-221.

Friend, G.R. 1985. Ecological studies of a population of *Antechinus bellus* (Marsupialia: Dasyuridae) in tropical northern Australia. *Aust. Wildl. Res.* 12: 151-162.

Friend, G.R. 1987. Population ecology of *Mesembriomys gouldii* (Rodentia: Muridae) in the wet-dry tropics of the Northern Territory. *Aust. Wildl. Res.* 14: 293-303.

Friend, G.R., M.L.Dudzinski & K.M.Cellier 1988. *Rattus colletti* (Rodentia: Muridae) in the Australian wet-dry tropics: seasonal habitat preferences, population dynamics and the effects of buffalo (*Bubalus bubalis*). *Aust. J. Ecol.* 13: 51-66.

Friend, G.R. (in press). Breeding and polulation dynamics of *Isoodon macrourus* (Marsupialia: Peramelidae): studies from the wet-dry tropics of northern Australia. In J.H.Seebeck, P.P.Brown, R.L.Wallis & C.M.Kemper (eds.), *Bandicoots and Bilbies.* Sydney: Surrey Beatty.

Gentilli, J. 1971. Australian climatic factors. In H.E.Landsberg (ed.), *World Survey of Climatology, Climates of Australia and New Zealand*, Vol. 13. Amsterdam: Elsevier.

Keith, K. 1968. Mammals of the Pellew Islands. *Mimag*, June 1968 (House Journal of Mount Isa Mines Ltd.).

Keith, K. 1970. The occurrence of a second species of naked-tailed rat, genus *Melomys*, in New South Wales. *CSIRO Wildl. Res.* 15: 19-26.

Kerle, J.A. 1983. The population biology of the northern brushtail possum. Ph.D. thesis, Macquarie University, North Ryde, New South Wales.

Kinnear, J.E., A.Cockson, P.Christensen & A.R.Main 1979. The nutritional biology of the ruminants and ruminant-like mammals – A new approach. *Comp. Biochem. Physiol.* 64A: 357-365.

Knox, E. 1976. Upper molar alveolar patterns of some muridae in Queensland and Papua New Guinea. *Mem. Ql. Mus.* 17: 457-458.

Knox, E. 1978. A note on the identification of *Melomys* species (Rodentia: Muridae) in Australia. *J. Zool. Lond.* 185: 276-277.

Lourensz, R.S. 1977. *Tropical cyclones in the Australian region July 1909 to June 1975.* Canberra: Australian Government Publishing Service.

McAlpine, J.R. 1969. The climate of the Adelaide-Alligator area. Lands of the Adelaide-Alligator Area, Northern Territory, IV. *CSIRO Aust. Land Res. Ser.* 25: 49-55.

McDougall, W.A. 1946a. An investigation of the rat pest problem in Queensland canefields. 4. Breeding and life histories. *Qld. J. Agric. Sci.* 3 (1): 1-43.

McDougall, W.A. 1946b. An investigation of the rat pest problem in Queensland canefields. 5. Populations. *Qld. J. Agric. Sci.* 3 (4): 1-81.

MacMillen, R.E. & J.E.Nelson 1969. Bioenergetics and body size in dasyurid marsupials. *Am. J. Physiol.* 217: 1246-1251.

Mahoney, J.A. 1965. The identity of *Mus burtoni* Ramsay, 1887 (Rodentia, Muridae, *Melomys*) from the neighbourhood of Derby, Western Australia. *J. Roy. Soc. West. Aust.* 48: 91-95.

Main, A.R. 1976. Adaptation of Australian vertebrates to desert conditions. In D.W.Goodall (ed.), *Evolution of Desert biota.* Austin & London: University of Texas Press.

Merchant, J.C. 1976. Breeding biology of the agile wallaby, *Macropus agilis* (Gould) (Marsupialia: Macropodidae) in captivity. *Aust. Wildl. Res.* 3: 93-103.

Merchant, J.C. & J.H.Calaby 1981. Reproductive biology of the red-necked wallaby (*Macropus rufogriseus banksianus*) and Bennett's wallaby (*M. r. rufogriseus*) in captivity. *J. Zool. Lond.* 194: 203-217.

Morton, S.R. 1980. Field and laboratory studies of water metabolism in *Sminthopsis crassicaudata* (Marsupialia: Dasyuridae). *Aust. J. Zool.* 28: 213-227.

Morton, S.R. & A.K.Lee 1978. Thermoregulation and metabolism in *Planigale maculata* (Marsupialia: Dasyuridae). *J. Thermal Biol.* 3: 117-120.

Newsome, A.E. 1964a. Anoestrus in the red kangaroo, *Megaleia rufa* (Desmarest). *Aust. J. Zool.* 12: 9-17.

Newsome, A.E. 1964b. Oestrus in the lactating red kangaroo, *Megaleia rufa* (Desmarest). *Aust. J. Zool.* 12: 315-321.

Newsome, A.E. 1965. Reproduction in natural populations of the red kangaroo, *Megaleia rufa* (Desmarest), in central Australia. *Aust. J. Zool.* 13: 735-759.

Newsome, A.E. 1983. The grazing marsupials. In F. Bourlière (ed.), *Tropical Savannas of the World.* Amsterdam: Elsevier.

Newsome, A.E., J.C.Merchant, B.L.Bolton & M.L.Dudzinski 1977. Sexual dimorphism in molar progression and eruption in the agile wallaby. *Aust. Wildl. Res.* 4: 1-5.

Nix, H.A. 1976. Environmental control of breeding, post-breeding dispersal and migration of birds in the Australian region. In H.J.Frith & J.H.Calaby (eds.), *Proc. 16th Intern. Ornith. Congr.*: 272-305. Canberra: Australian Academy of Science.

Poole, W.E. 1976. Breeding biology and current status of the grey kangaroo, *Macropus fuliginosus fuliginosus*, of Kangaroo Island, South Australia. *Aust. J. Zool.* 24: 169-187.

Redhead, T.D. 1973. A re-assessment of the rat problem in North Queensland cane fields, with emphasis on the ecology and reproduction of *Melomys littoralis* (Lönnberg). Unpubl. M.Sc. thesis, James Cook University of North Queensland.

Redhead, T.D. 1979. On the demography of *Rattus sordidus colletti* in monsoonal Australia. *Aust. J. Ecol.* 4: 115-136.

Russell, E.M. & B.J.Richardson 1971. Some observations on the breeding, age structure, dispersion and habitat of populations of *Macropus robustus* and *Macropus antilopinus* (Marsupialia). *J. Zool., Lond.* 165: 131-142.

Sadleir, R.M.F.S. & C.H.Tyndale-Biscoe 1977. Photoperiod and termination of diapause in the marsupial, *Macropus eugenii. Biol. Reprod.* 16: 605-608.

Sharman, G.B. 1955. Studies on marsupial reproduction. 2. The oestrous cycle of *Setonix brachyurus. Aust. J. Zool.* 3: 44-55.

Sharman, G.B. 1959. Marsupial reproduction. In A. Keast, R.L.Crooker & C.S. Christian (eds.), *Biogeography and Ecology in Australia. Monographiae Biologicae* 8: 332-368. The Hague: Junk.

Story, R. 1969. Grasses of the Adelaide-Alligator area and their utilization. Lands of the Adelaide-Alligator Area, Northern Territory, IX. *CSIRO Aust. Land Res. Ser.* 25: 131-139.

Taylor, J.M. & B.E.Horner 1970a. Reproduction in the mosaic-tailed rat, *Melomys cervinipes* (Rodentia: Muridae). *Aust. J. Zool.* 18: 171-184.

Taylor, J.M. & B.E.Horner 1970b. Observations on reproduction in *Leggadina* (Rodentia, Muridae). *J. Mamm.* 51: 10-17.

Taylor, J.M. & B.E.Horner 1971. Reproduction in the Australian tree-rat *Conilurus penicillatus* (Rodentia: Muridae). *CSIRO Wildl. Res.* 16: 1-9.

Taylor, J.M. & B.E.Horner 1972. Observations on the reproductive biology of *Pseudomys* (Rodentia: Muridae). *J. Mamm.* 53: 318-328.

Taylor, J.M. & B.E.Horner 1973a. Results of the Archbold Expeditions No. 98. Systematics of native Australian *Rattus* (Rodentia, Muridae). *Bull. Am. Mus. Nat. Hist.* 150: 1-130.

Taylor, J.M. & B.E.Horner 1973b. Reproductive characteristics of wild native Australian *Rattus* (Rodentia: Muridae). *Aust. J. Zool.* 21: 437-475.

Taylor, J.M., J.H.Calaby & T.D.Redhead 1982. Breeding in wild populations of the marsupial-mouse *Planigale maculata sinualis* (Dasyuridae, Marsupialia). In M.Archer (ed.), *Carnivorous Marsupials* Vol. 1: 83-87. Sydney: Royal Zool. Soc. NSW.

Tulloch, D.G. 1969. Home range in feral water buffalo, *Bubalus bubalis* Lydekker. *Aust. J. Zool.* 17: 143-152.

Tulloch, D.G. 1970. Seasonal movements and distribution of the sexes in the water buffalo, *Bubalus bubalis*, in the Northern Territory. *Aust. J. Zool.* 18: 399-414.

Tulloch, D.G. 1974. The feral swamp buffaloes of Australia's Northern Territory. In W.R.Cockrill (ed.), *The Husbandry and Health of the Domestic Buffalo*: 493-505. Rome: FAO.

Tulloch, D.G. 1978. The water buffalo, *Bubalus bubalis*, in Australia: Grouping and home range. *Aust. Wildl. Res.* 5: 327-354.

Tulloch, D.G. & A.Grassia 1981. A study of reproduction in water buffalo in the Northern Territory of Australia. *Aust. Wildl. Res.* 8: 335-348.

Tyndale-Biscoe, C.H. 1977. Environment and control of breeding in kangaroos and wallabies. In H.Messel & S.T.Butler (eds.), *Australian Animals and Their Environment*: 63-79. Sydney: Shakespeare Head Press.

Tyndale-Biscoe, C.H., J.P.Hearn & M.B.Renfree 1974. Control of reproduction in macropodid marsupials. *J. Endocr.* 63: 589-614.

Van Dyck, S. 1979. Behaviour in captive individuals of the dasyurid marsupial *Planigale maculata* (Gould 1851). *Mem. Qld. Mus.* 19: 413-429.

Watts, C.H.S. 1977. The foods eaten by some Australian rodents (Muridae). *Aust. Wildl. Res.* 4: 151-157.

Watts, C.H.S. & H.J.Aslin 1981. *The rodents of Australia.* London & Sydney: Angus & Robertson.

Wesley-Smith, R.N. 1972. Live weight gains of shorthorn steers on native and improved pastures at Adelaide River, Northern Territory. *Aust. J. Exp. Agric. and Anim. Husbandry* 12: 566-572.

Williams, C.K. 1987. Water physiology and nutrition in fluctuating populations of *Rattus colletti* in monsoonal Northern Territory, Australia. *Aust. Wildl. Res.* 14: 443-458.

Wood, D.H. 1971. The ecology of *Rattus fuscipes* and *Melomys cervinipes* (Rodentia: Muridae) in a south-east Queensland rain forest. *Aust. J. Zool.* 19: 371-392.

Woolley, P. 1973. Breeding patterns, and the breeding and laboratory maintenance of dasyurid marsupials. *Exp. Animals* 22 (suppl.): 161-172.

CHAPTER 11

Feral mammals and their environment

M.G. RIDPATH

THE FERAL STATE

Our ancestors took two giant strides when they began to domesticate animals and plants, some 9,000-12,000 years ago (Flannery 1969, Clutton-Brock 1981). By gradually assuming control over critical natural resources for hunting, food and fibre for clothing, and improving them, humans started to escape from their precarious dependance on the day-to-day vagaries of the environment. This technological change started in Asia, centred in its southwestern and central regions, and gathering momentum, it soon spread well beyond. Thus, some of the original domesticated animals were taken fairly soon to other parts of Asia, Africa and Europe, while domestication of a few more wild species in the new regions also took place. The process of domestication of mammals was largely over by the end of the Neolithic period (Zeuner 1963, Jewell 1969). The total of wild species involved was actually quite small. Nonetheless, so valuable (perhaps indispensable) were certain of them that European colonists subsequently took them to the Americas in the sixteenth century and to Australia at the beginning of the nineteenth. Today domesticated mammals are used in one way or another by the vast majority of the world's inhabitants.

Naturally individuals of domesticated species have escaped from time to time and lived wild. In some cases they and their descendants have bred successfully outside human control and maintained stable populations in the wild. The term 'feral' (as distinct from 'exotic', which refers to an unchanged wild species introduced into a region it has not occupied hitherto) is used to describe such animals which have reverted to life unsupported by human intervention (Hafez 1962, McKnight 1964). This must have happened repeatedly in human history. The various transitions which occur between the wild and domesticated states are summarized in Figure 61. One of the best documented is the interbreeding of the dingo with domestic dogs *Canis familiaris* in Australia and Asia (Newsome et al. 1980, Newsome & Corbett 1982, Corbett 1985).

Domestication probably began as a loose, often mutually beneficial ecological association (as Wharton 1968) between man and particular species, the behaviour and social organization of which predisposed its members to his subsequent dominance and advantage (as Clutton-Brock 1977, 1981). In these early stages merely tame but genetically unchanged individuals would often have withdrawn from the association with humans as a result of chance events. Once human control eventually had been established more positively by clear-cut husbandry and selective breeding, such reversions would have become progressively more difficult for the animals, and thus less frequent. To use an analogy with human history, domestic animals had by then been fully 'enslaved'.

Nonetheless these 'slaves' still escaped occasionally. This occurred particularly under the disturbed conditions when immigrant human populations were struggling to colonize parts of the world new to them, and their domesticated stock. The cattle of the North American prairies originated from those the early conquistadors brought with them in the sixteenth century, some of which had soon gone feral (Jennison 1929). Thereafter these were harvested, rather in the sense of hunting than of husbandry, and their economic value probably deteriorated. It was 200 years or more before they were redomesticated, although today a few still remain feral (McKnight 1975). To combine higher productivity and tractability with the capacity of feral cattle to withstand open-range conditions, especially in harsh western America, ranchers later crossed them with fresh imports of long domesticated cattle. On the vast, often little developed properties of Australia's harsh North, where pastoralism has often amounted to little more than annual cattle trapping, similar crossing has been tried. Both cases have something in common with that of the Naga tribes of Assam who still cross their domestic Gayal cattle (*Bos frontalis*) with the wild Gaur (*Bos gaurus*), from which the Gayal originate (Zeuner 1963). Likewise in India domestic buffaloes have been crossed occasionally with wild buffalo almost to this day (Epstein 1971).

Whatever the reasoning that led to their use, such

169

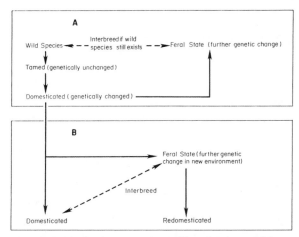

Figure 61. The feral state, and transitions between wild species and domesticated forms: A - in original range; B - in new range into which domesticated forms have been introduced by humans. Both domesticated and feral forms may undergo genetic change in the new range.

practices could well have some advantages to the owners/users. The process of domestication, whether wholly planned or not, involves selection for greater docility, size and productivity (i.e. higher reproductive rate, faster growth and earlier maturity). However such selection, especially for docility, also results in other characteristics in domesticated stock which loosely could be described as anomalous and apparently non-adaptive. They include what anthropomorphically might be termed 'stupidity' or 'insensitivity' to the environment, but which the animals' owners/users accept (or even favour). Conversely such behavioural and other side-effects disadvantage highly domesticated stock if they are ever run under open-range or semi-wild conditions, and then are to the owner's/user's economic detriment too. These side-effects even further disadvantage domesticated individuals which escape from human control and husbandry altogether and must try to survive in the wild. In both cases natural selection would tend to eliminate such side-effects and to favour traits better adapted to conditions closer to the wild state. Where the wild species and its domestic descendants co-exist, as occasionally still occurs with ungulates in Asia today, hybridization readily provides the possibility of new forms with a better mix of environmental adaptations and economic desiderata. However in new environments such as Australia the original wild ancestor is absent. Thus the only locally available reservoir of adaptations to a non-domestic environment resides in the feral population. Although this population has lost some wild genetic attributes as a result of its original domestication, it often exhibits (and may even have gained) a few new ones as a result of natural selection

since going feral. Many of the genes for such new traits were probably present among the original imports, but not substantially expressed in the population until favoured by the new selective forces in the new environment. Such genetic change in populations can occur quite rapidly in these situations, as has been shown in the case of rabbits imported into Australia (Barber 1954, Stodart 1965).

Whether for the study of natural selection in action or for the quest for environmental adaptations in stock, the dynamic relationships outlined above between feral mammals and new environments offer unique scope for research. Yet, with a few exceptions, this topic has been little studied. This is especially surprising in view of the fact that we start off with a much greater knowledge of the basic biology of these animals than about that of most wild species. Nonetheless, the evolutionary biology of colonizing wild species of animals, particularly insects, has begun to attract a good deal of attention recently (Parsons 1983), by comparison with that of colonizing feral animals.

This chapter describes the varying success with which feral mammals have occupied the tropical monsoonal lowlands, some of the characteristics they possess that may affect their success and how such characteristics may relate to the habitat of their wild ancestors. Special attention has been given to characteristics which appear unique to tropical animals. Discussion of the effects the feral mammals themselves have on their environment and on human concerns, positively and negatively, complete this over-view.

FERAL FORMS

Members of twelve families from six orders of mammals have been domesticated world-wide. Australia was the last continent to receive their descendants, apart from the dingo *Canis familiaris dingo* which probably arrived from Asia in association with Aboriginal immigrants around 3,500-4,000 BP (Millham & Thompson 1976). All the rest were imported by European colonists 170-190 years ago. Domesticated forms from four orders have gone feral after introduction into Australia: Lagomorpha (Leporidae: rabbits and hares), Carnivora (Felidae: cats, Canidae: dogs), Perissodactyla (Equidae: horses and asses), Artiodactyla (Suidae: pigs, Camelidae: camels, Bovidae: cattle, buffaloes and goats). All are present in the coastal monsoonal lowlands of tropical Australia, except for rabbits and camels. The introduction of domesticated mammals that have gone feral comprises one of the most dramatic human impacts on the pre-existing ecosystems of this region. By contrast only four wild species of mammals have been introduced: Fallow, Barasingha, Sambar and Rusa deer, *Dama dama, Cer-*

vus duvauceli, C. unicolor and *C. timorensis* (Bentley 1967, 1983). Of these 'exotic' species only some Sambar now survive, on Cobourg Peninsula (Calaby & Keith 1974).

Nine forms of feral mammals occur, eight being ungulates and one a carnivore (Table 39).

Because it arrived in Australia much earlier than the others, the dingo has had much more time in which to evolve in its new environment. In this sense its present status is indistinguishable from that of wild native species and it is treated here as such.

AVAILABLE NICHES AND COMPETITION

Australia with New Guinea became a separate continent in the Late Cretaceous when it broke away from the Gondwanaland supercontinent, originally including Africa, India, South America and Antarctica (Crook 1981, Powell et al. 1981, Williams: Chapter 1). Thus the non-marine flora and fauna of Australia have evolved in virtual isolation for the last 80 to 90 million years. This has resulted in an array of species and ecological relationships which differ greatly from those found elsewhere in the world. Such differences are particularly marked in those groups of the Australian biota for which the sea created an absolute barrier to invasion, including the plants and the terrestrial mammals.

In all other continents a significant part of the terrestrial mammalian fauna comprises large herbivores (ungulates) and large carnivores. The former account for 13-20% and the latter for 6-14% of the families present (McKnight 1975, for ungulates, and calculated similarly for the four families of large carnivora).

Table 39. Feral mammals of the monsoonal lowlands.

Feral form	Species	Domesticated form introduced from
Felidae		
Domestic cat	*Felis catus*	Britain
Equidae		
Brumby (and Timor Pony)	*Equus caballus*	Britain (and Timor)
Donkey	*E. asinus*	Europe (Britain)
Suidae		
Pig	*Sus scrofa*	Britain (and Timor)
Bovidae		
European cattle	*Bos taurus*	Britain
Brahman cattle (Zebu)	*B. indicus*	India and South Africa
Bali cattle	*B. javanicus*	Bali
Asian Water Buffalo	*Bubalus bubalis*	Timor
Goat	*Capra hircus*	Britain

Australia had no really large herbivorous or carnivorous marsupials when the first Europeans arrived. The largest were the Red and Grey Kangaroos *Macropus rufus* and *M. giganteus* and the Tasmanian wolf *Thylacinus cynocephalus*, already replaced on the mainland by the same-sized dingo which had come with Aborigines some 4,000 years earlier. Yet the presence until the Late Pleistocene of some very large herbivorous and carnivorous species (e.g. the huge *Diprotodon*, the 'marsupial lion' *Thylacaleo*, the giant kangaroos and wombats) shows that such ecological niches previously existed in Australia (Jones 1968, Merrilees 1968, Rich 1982). Exactly what they ate and how vegetation and prey had evolved in response to their pressure then and subsequent extinction, we do not know. Nor do we understand why they disappeared though this has been attributed variously to human activities such as burning and hunting, to climatic change or to some combination thereof (Archer 1981, Rich 1982). What is clear is that present Australian conditions have allowed several large domesticated herbivores and a medium-sized domesticated carnivore to establish themselves very successfully in the feral state. This has occurred in the presence of a number of native species, mostly abundant though somewhat smaller, with broadly similar needs for food and cover to those of the newcomers.

In such situations it is found that wild species partition their resources among themselves, often in quite subtle ways. The evidence is quite good that such partitioning results from competition and selection which have led each species to some degree of specialization, even though they share other components of the resource. A good illustration occurs in the six large herbivores which coexist on the Serengeti plains of East Africa (Jarman & Sinclair 1979). Likewise when a domesticated species becomes feral it is immediately in competition, to a greater or lesser degree, with any native species with similar demands on the environment. Its success or failure in becoming feral depends on the extent to which its adaptations to its original environment happen to fit its new environment and thus to competition with the native species there. At one extreme it may be so lacking in suitable attributes that it fails altogether to establish itself permanently in the wild. At the other extreme it may possess such well-suited attributes that it actually displaces native competitors. In either case it may take several generations for the new scenario to be played out. Furthermore this situation can be modified by changes the feral form may cause in its environment. Finally, as already mentioned, it may itself undergo genetic changes as a result of selection in the new environment. Ecosystems are dynamic and it is probable that in Australia those which include feral animals today are in a state of comparatively rapid change.

ECOLOGICAL FEATURES FACED IN THE ENVIRONMENT OF THE WET-DRY TROPICS

Five distinctive climatic features of this monsoonal environment are very important ecologically (Ridpath 1985).

1. The daily period of light varies only between 11.2 and 12.7 hours over the year.

2. Temperatures are high day and night, year round. For example the highest mean monthly maximum at Oenpelli, NT, 60 km from the north coast, is 37°C yet the lowest monthly minimum is still 18°C.

3. Rainfall is highly seasonal resulting in very great differences in soil moisture and thus in plant growth during the year (see McDonald & McAlpine, Chapter 2, Williams & Newsome, Chapter 10, for further details).

4. The pattern of rainfall during the 'wet season', especially during the first half, varies appreciably both within the season and between years. This variability is reflected in soil moisture and thus in the conditions for plant growth.

5. Almost all tropical soils are very infertile except those of the sub-coastal plains.

In view of the uniformity of light and temperature regimes it can be seen that plant growth is largely determined by the marked annual alternation between a wet season and a very dry one. The final determinant of productivity is the significant variability of soil moisture within and between the wet season(s). Fires, of fairly erratic incidence nowadays, further vary the availability of vegetation in the dry season. Thus each year native and feral herbivorous mammals, and hence often carnivores too, face a flush of food of rather uncertain time of commencement, composition, magnitude and duration. This is followed by a period of drought and a shortage of food, occasionally severe. Their survival, growth and reproduction depend on their capacity to respond successfully to these major fluctuations in resources. One widespread strategy depends on the capacity to move in harsh times to where resources have persisted best and to spread out again as things improve. Another important capacity is to be able to fine-tune reproduction to coincide with the somewhat variable availability of resources for the production and nourishment of the young. Likewise reproductive rates (i.e. the size and frequency of litters) need to be in keeping with the likely size and duration of these resources.

Another important set of ecological features arises from the permanently high temperature in the tropics. These cause high process rates which lead to the rapid release of nutrients from litter and their frequent loss by leaching in the heavy monsoonal rains (Ridpath et al., 1985). Hence the only fertile tropical soils are generally those of recent origin. In this respect a capacity to utilize alluvial situations, such as wetlands, has several advantages. The absence of low temperatures in the tropics favours poikilothermic animals such as invertebrates because it allows activity year-round. As a result herbivorous insects such as termites and grasshoppers are varied, widespread and abundant. They probably are in competition with herbivorous mammals and this could place nutritional efficiency at a premium. Finally, the capacity to lose heat and/or to avoid gaining it is very important for all homeothermic animals in the permanently high temperatures of the tropics.

NICHES IN A NEW ENVIRONMENT

Systematic scientific research on the ecology of feral mammals in this region began in the late 1950s with D.G.Tulloch's pioneering work on the Water Buffalo (Tulloch 1967, 1969, 1970, 1978, 1979b, 1981). His and subsequent investigations, especially by CSIRO in the Kapalga area (see Fig. 1), provide an outline of the ecological niches occupied by buffaloes, pigs and, to a lesser extent, the three kinds of feral cattle, brumbies and cats. Less information is available on donkeys and goats and it has been necessary to draw partly on what is known about them elsewhere in northern and southern Australia.

Cat (Felis catus)

The domestic cat originated from the North African race of the wild *F. silvestris*, possibly around 3500 BP in Egypt (Bodenheimer 1935, Zeuner 1963, Clutton-Brock 1981). The Romans introduced it into central and northern Europe including Britain where some interbreeding with the wild European race of *F. silvestris* took place. Thus the domestic cat brought by British settlers to Australia, which subsequently went feral, had a mixture of North African and temperate origins. It is not known when it reached the northern coast though presumably this occurred sometime after permanent settlement began in 1869. They were at Kapalga by 1920 (Warburton 1934). Feral cats occur sparsely in open forest and woodland and less commonly on the sedgelands of the monsoonal lowlands but nowhere appear abundant. On 600 km² at Kapalga the population has been estimated at 16 cats (L.K.Corbett, pers. comm.). Their diet at Kapalga consists mainly of the swamp-dwelling Dusky Rat *Rattus colletti*, with some Northern Brushtailed Possum *Trichosurus arnhemensis*, Magpie Geese *Anseranas semipalmata* and small quantities of carrion and other birds (L.K.Corbett, pers. comm.). This accords with data from southern Australia, North America, New Zealand and Britain where feral cats were found to prey very largely on the most commonly available small mammals, up to the size of a rabbit (Eberhard 1954, Coman & Brunner 1972, Gibb et al. 1978, Corbett 1979, MacDonald 1981). The rest of their diet comprises small quantities of other items such as birds, reptiles, frogs, carrion, grasshoppers, beetles and other insects. Feral cats are solitary and probably territorial. Elsewhere the litter usually comprises three to five, produced sometimes twice a year (Jones 1983).

Brumby or feral horse (Equus caballus)

The Wild Horse *Equus ferus* was an inhabitant of the steppes, forests and tundra of Europe and northern and central Asia. It was probably first domesticated in the well-watered Eurasian grasslands of the Ukraine and Caucasus about 6,000 BP (Epstein 1971, Protsch & Berger 1973, Clutton-Brock 1981). Fairly soon it was taken to southwest Asia and the Middle East, but only much later into Europe. There, domesticated horses may have been crossed with local wild *E. ferus* of the forests in the early days. Much later they received an admixture of the domesticated strains from the Middle East ('Arab blood'), which had undergone further breeding for hotter, often drier conditions, and speed. Thus the domestic horses brought to Australia by British settlers were basically derived from ancestors from the steppes but included some partly adapted to hot and mediterranean conditions and others to temperate, more forested habitats. The first of such horses were imported into New South Wales in the early 1800s and some soon became feral (McKnight 1976). By the 1880s there was evidence of their presence in the northern coastal region where now they are not uncommon (McKnight 1976, Letts et al. 1979). They occur at around 6 per 10 km^2 in a typical area of the monsoonal lowlands at Kapalga (Table 40), although they may be at slightly higher densities here and there elsewhere in the region (Graham et al. 1982). All these brumbies at Kapalga probably originated from 300 horses with which the location was stocked by 1920 (Warburton 1934). Brumbies are normally seen in mobs of five to ten mares with one stallion. Observations at Kapalga suggest they range quite widely to feed during the day, and further still during the dry season.

In fact the earliest occurrence of feral horses in the monsoonal lowlands resulted from two unsuccessful attempts at settlement on the Cobourg Peninsula, 1826-1849 (Calaby & Keith 1974). Timor Ponies were imported on both occasions and released on departure. Their descendants now number about 400 almost entirely confined to the 120 km long peninsula (Letts et al. 1979). Only a few occur 30-40 km beyond its narrow neck (5 km across) where they interbreed with

Table 40. Approximate densities of feral ungulates and Agile Wallabies at Kapalga NT in dry season 1981 (Ridpath et al. 1983 and unpubl. data).

Species	Number per 10 km^2		
	Sedge-lands	Fringing forest bordering sedge-lands, water-holes, creeks and rivers, and monsoon forest	Open forest and woodland
Feral			
Brumby*	<1	6	<1
Pig**	<1	12-45	<1
Cattle***	<1	3	<1
Water buffalo	148	339	36
Total	148	393	36
Native			
Agile Wallaby	<1	18	4

* Includes small number of donkeys and mules.
** Two different counting techniques were used.
*** Includes *Bos taurus* and *B. indicus*.

other feral horses. The Timor Pony has a stature of only c. 1.1 m and its morphology is reminiscent of the mongolian form of the domesticated horse (J.A.Calaby, pers. comm.). This and its presence on Timor before Portuguese colonization might both be explained by ancient Chinese, Indian and Javanese trade links (Ormeling 1957). This pony in tropical mountainous Timor is reputed to be agile, strong, hardy, resistant to high temperatures and able to survive exclusively on native vegetation (Valdes 1926, Silver 1976). At first sight this makes its failure to spread in the monsoonal lowlands of nearby northern Australia surprising, and contrasts with the greater success there of feral horses without a tropical history.

As a rule members of the horse family are able to utilize fairly rank, coarse grass, poor in protein but rich in cellulose. They manage this by ingesting very large quantities daily and also by virtue of the efficiency of the cellulytic processes in their non-ruminant caecum (Gwynne & Bell 1968, Bell 1971). However the nutritional content of most grasses in the monsoonal lowlands is remarkably low in protein and in usable carbohydrates too, even in the growing season (Norman 1963, 1966, Minson 1981, Andrew et al. 1985). Furthermore, quality and quantity of both decline sharply in the dry season. Thus their striking ease of mobility puts brumbies at a great advantage in seeking out and using the best of what fodder remains at this extremely lean time of year. Perhaps that is why the much smaller Timor Pony, presumably with a correspondingly lessened locomotory ability to roam, is handicapped in this wide flat environment – agile and hardy though it is among tropical hills and mountains.

Donkey (Equus asinus)

The donkey was first domesticated from the hill-dwelling African Wild Ass *Equus africanus*, which still survives today. This may have taken place in Egypt or the Sudan, at least 5,500 BP, although the original domesticated form probably was crossed later with the larger wild Somali race, also still extant (Epstein 1971, Clutton-Brock 1981). Donkeys were introduced into Australia in 1866 from Europe (Gooding 1983). They were used in northern Australia for haulage, pack animals, commercial breeding and as a substitute for horses because of their immunity to certain poisonous shrubs which kill horses (McKnight 1976, Letts et al. 1979). They became feral around the turn of the century. They are uncommon on the monsoonal lowlands, although very abundant in semi-arid hilly areas further inland and to the west (McKnight 1976, McCool et al. 1981). Their density is very low at Kapalga. Their distribution is in keeping with the fact that their ancestors inhabited semi-arid hilly country in North Africa. No study of their diet has been made but elsewhere they are known to graze and browse less selectively than brumbies. Nonetheless, as in the case of Timor Ponies, the donkey's small size and correspondingly reduced capacity to roam far in search of adequate food in times of scarcity may explain its lack of success. They mate with brumbies not uncommonly where their ranges overlap, giving rise to feral mules.

Pig (Sus scrofa)

Pigs were first domesticated 9,000 BP in southeast Europe and/or southwest Asia, and were introduced gradually into central and northern Europe in Neolithic times (Epstein 1971, Protsch & Berger 1973, Clutton-Brock 1981). The wild ancestral species extends as a number of races from northern Africa and western Europe throughout the Palaearctic region to Asia, and the Oriental region to India, Malaysia and Indonesia. After the domesticated form arrived in new areas in early times it appears to have been crossed often with these local wild races. This, together with selective breeding, has resulted in the occurrence of a variety of breeds of domesticated pigs in Europe, Asia and Africa. It was the domesticated form of northern Europe that the first British settlers brought with them to southern Australia, and it had already gone feral by the early 1800s (Tisdell 1982). The first domestic pigs to go feral in the monsoonal lowlands became established on the Cobourg Peninsula between 1829 and 1849. Descendants were still there in 1866 but are not common today (Howard & Edmunds 1969, Calaby & Keith 1974). They originated from a mixture of domestic pigs imported from tropical Timor and temperate southern Australia. Another significant release of domestic pigs, possibly of Asian origin, occurred in 1866 near the mouth of the Adelaide River (Letts et al. 1979). In summary it can be seen that although the domestic pig originated in southwestern Asia the wild species is very cosmopolitan, ranging from northern Europe to the tropics. Furthermore the feral population of the monsoonal lowlands probably comprises a mixture of ancestors from later local breeding in Europe, Asia and Timor.

By the 1920s feral pigs had been reported as common from Oenpelli and the Kapalga area in the east to the Finniss river in the west (see Fig. 1) (Warburton 1934, Kelsey 1975, Letts et al. 1979). They are restricted to fairly thick vegetation in damp places with water nearby, especially the swampy and forested margins of the sedgelands, creeks, rivers and waterholes, and also monsoon forest patches. They reach densities in such places of 10-45 per 10 km^2 (Table 40). Pigs are susceptible to extremes of temperature (Pond & Houpt 1978). They avoid heat stress by spending a part of the hot tropical day resting in the shade of these habitats, and/or in wallows. A good deal of feeding is nocturnal, when they may also range out onto the edges of the sedgelands.

Feral pigs are basically vegetarian but supplement their diet with some invertebrates and vertebrates, including carrion. Their main food items in the monsoonal lowlands are grasses, forbs, and the shoots and nuts of *Pandanus* palms (R. H. Barrett, unpubl. data). All these are important sources of protein, the availability of which has been argued to limit populations in temperate Australia (Giles 1980). When these foods become scarce towards the end of the dry season adult pigs can tide themselves over by turning to roots, for instance of spike rushes *Eleocharis* spp. and of lilies. These contain adequate carbohydrates but so little protein that young pigs may have difficulty in surviving during this period. There are some indications that where there is a high density of buffaloes the density of pigs is often low, and vice versa. If confirmed, this could well be related in some way to the effect of buffaloes on the food supply of pigs. For instance the known compaction of the soil by the trampling of buffaloes and its subsequent hardening (Ruskin 1981), as reported world-wide for ungulates (Gifford & Hawkins 1978), might make it difficult for pigs to obtain roots during critical periods of food shortage in the dry season.

Pigs in the monsoonal lowlands appear to be able to breed twice a year and in most months (R. H. Barrett, unpubl. data). However the survival of the young seems to fluctuate according to their season of birth. Litters of young piglets have mainly been recorded in the dry and early wet season during regular transects spanning the year at Kapalga (Ridpath, unpubl. data).

Likewise piglets only occur in the diet of the dingo during the dry period (L.K.Corbett, unpubl. data). Pigs' litters vary in size from four to six and can reach eight or more in good conditions. Given the variability of the onset, duration and intensity of plant growth in the wet-dry tropics the environmental conditions at conception are often a poor guide to those that mother and offspring will face at birth nearly four months later. Thus variations in recruitment are likely to occur and are capable of substantial effects on population levels. Lack of protein and predation by dingoes are probably responsible for frequent high mortality of juveniles in their first year of life.

Pigs are often seen singly and otherwise in groups which rarely exceed 20 individuals. In times of severe drought concentrations of over 100 occur round remaining waterholes. Normally they are sedentary, the longest distance travelled by a trapped animal being 15 km at Kapalga.

Cattle: European (British) breeds (Bos taurus) and Brahmans (B. indicus)

The wild Uru *Bos primigenius*, now extinct, originally ranged from Europe to North Africa and most of Asia. Cattle were first domesticated from the European and/ or Asiatic race(s) of wild Urus in southeast Europe and/or southwest Asia around 9,000 BP (Epstein 1971, Protsch & Berger 1973, Clutton-Brock 1981). The original domesticated longhorned form gave rise both to the predominantly shorthorned *Bos taurus* and the humped *B. indicus* cattle of today. On the one hand *B. taurus* from SE Europe or SW Asia spread across Europe around 5,000-6,000 BP, probably being crossed in early times with local temperate, wild long-horned Urus. On the other hand the Brahman (or Zebu) *B. indicus* form with its humped back and baggy dewlap from the throat, appears to have been developed in a particularly hot dry region of SW Asia (Epstein 1971) or possibly of India (Clutton-Brock 1981). There it acquired marked physiological resistance to high temperatures, drought and poor and fluctuating fodder (Epstein 1971). On the first view it was introduced into India around 4,000-5,000 BP, and there is general agreement that it was taken to Africa a little later.

When Australia was settled the first imported cattle were almost entirely *B. indicus*, mainly Afrikander (Phillip 1789), until about 1820 (J.A.Calaby, pers. comm.). Thereafter most imports were of British breeds, *B. taurus*. Later still, Brahman cattle and other breeds containing *B. indicus* blood were introduced increasingly into northern herds because of their adaptations to tropical conditions. The Australian beef industry increased over 150 years to the point where the cattle population in 1976 stood at 19 million distri-buted throughout the continent (McKnight 1976). Understandably, beasts sometimes strayed and became feral, particularly in the poorly fenced or open range conditions of northern Australia. There are estimated to be roughly 75,000 feral cattle from British breeds and Brahmans in the Top End of the Northern Territory (Letts et al. 1979). They originate from cattle such as the British breeds introduced by F.A.Smith to un-fenced land at Kapalga between 1913 and 1920 (Warburton 1934). British breeds introduced onto the Cobourg Peninsula between 1838 and 1849 went feral there but did poorly and died out, apparently within fifty years of the abandonment of the settlement (Calaby & Keith 1974). Most feral cattle occur in the western part of the monsoonal lowlands, in the vicinity of the Daly River, and in drier habitats further inland, originating from the substantial managed herds of domestic cattle in that region (Aldrick et al. 1978, Graham et al. 1982). Nonetheless they occur at around 3 per 10 km^2 in the monsoonal lowlands at Kapalga (Table 40) and at similar densities in most of the rest of these lowlands (Graham et al. 1982). They are largely confined to damp places with water nearby, especially the forested margins to the sedgelands, creeks, rivers and waterholes, and also monsoon forest patches. This generally low density is no doubt due to the low nutritional value of northern forage already mentioned when discussing brumbies. Cattle rely on building up a reserve of fat during the annual wet season on which to survive in the dry season when food is short. The severity of this annual stress is reflected in a loss of 15-19% of body weight by domestic cattle each dry season (Story 1969). This limitation is particularly marked in the monsoonal lowlands. Under these conditions the lower metabolic rate of Brahman cattle allows them to tolerate greater heat stress and a lower plane of nutrition than temperate *B. taurus* genotypes (Frisch 1981, Frisch & Vercoe 1982). Even though they pay for this environmental advantage at the cost of lower fertility and growth rates, feral Brahmans have spread with disproportionate success in the monsoonal lowlands, when their later arrival is taken into consideration.

Feral cattle are usually seen singly or in parties of up to ten, often a mixture of Brahmans and British breeds. They are normally sedentary.

Bali cattle (Bos javanicus)

Bali cattle were domesticated from the wild Banteng of southeast Asia, probably in prehistoric times (Calaby 1975). The Banteng occurs in savanna with dry deciduous and mixed evergreen (monsoon) forest from Burma east and south as far as the Indonesian islands of the Sunda Shelf and Bali (Wharton 1968, Hoogerwerf 1970, Lekagul & McNeely 1977). Its original habitat must have been much expanded first by burning in the Palaeolithic period and later by shifting cultivation (Wharton 1968). Although predominantly a grazer this species does a limited amount of browsing too (Hoogerwerf 1970, Calaby & Keith 1974, Sinclair 1977).

Domestic Bali cattle were introduced from Bali onto the Cobourg Peninsula between 1838 and 1849 and went feral after the settlement at Port Essington was abandoned (Calaby & Keith 1974, Calaby 1975, Letts et al. 1979). After 130 years the population is around 1000 beasts which remain restricted to the peninsula. No doubt this is because there is no other area which they can reach which provides a combination of monsoon forest with well-watered plains that sufficiently resembles the habitat of their wild ancestors. Bali cattle rely more on cooling by sweating and less on respiratory cooling than do Brahmans and they are also more easily stressed by activity under very hot conditions (Moran 1973). This is in spite of their short sleek coat which reflects the sun's rays but allows the dissipation of internal heat from its skin. The differences may be explained by Bali cattle's reliance, when conditions are hottest, on the greater amount of shade available in their monsoon forest habitat. Bali cattle and Brahmans appear to be equally efficient in their utilization of tropical forage of low nutritive value (Kirby 1979). However Bali cattle do not lose weight and have higher fertility than Brahmans. Thus in certain respects they show superior tropical adaptations.

Feral Bali cattle are smaller than British breeds or Brahmans and are not unlike deer in appearance. Bulls are dark brown and cows much lighter, both having striking white patches on the buttocks. They are very shy. Solitary bulls are often seen; otherwise parties of up to sixty occur, mixed or of the same sex (Calaby & Keith 1974).

Asian Water Buffalo (Bubalus bubalis) (Photograph 12)

The buffalo was first domesticated in India, or possibly Indo-China, from the wild Asian Water Buffalo, or Arnee, around 5,000 BP (Epstein 1971, Clutton-Brock 1981). The wild species inhabits wetlands and riverine areas and is restricted now to remote parts of central and northeastern India, and to the Celebes and the Philippines, in both of which relicts of dwarf races still occur (Mason 1974a). Wild Water Buffaloes originally were widely distributed over the Indian sub-continent, southeast Asia and even Mesopotamia and Iran (Epstein 1971, Mason 1974a). Domesticated Water Buffaloes probably were brought to southeast Asia including Indonesia around 4,000 BP with the proto-Malays (Loofs 1974). Their long association with people in Indonesia and Bali is also attested by the complexity, intimacy and apparent antiquity of the cultural bonds with buffaloes (Cockrill 1974).

Domesticated Water Buffaloes were first brought to northern Australia from Timor and, in a few cases, other Indonesian islands. Introductions, probably totalling no more than 100, took place at Melville Island in 1826, the Cobourg Peninsula in 1828, 1838 and 1843 and Escape Cliffs near the mouth of the Adelaide River between 1864 and 1866 (Letts et al. 1979, Tulloch 1979a). All these settlements were abandoned within a few years and their buffaloes turned loose, joining those which had already escaped. Over the following 150 years the feral descendants from these few small populations have occupied most of the monsoonal lowlands east to the border with Queensland on the Gulf of Carpentaria and west to the Ord River on the Joseph Bonaparte Gulf in the Kimberley region of Western Australia (Tulloch 1983). At first they spread fairly slowly through the extensive areas of wetlands. In 1883-1885 buffaloes were common near the Liverpool River in Arnhem Land and were present in large numbers in the East Alligator area, but 50 km further west the population at Kapalga still seems to have been low (Lindsay 1884, Australian Archives a). By 1913 they had become plentiful enough at Kapalga to be harvested by F.A.Smith, the first European settler there. Buffaloes have also penetrated up to 350 km inland into generally drier country, wherever moister locations occur. Lone bulls occasionally wander up to 700 km beyond these limits (McKnight 1971). The buffalo's success can be gauged by the fact that the population stood at over 280,000 in 1981 (Graham et al. 1982), in spite of the removal of over 700,000 since shooting for profit began in 1886 (Fig. 62). In fact the population once had probably been a good deal higher than that but there had been a major increase in the rate of removal of buffaloes over most of the monsoonal lowlands since 1963 (Fig. 62). By far the greatest part of the population occurs in the lowlands within 100 km of the coast.

At Kapalga in 1981 they were at a density of c. 340 per 10 km^2 in damp places with water nearby, especially the swampy and forested margins of the sedgelands, creeks, rivers and waterholes, and also monsoon forest patches (Table 40). On the open sedgelands density was slightly under half that figure, while in the drier open forest and woodlands it was only about a ninth of that of the most favoured habitats. With one exception (below) the densities at Kapalga in 1981 were much higher than elsewhere because its population had been untouched for eight years; whereas most of the surrounding region had been regularly harvested during the same period.

Photograph 12. Feral buffaloes in dry eroding channel during severe drought, at Murrabibbi Springs near Kapalga, October 1972 (Photograph: M.G. Ridpath).

Photograph 13. Buffalo bull in a wallow at Kapalga, January 1982 (Photograph: D.G. Tulloch).

Photograph 14. Soil rooted by feral pigs on the edge of retreating water level at Humpty Doo, July 1971 (Photograph: M.G.Ridpath).

Photograph 15. Plot from which feral buffaloes and pigs had been experimentally excluded for two years at Kapalga, January 1980. The effects of grazing and trampling by buffaloes outside the enclosure can be seen, by comparison. The area of floating aquatic plants in the background had just started to grow, due to the commencing wet season (Photograph: M.G. Ridpath).

Photograph 16. Feral buffalo 'swim channel', in foreground, and narrow trails through sedgelands at Kapalga, March 1982. Many wallows can be seen between the sedgelands and wooded area (Photograph: J.A.Taylor).

Photograph 17. Feral buffaloes on heavily compacted, much pugged and largely bare sedgelands in the dry season 1980, at Kapalga (Photograph: D.G.Tulloch).

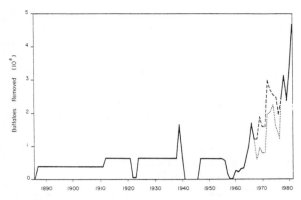

Figure 62. The annual totals of buffaloes removed from the monsoonal lowlands in the Northern Territory 1886-1980 for hides, human consumption, pet meat, domestication, live export and eradication (calculated from Christian & Stewart 1953, McKnight 1971, 1976, Tulloch 1974, Letts 1977, Rideout 1977, Letts et al. 1979, Ford 1982, Australian Archives b, Conservation Commission NT records). Only means could be obtained for 1886-1910, 1911-1937, 1946-1954. Figures used ran from July to the following June and have been plotted midway, by when most had already been taken.

Annual totals 1967-1975 are shown by (a) dotted line for all purposes except pet meat, and (b) broken line for all purposes, including numbers for pet meat estimated as below. Detailed records of the numbers taken for pet meat (as distinct from those for other purposes – mainly human consumption and live export) were not available for 1967-1975 and have been estimated as follows: 1967-1970, on the basis of departmental information (P. Sullivan, pers. comm.), the conservative assumption was made that as many were taken for pet meat as were known to have been taken for other purposes; 1971-1972 and 1974, estimated from allocations on licenses issued by Dept of NT, and calculated by proportion when copies of licenses were only available for part of year; 1975, calculated from number taken for other purposes on the assumption that the proportion for pet meat was as estimated for 1974; 1973 estimate as quoted by Letts et al. (1979).

The highest concentration of buffaloes on the monsoonal lowlands in 1981 occurred in the area of the Finniss and Reynolds Rivers, where the overall density was over twice that at Kapalga (Graham et al. 1982). Like Kapalga, its population had been relatively untouched for many years. The difference in the density of buffaloes between the two areas could be due to natural differences in their carrying capacities. On the other hand it is also possible that the difference has to do with the fact that the Finniss area was not fully colonized by buffaloes till the early 1960s (Hill & Webb 1982), sixty to eighty years later than Kapalga. Evidence that buffaloes can cause changes in vegetation is given later, and this may alter carrying capacity in the long term. The history of the populations of feral

buffaloes on the Cobourg Peninsula and Melville Island, where they have been present longest, provides examples of a large increase followed by a decline to a permanently lower density. This is in spite of the fact that shooting pressures in both areas have been much reduced in recent times. Information on numbers and harvesting shows they had increased greatly during the first thirty to forty years after they went feral on Cobourg Peninsula (Calaby 1974). However the density in 1965-1966, when buffaloes were outnumbered by Bali cattle, seems well below that indicated for the period seventy to ninety years previously (Calaby 1974, Calaby & Keith 1974). On Melville Island high densities built up from the handful released in 1829 and supported a thriving local harvesting industry by the 1890s, which lasted well into the 1900s (Hill 1951, McKnight 1971, Pye, n.d.). A total of 6,000 were shot in a single year. Harvesting collapsed some time later on and by 1972-1974 the density of feral buffaloes on the island was low (Ridpath, unpubl. data).

The Water Buffalo is well adapted to a semi-aquatic environment, filling a somewhat similar ecological niche in Australia and Asia to that of the Lechwe *Kobus lechwe*, a semi-aquatic antelope in tropical Africa (as described by Rees 1978a, b, and Howard 1985). The buffalo can move easily in muddy situations because of its short legs with their very flexible lower joints and large spreading hooves. Its compact balloon-like body is good for part sliding, part floating in slushy places. It swims and dives efficiently and can graze under water. All these adaptations contrast strongly with thinner, long-legged cattle which bog easily and are not adept in water. In keeping with the high humidity of the wetland environment, which reduces the efficacy of evaporative cooling by sweating, buffaloes rely less on sweating than cattle or Bali cattle, having only one-third to one-sixth the density of the sweat glands of cattle (Hafez et al. 1955, Ruskin 1981). Instead buffaloes have two other adaptations. Firstly, they breathe faster at higher temperatures (Mason 1974b, Ruskin 1981). Hence the increased flow of air and associated cooling in the respiratory tract partly compensates for the reduction in the rate of evaporation due to high humidity (Schmidt-Nielsen 1979). This partly explains why buffaloes have to drink as much as 20 litres a day in the wild, in spite of sweating little (Williams & Ridpath 1982, Williams & Green 1982). The second and equally important adaptation of buffaloes to this environment is their avoidance of the midday heat of the monsoonal summer. This is achieved by wallowing (Photograph 13), where they lose heat by convection, or by being in the shade where they lose much heat by radiation (Camoens 1976). Their almost hairless, highly vascularized skin maximizes the efficiency of cooling by radiation, which can account for 90% of the heat lost over 24 hours. No doubt these attributes of the skin are equally important for loss of heat by convection since wallowing is as efficient in keeping cool as being in the shade (Tulloch & Litchfield 1981). Wallowing has the added advantage of providing a protective coating of mud against the extremely high densities of insects which bite buffaloes in tropical wetlands, where biting rates of 34,000 per hour per animal have been recorded (Dyce & Standfast 1979). Unlike sweating, the buffalo's cooling adaptations are well suited to its humid habitat. They have one drawback, however. The buffalo cannot quickly dissipate heat gained. Thus it suffers serious heat stress if exposed to the sun in the hottest part of the day in the hot humid summer especially if obliged to be active (Mason 1974b, Ford 1982). By contrast, in the cooler dry season (winter) buffaloes spend the night on the wooded ridges, where it is warmest, and descend during the morning to graze (Ridpath 1981). Those which graze on the open sedgelands spend the warmest, middle part of the day there. Although the tropical winter seems mild by the standards of a temperate winter this routine is the opposite of that for summer, when their movements are geared to the coolest place in which to sleep and the coolest period of the day in which to expose themselves to the sun (i.e. early).

Buffaloes graze and browse in a rather unselective manner. Their main forage includes grasses, sedges, other aquatic plants and fibrous material such as leaves of the palm *Pandanus spiralis* (Tulloch 1974, Ruskin 1981). Highly favoured items eventually may be suppressed, for instance, the grass *Hymenachne acutigluma*, the reed *Phragmites karka* and the Red Water Lily *Nelumbo nucifera*. Buffaloes can obtain green fodder readily throughout most of their habitat in the wet season, although at its height extensive flooding often prevents them reaching all the sedgelands. As the dry season progresses at Kapalga green forage becomes increasingly restricted to the forested, swampy margins of the sedgelands and other damp habitats where most of the population then feeds. At the very driest time of year, usually between September and November, the almost sole sources of forage are often the few remaining deep swamps containing a spike rush *Eleocharis sphacelata* and other sedges and water plants. Because of their adaptations for grazing in an aquatic environment buffaloes are able to concentrate and survive there on around 6 kg (dry weight) each of this food per day (Williams & Ridpath 1982). Thus they normally do not lose much weight when food is shortest whereas cattle lose 15-19% (Story 1969, Vercoe & Frisch 1977, Letts et al. 1979, Ruskin 1981, Ford 1981, 1982). However, if the dry season is unusually prolonged even this resource becomes exhausted eventually and buffaloes starve, bog and some may die. Their mortality can be as high

as c. 30% of the population, many under one year of age or very old (Ridpath, unpubl. data).

Buffalo bulls appear to reach an adult level of fertility at around four to five years of age while cows normally first conceive at two and a half to three years (Tulloch & Grassia 1981). The pregnancy rate of cows of three and a half years and older is 75-90%, and gestation lasts c. 10.5 months (Ford 1977, Ruskin 1981). The calves are born between October and July; i.e. throughout the wet season and during the beginning of the dry season (Tulloch & Grassia 1981). The peak of births is in the middle of the wet season during the months of January to March (derived from conceptions 10.5 months earlier; see Chapter 10, Fig. 58). Thus the period when calves are born, during and after which they and their mothers require the best resources, extends over ten months and its peak over three. This pattern of calving accords broadly with the somewhat variable availability of resources which themselves are dependent on a wet season, the commencement and duration of which vary between years. Apart from mortality due to starvation, as above, calves up to the age of about six months are also taken by dingoes, not uncommonly at the moment of birth.

Cows and their calves live in groups with a well-developed dominance hierarchy; while bulls over three years old live in looser groups and also singly (Tulloch 1978). Groups vary in size between 30 and 250 animals. Parties of up to ten calves of four to ten months from different mothers are sometimes left in the forested margins with one adult cow (Tulloch 1979b). While the real mothers graze out on the open sedgelands, the maternal substitute temporarily guards them, especially against dingoes. In the wet season one bull takes temporary possession of a group of cows to mate. It ejects any young bulls present which have reached the age of three years and fights any intruding bulls, sometimes to the death. The bull's sharp but massive horns are well adapted to charging head-on and to close-up lunging in such combats. The cow has longer narrower horns better suited to stationary protection of her sheltering calf, with wide sweeps against predators (which in Asia included tigers and other large cats). Buffaloes are normally very sedentary with a home range over the year of 4-12 km in diameter (Tulloch 1969, 1978, Ridpath 1981). However bulls, especially younger ones, sometimes wander further. There is a tendency for bulls to feed furthest out on the sedgelands and also up on the wooded ridges, in both of which situations the forage is only temporarily abundant and probably least nutritious (Tulloch 1970, 1978, Williams & Ridpath 1982). The group's movements within its home range are regular and stereotyped over each season, resulting in the formation of long, bare, heavily compacted pads. This is not surprising in view of an average weight of 400-500 kg, with very large bulls perhaps even reaching 1000 kg (Ford 1981, 1982).

Buffaloes are known to live up to twenty years in captivity (Tulloch 1979a). Published data suggest most feral buffaloes do not live longer than five to six years with only a few surviving to nine to ten years (Tulloch & Grassia 1981). This is surprising in view of the fact that many African buffaloes *Syncercus caffer*, of about the same size as the Water Buffalo, reach ten to twelve years and a few survive to 16-17 years in the wild (Sinclair 1977). On the face of it feral Water Buffaloes in northern Australia survive no better than a much smaller ungulate, the introduced goat-like Himalayan Thar *Hemitragus jemlahicus* in New Zealand (Caughley 1966). It could be argued that, when assessed, the comparatively short apparent lifespan of buffaloes had resulted from previous shooting and/or earlier mortality from starvation in 1961 and 1965 (Tulloch 1970). On the other hand the absence of a reliably calibrated aging technique beyond four to five years (Tulloch 1969) suggests more research on demography is called for.

Goat (Capra hircus)

Goats were first domesticated from the Wild Goat in southwest Asia around 10,000 BP, possibly soon followed by some hybridization with the larger wild Markhor *C. falconeri* of the mountains (Epstein 1971, Protsch & Berger 1973, Clutton-Brock 1981). The wild ancestral species occurred in mountainous and hilly areas and wooded uplands from the Balkans to Pakistan, north to the Caucasus and south to Arabia. A few still exist in parts of this region where they live as browsers. The domesticated form occurred in Greece by 9,000 BP. Goats were brought to Australia from Britain by the very first settlers and were soon taken to all parts of the continent (McKnight 1976, Mahood 1983). They now occur in the feral state widely in arid and semi-arid areas. Their only occurrence in the monsoonal lowlands was in small numbers in the Point Stuart area but these have now disappeared (Letts et al. 1979). However they do occur in large numbers on two small offshore islands in the Gulf of Carpentaria. Their failure to colonize the monsoonal lowlands successfully can be attributed to the fact that their ancestors were adapted to life in hot dry savanna and woodlands in hilly areas with plenty of shrubby vegetation on which to survive. Only on small islands, where maritime conditions favour shrubs, bushes and little trees, does their ecological niche exist in the immediate vicinity of the monsoonal lowlands.

Success and failure in a new environment

We can now summarize broadly by saying that buffaloes and pigs have done well after going feral in this new environment; feral Bali cattle have managed reasonably well in one particular but restricted habitat; while brumbies, feral cattle and cats have done poorly, donkeys worse and goats have failed (Table 41). Thus the three most successful feral mammals are those originating in the tropics. The remainder, which originated in the hot semi-arid regions of SW Asia and North Africa and the Eurasian steppe, have been much less successful in the monsoonal lowlands (Table 41). This, of course, is in contrast with their performance in the hot, semi-arid regions further inland in northern Australia where feral cattle, brumbies, cats and donkeys all do well.

Success or failure in the monsoonal lowlands depends on the invader's capacity to respond adequately to the special features of this environment. At one extreme, the absence of enough of the right kind of shrubs and bushes is probably responsible for the failure of feral goats to become established. The seven other ungulates have a range of physiological adaptations which allows them, with varying degrees of success, to survive on the forage of generally low nutritional value characteristic of the tropics. Two of the most severe stresses which have to be met are the high temperatures with high humidity, especially in the wet season; and the shortage of food in the dry season accompanied by deterioration in its quality, already low anyway. The three tropical ungulates rely most on staying in the shade or wallowing in the middle of the day to avoid these high temperatures (Table 41). The buffalo's reduced capacity to sweat, compared with that of Bali cattle, may be associated with its use of wallows and with the exceptionally high humidity and windlessness of the damp surroundings of swamps in which it shelters.

The three tropical ungulates overcome the severe shortage of forage in the dry season by their capacity to exploit various alternative sources of food (Table 41), sometimes of lower nutritional value. Thus buffaloes do not need to build up reserves of fat to tide them over the harsh times, as cattle do (Williams & Dudzinski 1982). An overall measure of the relative success of feral buffaloes' tropical adaptations can be gleaned from their density, which is now over 100 times greater than that of feral cattle, after about seventy years coexistence in the monsoonal lowlands at Kapalga. This strong preponderance of feral buffaloes over feral cattle is typical of the situation over the monsoonal lowlands as a whole. However the density of managed herds of domestic cattle equals that of feral buffaloes in some of the western monsoonal lowlands, where both exist at high densities (Graham et al. 1982) and the buffalo has only been present for twenty to thirty years. Nonetheless, even there a tendency exists for

Table 41. Adaptive responses of feral mammals to two major environmental stresses in the monsoonal lowlands.

Feral form	Origin of wild ancestor(s)	Decreasing order of abundance	Adaptive responses to: High temperatures, especially in wet season	Adaptive responses to: Shortage of food in dry season (including further reduction in its quality)	Comments
Water Buffalo	Wet-dry tropics	Very abundant	Seeks shade and/or wallows	Feeds in swamps	
Pig	Cosmopolitan including tropics	Locally abundant	Seeks shade and/or wallows; nocturnal	'Roots' for underground foods	
Bali cattle	Wet-dry tropics	Common in one area	Seeks shade and sweats more	Browses more in monsoon forest?	Very restricted habitat and distribution
Brumby	Eurasian steppe	Sparse	?	Ranges widely to find enough suitable food	
Cattle (*B. taurus* and *B. indicus*)	Hot, dry, SW Asia mixed with genotypes of temperate Europe	Sparse	Metabolic rate lowered in *B. indicus* genotypes	Uses fat reserves accumulated in wet season	
Cat	Hot, dry, N Africa mixed with genotypes of temperate Europe	Sparse	(Nocturnal)	?	Nocturnal even in temperate regions
Donkey	Hot, dry, hilly north Africa	Rare	?	Poorly adapted	
Goat	Hot, dry, hilly SW Asia	Absent – failed to establish	?	Poorly adapted	Only on two small islands now

the domestic cattle to do best in those parts where feral buffaloes are not at maximum density at present (Graham et al. 1982). This could mean that feral buffaloes are even superior in competition with managed domestic cattle too.

Another important environmental feature which faces these invaders of the monsoonal lowlands is the variable onset, intensity and duration of the annual wet season and associated plant growth. Such variability in the wet season can also affect conditions in the following dry season. Thus the best period in which to raise young may vary appreciably each year, in whichever season a species breeds. Of the eight feral mammals which have established themselves six are large and long-lived, take two to three years to become sexually mature, and have a gestation period of at least nine months. They can be thought of as conservative in the sense that they spread their life's reproductive output over many years, one offspring at a time, and thereby with least strain on the mother. Thus they have many chances to breed, albeit at the price of low productivity on each occasion. This may compensate for the fact, just mentioned, that in the monsoonal lowlands the environmental conditions when they mate are poor predictors of those that mother and young will face many months later. In other words, the timing of their

breeding, though bound to be ill-suited to the prevailing conditions in some years, will be well suited to them in others. Nonetheless, given their long life they should be able to breed successfully often enough to leave behind sufficient young to replace their parents when they die.

By contrast cats and pigs live shorter lives than these large ungulates but become sexually mature within their first year, have gestation periods of only nine to sixteen weeks and litters of three to eight young, and may even breed twice a year. Their bigger litters and their capacity to breed more often compensate for a shorter lifetime over which to breed with its fewer chances of breeding successfully in the prevailing conditions of variable wet or dry seasons. We can think of these two smaller animals as risking more of their life's reproductive effort each time they breed than the larger conservative ungulates. However, whenever prevailing conditions do allow their success, the productivity of these small-sized risk-takers is greater than that of the larger conservative forms. As buffaloes and pigs are the two most successful colonizers of the monsoonal lowlands it appears both strategies can work well there. This leaves unanswered the question of why these differences exist.

A comparison between the breeding regimes of the

two largest feral ungulates and the Agile Wallaby *Macropus agilis* has been made earlier (Williams & Newsome, Chapter 10). It showed that the largest common native mammalian herbivore combines the conservatism of the large ungulates (single offspring) with a capacity to respond quickly and flexibly to this partly unpredictable environment. This flexibility is achieved as a result of a gestation period of only four weeks and the other unique features which, compared with placentals, give some marsupial mothers, especially in the kangaroo family, much more control over their reproductive mechanisms and at a lower daily cost (Parker 1977, Morton et al. 1982). Ungulates, as large placentals, lack the ability to fine-tune their reproductive output to meet prevailing conditions. Nonetheless this has not prevented feral buffaloes and pigs from reaching much higher densities than the Agile Wallaby in the damp places it favours most, also preferred by the other ungulates (Table 40). These places are also the main habitat of grass-eating termites, the mounds of which are often common there. We do not know whether the feral ungulates compete with wallabies and/or termites in this habitat so vital to them all, but it is at least a possibility. In fact Agile Wallabies are still relatively successful in the drier, higher open forest and woodland at Kapalga (Table 40). Their density in this extensive zone is just under a quarter of that in the favoured damp places, whereas the density of buffaloes is only a ninth of what it is in the damp places. Nonetheless, the actual density of buffaloes in the open forest and woodland still exceeds that of wallabies by nine times, and they may still compete there.

THE EFFECTS OF FERAL MAMMALS ON THEIR ENVIRONMENT

Measuring and assessing the effects of feral mammals on their environment is not a simple matter. Some effects, such as the wallow of a buffalo, are immediately obvious. However, many other effects, such as the complete absence of a once common plant or an enforced change in an animal's breeding regime or the disappearance of a subterranean organism, all may easily fail to impress themselves on the notice of the observer on his early acquaintance with the ecosystem involved. The consequences arising from the prime effects may be even less clearly apparent to him. These and other complexities are best resolved by experimentation. It will be noted that the word 'damage', often used in discussion of environmental effects, is eschewed here. This is because the definition of damage depends on the particular purposes for which particular areas are managed or dedicated, and these vary. This does not imply that many effects described below are not regarded as highly undesirable by many, according to the particular purposes they have as objectives.

The effect of feral mammals on their environment in the monsoonal lowlands depends largely on their density.

Rare or sparse feral mammals of non-tropical origin

Goats are absent from the monsoonal lowlands but have caused severe denudation of vegetation on two small offshore islands (Letts et al. 1979). Donkeys are too uncommon on the monsoonal lowlands to have significant effects. Brumbies and feral cattle are sparse in the wettest areas closest to the sea. Whatever effects they may have there are masked by those of buffaloes (see below). The density of feral cattle in the monsoonal lowlands increases with increasing distance from the sea but little data exist on their environmental effects. However further inland still and westwards they are common in hilly areas of semi-arid regions where they cause erosion and affect vegetation, as is also the case with donkeys (Kabay et al. 1977, Aldrick et al. 1978, McCool et al. 1981). Further investigation of their environmental effects is needed, especially as regards fauna (see also Haynes, Chapter 5).

There are not sufficient data on which to assess the effects of cats on their environment in the monsoonal lowlands. The effects of their predation on the density of uncommon species of small mammals also needs further investigation.

Abundant or locally abundant feral mammals of tropical origin

On the Cobourg peninsula, Bali cattle graze and trample ground vegetation on the sandy plains with appreciable intensity, aggravate the erosion of sand dunes and destroy the burrows of rodents there (Calaby & Keith 1974, Letts et al. 1979). However, their browsing in the very restricted monsoon forest association may be of greater significance.

The most important effect pigs have on their environment occurs in the dry season in the monsoonal lowlands. Then they root in soft ground on the edge of the sedgelands (Photograph 14), their forested margins and similar damp places, including pockets of monsoon forest. However, such rooted patches, churned to a depth of 10-20 cm, rarely exceed 70 m in diameter. The consequences for the vegetation are unknown.

Feral buffaloes have by far the biggest impact on the monsoonal lowlands, in keeping with their high biomass there. Taking all habitats into account at Kapalga, the buffalo averages 5,200 kg per km^2 overall (Ridpath et al. 1983). In fact that biomass rises by over 2.5 times in their preferred habitat, where buffaloes

Table 42. Approximate biomass of feral ungulates and Agile Wallabies in fringing forest bordering sedgelands, waterholes, creeks and rivers, and monsoon forest patches at Kapalga, NT, in dry season 1981 (Ridpath et al. 1983 and unpubl. data).

Species	Biomass* kg/km²
Feral	
Brumby (300 kg)	190
Pig (70 kg)	80-360**
Cattle (290 kg)	100
Water Buffalo (400 kg)	13,570
Total	14,220
Native	
Agile Wallaby (20 kg)	36

* Based on the conservatively estimated individual live weights shown in brackets against each species.
** According to counting technique.

vastly outweigh other feral mammalian herbivores and the herbivorous native Agile Wallaby (Table 42). The overall biomass of feral buffalo at Kapalga stands at the lower end of the range of that of all wild ungulates taken together in equivalent situations in tropical Africa, and is higher than that in southern Asia (Ridpath et al. 1983). In Africa and Asia the vegetation has developed mutualistic adaptations to heavy grazing by indigenous herbivores over a long evolutionary span (Parsons 1980). Australia had more and some larger marsupial herbivores until the end of the Pleistocene than it has today, although we know little about their abundance and ecological niches (Archer 1981). The persistence of whatever such adaptations may have existed once in the vegetation of the monsoonal lowlands

is unknown, but their present existence seems dubious (Ridpath et al. 1983). Elsewhere, heavy grazing by introduced ungulates has been shown to cause substantial changes in the species composition and physiognomy of ground vegetation (Wraight 1964, Snayden 1981, Riney 1982). This also appears to have occurred in the wetlands of the monsoonal lowlands (Fox et al. 1977, Williams & Ridpath 1982). Some evidence has already been given of a major decline in the carrying capacity of the two areas longest colonized by buffaloes, which may even be irreversible there. Long-term changes induced in vegetation are a likely cause of such declines.

Buffaloes have affected vegetation, soils, hydrology, water quality and thereby the habitats of a variety of native fauna, as summarized in Figure 63 (from Fogarty 1982). Much of the evidence for their effects depends on correlations with their first arrival and/or density, but some is experimental (Photograph 15). The biggest changes seem to have occurred in the wetlands. Reeds such as *Phragmites karka*, Bullrushes *Typha* sp. and Red Water Lilies *Nelumbo nucifera* (see Chapter 4, Plate 4) appear to have been common in the wetlands between the East Alligator and Adelaide Rivers before buffaloes became abundant (Leichhardt 1847, Brackenbury 1896, Lewis 1922, Letts et al. 1979). These species virtually disappeared thereafter but began to reappear in recent times (e.g. Williams 1979) as the rate of removal of buffaloes increased (Fig. 62). All three species have remained abundant 250 km further east in the Arafura Swamp far into the Aboriginal lands of Arnhem Land (J.A.Taylor, pers. comm.), where buffaloes are still scarce – for reasons that have never been investigated. Likewise, thick floating mats of aquatic vegetation (see Chapter 4, Plate 6 and Photograph 3) were still present as late as

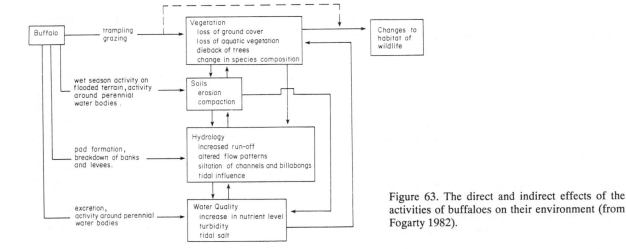

Figure 63. The direct and indirect effects of the activities of buffaloes on their environment (from Fogarty 1982).

1920 in the region between the East Alligator and Adelaide Rivers (Warburton 1934) and became very rare later, although they began to reappear in a few places more recently (Williams 1979). However about 200 km further west aerial photography in 1963-1964 shows similar mats still covered 30% of a lagoon between the Finniss and Reynolds Rivers, and this was typical of the lagoons throughout the region (Hill & Webb 1982) (as in the case in the Arafura Swamp too). Buffaloes first attained high densities in the region about that time. Fifteen years later further aerial photography showed the cover of floating mats had been reduced to 5% on that and the other lagoons of the region. The loss of mats was largely attributable to observed heavy grazing by buffaloes, which were then at maximum density in this region, with little or no removal. Such mats are composed of a diverse array of aquatic plants and were important as nesting places for Saltwater Crocodiles *Crocodylus porosus* and Magpie Geese (Thompson 1949, Webb et al. 1983, Webb, Chapter 8).

Some plants may not always disappear altogether under heavy grazing, but may be suppressed for an unknown period. This has been illustrated at Kapalga where the buffalo population was left untouched for nine years allowing it to reach maximum density. Then all the buffaloes were removed from half the area and reinvasion from the other half prevented by fencing. Within 12-18 months Red Water Lilies, which had not been recorded as emergent vegetation during the eight years since observations began, grew very thickly in some lagoons (see also Chapter 4, Plate 4); while the perennial semi-aquatic grass *Hymenachne acutigluma*, which had always been extremely sparse, suddenly grew luxuriant swards, in which Magpie Geese were recorded as nesting for the first time (D.G.Tulloch, pers. comm.). This grass has remarkably high protein levels in nature (Calder 1981) which no doubt explains buffaloes' preference for it. If ungrazed it also forms a mulch which causes soils to retain their moisture longer in the dry season than when it is absent. Apart from their effects on water lilies, buffaloes also can cause reductions in other floating and submerged macrophytes by grazing and by creating deeper channels along which they swim (Forbes 1980). 'Swim channels' (Photograph 16) are the continuation of well-defined trails in the woodlands, down which silt also passes from erosion caused by buffaloes (Williams 1976). This results in significant siltation, turbidity and other effects in swamps and billabongs with important consequences for the aquatic biota (McKnight 1976, Forbes 1980, Bishop & Forbes, Chapter 6). Experimental measurements at waterholes with and without buffaloes have shown that their faeces and urine cause eutrophication and algal blooms. Many of their effects on aquatic

ecosystems have been most intense in the dry season when natural stresses on the aquatic biota are also most severe.

Some of the most important effects buffaloes have had on the wetlands are due to their physical activities of wallowing (Photograph 13), pugging, slithering on banks and progressing along channels. Widespread pugging causes bare areas of soil in the wet season, whilst in the dry season it compresses the soils which then harden and often remain bare (Photograph 17) (Ruskin 1981, Fogarty 1982, Taylor & Friend 1984). Slithering tends to break down the levee banks of the rivers and banks of channels and, together with movement along them, deepens channels (Fogarty 1982). Intrusion by tidal saltwater into freshwater swamps resulting in the death of their paperbark *Melaleuca* forests is a feature of parts of the wetlands. It was caused by the deepening and lengthening of channels from the tidal rivers across the sedgelands to the deeper swamps. Between the Finniss and Reynolds Rivers there is observational evidence that this occurred because of the activities of buffaloes in and along the channels (Fogarty 1982). Aerial photography shows that the intrusion of saltwater and first death of paperbark forest took place between 1963 and 1978 just after buffaloes had first attained high densities there. Similar intrusion of tidal saltwater appears to have begun ten to twenty years earlier between the Adelaide and East Alligator Rivers (Fogarty 1982), even though buffaloes had been at high densities there for about sixty to eighty years longer than in the Finniss-Reynolds area. In this region some of the intrusion had been natural (Tulloch 1977) but accelerated by the activities of buffaloes (Stocker 1971, Fogarty 1982 but see also Williams 1976 and Chapter 1). In the lower Mary River system such deaths of paperbark forest were extensive.

Shooting of buffaloes for their hides in early times (Fig. 62) may well have slowed down the growth of the population. Nevertheless it is clear that this toll did not prevent their eventual increase, spread or effects. In 1940 the rate of removal of buffaloes from the monsoonal lowlands (Fig. 64) dropped to zero for five years, after which it rose and remained for ten years at just below its pre-war rate, then dropped again to very low levels for another seven years. From 1963 onwards the rate of removal accelerated fairly steadily till 1980 when it was more than seven times higher than it had ever been before 1963 (except in one year). The relaxation in the removal of buffaloes between 1940 and 1963, and especially in the later years, must have allowed a good deal more rapid increase in the population during this period (it was then they colonized the Finniss River area). However the accelerating rate of removal after 1963 would have reversed such a trend with increasing stringency. What happened to the buf-

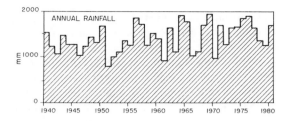

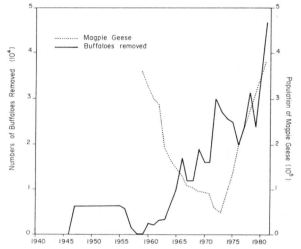

Figure 64. Annual totals of buffaloes removed (for all purposes) from the monsoonal lowlands 1940-1980 (as Fig. 62), annual aerial counts of Magpie Geese in the monsoonal lowlands 1958-1980 (redrawn from Tulloch & McKean 1983) and annual rainfall 1940-1980 (Commonwealth Bureau of Meteorology).

faloes' major habitat, the wetlands, during such increases and decreases in its density? Even though we could do with more experiments to understand better the processes involved, it can be seen from the summary just given that buffaloes have a wide variety of effects on the microtopography and vegetation of the wetlands when and where they occur at high densities. That evidence also indicated that usually the environment gradually changes back towards its original condition after densities have been reduced. By 1987, this process of recovery of much of the natural vegetation of the coastal lowlands had gained considerable momentum (Braithwaite & Werner, 1987).

The Magpie Goose is a conspicuous and common species, the wetland habitat of which largely overlaps that of the buffalo, when they occur together (Frith 1982). The disappearance of Magpie Geese throughout southern and eastern Australia as far north as the tropics has been attributed to the effects of stock on its wetland habitat and killing by man (Frith 1982). In the monsoonal lowlands the population of this waterbird underwent a dramatic decline between 1958 and 1972

followed by an equally striking return to its former abundance over the following ten years (Fig. 64). A variety of causes has been postulated for these changes, including the effects of fires, of hunting, of alleged commercial exploitation of eggs by Aborigines, of emigration, of irregularities in annual rainfall and thereby in water levels, and of changes in their habitat. For such a marked and consistent pattern of change as did occur, a cause of equal magnitude, consistency and amplitude, and related timing of reversal, must be sought. No evidence exists for such consistent trends in fires or hunting. Although significant mortality from such events may have occurred occasionally there is no reason to believe it was restricted to the period when the numbers of geese were declining. Large-scale banding in these wetlands and subsequent recoveries show that Magpie Geese were in fact rather sedentary, with only seasonal movements within the Top End during the period of their decline, so emigration too can hardly have been the cause of the decline (Frith 1977). Nor is asserted commercial exploitation of the eggs of Magpie Geese a realistic possible cause of this long-term decline of the adults, given the inherently vast logistic difficulties in procuring and marketing substantial numbers of eggs, and the occurrence anyway of heavy natural predation on the eggs of Magpie Geese (Frith & Davies 1961), as is normal for waterfowl (Lack 1954). Breeding by Magpie Geese in the monsoonal lowlands depends mainly on the level of water in the swamps each year, which thus affects their productivity and population size (Frith & Davies 1961). However the water level ultimately depends on the annual rainfall, almost all of which falls within the four to six months of the wet season. It can be seen that annual rainfall was clearly not correlated with the decline or increase of the population of Magpie Geese from 1958 to 1980 (Fig. 64).

In brief, this discussion suggests that of the possible causes only changes in the habitat shared with buffaloes are likely to have had a pattern consistent with that of this decline and rise of the population of Magpie Geese over 23 years (see also Morton & Brennan, Chapter 9). Thus the long period during which few or no buffaloes were removed from a probably expanding population was followed by a severe decline in that of Magpie Geese (Fig. 64). As removal of buffaloes subsequently increased, the decline of geese eventually slowed down (somewhat later as might be expected). Finally a major increase in the population of Magpie Geese followed (again with a lag of a few years) the further large increases in the rate of removal of buffaloes. This overall explanation still requires more corroboration. It is now being investigated experimentally at Kapalga by measuring and comparing the densities of the nests of Magpie Geese where there is a high population of buffaloes, and where all buffaloes have been removed (Anon. 1982a). A survey of a 1000 ha. site carried out there from 1980 to 1983 established the preferences of Magpie Geese for the depth of water and species of aquatic vegetation in which to breed (Tulloch et al. 1988). However, it did not measure the effect of buffaloes on the density of vegetation (shown by Davies 1961 to be important for breeding) and failed to indicate the effects of buffalo on the density of the nests of breeding geese.

Buffaloes also affect the rather drier wooded lands in the vicinity of the wetlands. This is especially marked in the narrow largely forested marginal zone beside the sedgelands where they scar the trunks of trees, especially White Gums *Eucalyptus alba*, Ghost Gums *E. papuana* and Paperbarks *Melaleuca* spp., on which they regularly rub their horns. They also create small crater-like wallows there. Exclosures at Kapalga show buffaloes prevent the regeneration of the characteristic thickets of the palm *Pandanus spiralis* in this margin. Relict patches of monsoon forest, most of which also occur in this narrow zone, are particularly susceptible to the effects of buffaloes which seek their shade and so are their fauna (Calaby 1980, Russell-Smith 1985). Surveys between 1885 and 1895 when buffaloes were beginning to get established between the East and West Alligator Rivers describe such patches of monsoon forest. They were said to be luxuriant, impenetrable, interlaced with many creepers, having conspicuous palms and plentiful orchids and with dense undergrowth in the dry season (Aust. Archives a, Brackenbury 1896). Today these features are either absent or no longer prominent. High densities of buffaloes have been shown to compact their soils and are associated with the death of large trees resulting in the lowering and opening of the canopy causing more ground and shrubby growth in the wet season (Braithwaite et al. 1984). This effect is probably because the hardening of the soils reduces the recharge of groundwater. Nearly three-quarters of the species of vertebrate inhabitants of monsoon forest, for which adequate data could be obtained, had their abundance affected in one way or another by such changes in their environment. In some cases small patches of monsoon forest or rainforest have almost totally disappeared (Chapter 4, Photograph 2).

Both in monsoon forest and in the wetlands a number of species of animals are disadvantaged by the environmental effects of buffaloes. However it is understandable that a number of others are actually advantaged by extension of their particular niches (Braithwaite et al. 1984, Friend & Taylor 1984). Nonetheless these two opposite categories cannot be equated ecologically with each other in terms of the survival of their populations, or even the species. The advantaged species or populations probably were never in danger of extinction there in the first place. The disadvantaged have been pushed a step or more in that direction.

Not only do buffaloes affect the indigenous biota but probably introduced plants and animals too, as discussed earlier. The establishment of introduced weeds may be influenced by the presence of feral buffaloes. For instance, in the forested margins of the wetlands, dense stands of the common annual weeds *Hyptis suaveolens* and *Cassia obtusifolia* are often associated with high densities of feral buffaloes now, or in the past. The relative unpalatability and/or other advantages of some introduced plants in heavily grazed situations could be important in competition with more vulnerable native vegetation (C.K. Williams, pers. comm. 1983). Thus it may be significant that the prickly invasive tropical shrub *Mimosa pigra*, a serious weed originally from South America, first became established in the Australian tropical wetlands some time between 1940 and 1947 (Miller 1983), while heavy grazing pressure must have existed. There would have been an appreciable increase in the population of feral buffaloes during that period resulting from the lengthy major relaxation of shooting then (Fig. 62). Nonetheless the shrub took another fifteen to twenty years to become abundant where it had started in the monsoonal lowlands (Miller 1983, Anon. 1984). Ungulates including cattle will graze *Mimosa* under some circumstances, for instance in southeast Asia and in the dry season in West Africa under comparable conditions with those of tropical Australia (K. Harley, pers. comm., P. Hiernaux, pers. comm.). It may be no coincidence that this shrub started to become abundant in the mid 1970s, a few years after heavy reduction of the population of feral buffaloes had begun (Fig. 62). Thus *Mimosa*'s progress may still have been slowed down to some extent by grazing and trampling, especially of seedlings, whatever its competitive advantage(s) over more susceptible native vegetation. However, at the same time, buffalo may also spread *Mimosa* seeds and facilitate their initial establishment (Braithwaite et al. 1989). None of this discussion is to deny the possible importance of other factors as well in *Mimosa*'s spread and establishment, such as flooding and climatic variation between years (as Miller 1983, M. Lonsdale, pers. comm.).

DISEASES

Feral mammals are susceptible to a number of diseases, especially exotic diseases (Letts et al. 1979). Whereas certain exotic diseases have decimated many native ungulates in Africa (Sinclair 1979), exotic diseases cause only low mortality in the feral ungulates of the monsoonal lowlands, on present evidence. No doubt this is because of some degree of immunity, long evolved in their place of origin. Unfortunately this also allows them to act as reservoirs of exotic diseases which affect stock, and humans.

Bovine tuberculosis occurred in c. 10% of feral buffaloes in 1972 but infection dropped to c. 3% by 1980 (Letts et al. 1979, Carpenter 1981). Its incidence in feral cattle is variable but probably comparable with that in buffaloes. Feral pigs in the monsoonal lowlands appear to be only an end host for bovine tuberculosis and thus are not important as an infective reservoir (Corner et al. 1981). Ephemeral fever and Akabane disease, both caused by viruses, occur widely among buffaloes and cattle in the monsoonal lowlands (Letts et al. 1979). Brucellosis is absent from the monsoonal lowlands, though present further inland (Williams & McPherson 1981). Cattle ticks and associated tick fever occur in feral cattle. Although feral buffaloes are capable of carrying ticks they are rarely infected, and then only lightly, in the monsoonal lowlands or elsewhere (Griffiths 1974, Bainbridge 1977). Melioidosis is an often fatal, tropical human disease associated with soil, which now occurs more frequently than in the past in the monsoonal lowlands (Ashdown et al. 1980, Rode & Webling 1981). Elsewhere it has been found in cattle and pigs (Glazebrook 1981), thus the frequent rooting of soil by the latter may be an important activity in relation to this disease. Feral pigs in the monsoonal lowlands are not parasitized by the Common Roundworm *Ascaris lumbricoides* (R.H.Barrett, pers. comm.), but they probably have sparganosis, which is common elsewhere in tropical Australia, and also Leptospirosis and Murray Valley Encephalitis (Letts et al. 1979, Pavlov 1983). Murray Valley Encephalitis also occurs in cattle in the monsoonal lowlands where this disease and other arboviruses show evidence of significant incidence in the human population (Stanley 1979, Whelan 1981).

Finally, feral mammals in the monsoonal lowlands are close to Indonesia and are potential reservoirs for serious exotic diseases which have not yet reached Australia (Letts et al. 1979). African horse sickness could be transmitted by brumbies and feral donkeys; African swine fever by feral pigs; Foot-and-mouth disease (from SE Asia) by feral buffalo, cattle and pigs; Jembrana disease (from Bali) by feral buffalo and cattle; and virulent strains of Blue tongue (from Asia) by feral buffalo and cattle as non-apparent carriers (St George & McCaughan 1979, Snowdon 1979, St George et al. 1980).

MANAGEMENT

Control

Their effects on the environment and the increased risks of outbreaks of exotic diseases have led to the need to control the three most common feral mammals: buffaloes, pigs and Bali cattle.

Pigs' reproductive rates are so high that a population in temperate Australia returned to its previous density within a year of being reduced by 58% (Hone & Pedersen 1980). Eradication by repeated poisoning followed by shooting is possible but lengthy, costly in money and effort and unlikely to be complete even then (Hone & Robards 1980, Tisdell 1982, Hone 1983). Short-term reduction of populations by 70% can be achieved by poisoning alone. More research is needed on anti-coagulant poisons which are less hazardous to many other vertebrates than to pigs and rodents. Pigs are most easily poisoned in the monsoonal lowlands from September to November when they are readiest to take bait.

Bali cattle are restricted to the Cobourg Peninsula and therefore can be controlled and/or eliminated without enormous cost by shooting or capture. However they should be retained in captivity elsewhere as a genetic resource (Kirby 1979).

The Commonwealth and Northern Territory Governments plan to bring under control and eliminate disease from some of the population of feral buffaloes in the monsoonal lowlands, and to remove the remainder by 1992 (Digby 1981, Forbes 1982, Anon. 1982b). Long-term reduction is feasible, as already indicated, resulting from the buffalo's low reproductive rate and the fact that it is large and conspicuous enough to shoot or capture quite easily. Large numbers can be mustered and/or shot, initially for profit, using a helicopter, and more slowly using vehicles (McCool 1981, Hill 1981, Lasry 1983a, b). Such concentrated catching and shooting over five months succeeded in removing c. 97% of the population in a 389 km^2 portion of Kapalga, under ideal conditions (Ridpath & Waithman 1988). The small residue increased by about twice over the following year because of immigration from an adjacent less heavily controlled population outside Kapalga (internal fencing prevented reinvasion from within); and likewise again a year after a second control operation. The results suggest that total eradication over the whole 245,000 km^2 often trackless habitat of the buffalo in the north would not be possible and, even were it so, prohibitively costly.

Redomestication and control

It is fitting that this chapter, which began by describing the start of domestication many thousands of years ago, should end by suggesting redomestication might be one effective and profitable way, in certain areas, of solving the problems some domestic stock have created since they escaped and became feral. Once feral stock become again the property of an owner it also becomes immediately in his best interests to keep them under control and eliminate disease from his herd. This then obviates the need for costly final clearance of a feral residue at government expense.

The superiority of the buffalo's adaptations, including its much smaller loss of weight in the dry season, indicates it may have greater economic potential than cattle in the wetter parts of the monsoonal lowlands. Furthermore it has been shown to be readily redomesticated (Tulloch 1979a, 1981). The dressed weight of buffalo carcasses as a proportion of gross weight is 3-5% lower than it is in cattle but buffalo meat has the commercial advantage of being leaner (Ruskin 1981, Ford 1982). In fact the value of all products of feral buffaloes in 1981/82 in the Northern Territory was not far short of $9 million (Anon. 1982c, plus approximation for pet meat value). Properly controlled, disease-free farming of redomesticated buffaloes should find good future markets if adequately promoted, especially for the export of meat and breeding stock (Letts et al. 1979, Lemcke 1981, Simpson 1981, Anon. 1982c, d, Ford 1982). Technical problems such as fencing are now soluble, especially with electrification (Wesley-Smith et al. 1981, Calder 1981). Advantage can be taken of the buffalo's differential seasonal use of its environment which tends to even out grazing pressure over the year. This could be achieved by long narrow paddocks extending across the sedgelands up into the adjacent drier wooded lands (as Andrew & Taylor, 1986). Such development requires capital for the necessary infrastructure (Anon. 1983). However, a profitable return on investment is possibly more likely than on backing stock ill-adapted to the monsoonal lowlands.

The proportion of the monsoonal lowlands economically suited to such use is limited for various reasons; while some of the region, in any case, has been allocated for other purposes such as national parks. Any commercial buffalo farming in the remainder would need to be carefully planned, especially with regard to the environment. The most serious worry concerns the density of stock in any such enterprise. If the density of stocking of redomesticated buffaloes were to be as high as the maximum feral densities before serious trimming began, then significant environmental changes would continue or recur. This would jeopardize the very resource on which such an industry would depend. It would also be of major public concern because of the effects on the native fauna with which the buffalo shares the monsoonal lowlands. Thus it would be critical for production to be set at a level that took into account not only the sustainable use but also the survival of the existing native vegetation and animals. Experience elsewhere in Australia indicates the need for the careful setting and enforcement of such upper limits to the density of stock in any such enterprise. The tropics can least afford a repetition of the deterioration of rangelands which has occurred elsewhere in Australia.

The present comprehensive campaign of control and reduction of feral stock in the monsoonal lowlands, perhaps in harness with some prudent and sustainable commercial utilization, must be seen as a particularly important step in the conservation of Australia's monsoonal heritage.

ACKNOWLEDGEMENTS

I am indebted to Peter Pangquee for the concept of very longterm and ultimately irreversible changes in carrying capacity, as I am to him for many other original insights into the ecology of the monsoonal lowlands. I am also grateful to Don Tulloch for much information about buffaloes; to Reg Barrett and Laurie Corbett for letting me examine and refer to unpublished data; to John Calaby for faunistic and historical advice; to Kent Williams for illuminating discussion about tropical ecology; to both the latter and Chris Haynes for their comments on the manuscript and to Rosalie Hall who so patiently typed it.

REFERENCES

Aldrick, J.M., D.F.Howe & C.R.Dunlop 1978. Report on the lands of the Ord River catchment, Northern Territory. *Tech. Bull.* 24. Darwin: NT Dept. Primary Production.

Andrew, M.H., P.N.Gowland, J.A.Holt, J.J.Mott & G.R.I.N.Strickland 1985. Constraints to agricultural development: Vegetation and Fauna. In R.C.Muchow (ed.), *Agroresearch for the Semi-arid Tropics: North West Australia:* 93-111. St.Lucia: University Queensland Press.

Andrew, M.H. & J.A.Taylor 1986. Intensification of the buffalo and cattle industries in the Australian wet-dry tropics – implications for the rangelands resource and their management in the Northern Territory. In P.J.Ross, P.W.Lynch & O.B.Williams (eds.), *Proc. 2nd International Rangelands Congress:* 228-229. Canberra: Aust. Acad. Sci.

Anon. 1982a. Ecological studies of Tropical sub-coastal wetlands. *Report 1980-82:* 22-28. CSIRO Division Wildl. Res.

Anon. 1982b. BTB NT Government policy and its implications for cattle producers. *N.T. Rural News* 7: 9-27.

Anon. 1982c. Northern Territory Primary Production Statistics, 1980/81. *Tech. Bull.* 59. NT Dept. Primary Production.

Anon. 1982d. Buffaloes go to PNG. *N.T. Rural News* 7: 9.

Anon. 1983. Annaburroo station moves ahead in buffalo domestication. *N.T. Rural News* 9: 4-5.

Anon. 1984. Joint research to control *Mimosa pigra. N.T. Rural News* 10 (3): 14-15.

Archer, M. 1981. A review of the origins and radiation of Australian mammals. In A.Keast (ed.), *Ecological biogeography in Australia:* 1437-1488. The Hague: Junk.

Ashdown, L.R., V.A.Duffy & R.A.Douglas 1980. Melioidosis. *Med. J. Aust.* 314-316.

Australian Archives (a) CRS A1640. Report on coast from Roper River to Castlereagh Bay by Captain Carrington 1885-87. Item 85/1251.

Australian Archives (b). Buffalo license to shoot or take, and returns. Dept. of NT Forestry, Fisheries, Wildlife and National Parks Branch 2005/I-III; Correspondence files.

Bainbridge, M.H. 1977. Parasites of Buffaloes in the Northern Territory of Australia. In B.D.Ford & D.G.Tulloch (eds.), *The Australian Buffalo − a collection of papers: 43-49. Tech. Bull.* 18. Darwin: Dept. NT Animal Industry and Agric. Branch.

Barber, H.N. 1954. Genetic polymorphism in the rabbit in Tasmania. *Nature* 173: 1227-1229.

Bell, R.H.V. 1971. A grazing ecosystem in the Serengeti. *Sci. Amer.* 255: 86-93.

Bentley, A. 1967. *An introduction to the deer of Australia.* Melbourne: Hawthorne Press.

Bentley, A. 1983. Deer. In R.Strahan (ed.), *The Australian Museum Complete book of Australian Mammals*: 500-512. Sydney: Angus & Robertson.

Bodenheimer, F.S. 1935. *Animal life in Palestine.*Jerusalem.

Brackenbury, L.J. 1896. *South Australia. Northern Territory. Report on Agriculture and other Lands.* Adelaide: Govt. Printer.

Braithwaite, R.W., M.L.Dudzinski, M.G.Ridpath & B.S.Parker 1984. The impact of Water Buffalo on the Monsoon Forest Ecosystem in Kakadu National Park. *Aust. J. Ecol.* 9: 309-322.

Braithwaite, R.W. & P.A.Werner 1987. The biological value of Kakadu National Park. *Search* 18: 296-301.

Braithwaite, R.W., W.M. Lonsdale & J.A. Estbergs 1989. Alien vegetation and native biota in tropical Austrialia: the impact of *Mimosa pigra. Biol. Conser.* 48: 189-210.

Calaby, J.H. 1974. Historical background. In H.J.Frith & J.H.Calaby (eds.), *Fauna survey of the Port Essington District, Cobourg Peninsula, Northern Territory of Australia*: 7-19. CSIRO Div. Wildlife Res., Tech. Paper 28.

Calaby, J.H. 1975. Introduction of Bali Cattle into northern Australia. *Aust. Vet.* 51: 108.

Calaby, J.H. 1980. Ecology and Human Use of the Australian Savanna Environment. In D.R.Harris (ed.), *Human Ecology in Savanna Environments*: 321-337. London: Academic Press.

Calaby, J.H. & K.Keith 1974. Mammals. In H.J.Frith & J.H.Calaby (eds.), *Fauna Survey of the Port Essington District, Cobourg Peninsula, Northern Territory, Australia*: 179-208. CSIRO Div. Wildl. Res., Tech. Paper 28.

Calder, G.J. 1981. *Hymenachne acutigluma* in the Northern Territory. *Tech. Bull.* 46. Darwin: NT Dept. of Primary Production.

Camoens, J.K. 1976. *The Buffalo in Malaysia.* Kuala Lumpur: Ministry of Agriculture.

Carpenter, M. 1981. T.B. in Buffalo − A Territory view. In F.G.Best (ed.), *Proc. Buffalo Industry Symp.*: 22-26. Darwin: NT Dept. Primary Production.

Caughley, G. 1966. Mortality patterns in mammals. *Ecology* 47: 906-918.

Christian, C.S. & G.A.Stewart 1953. General Report on survey of the Katherine-Darwin Region, 1946. *Land Res. Ser.* 1. Melbourne: CSIRO.

Clutton-Brock, J. 1977. Man-Made Dogs. *Science* 197: 1340-1342.

Clutton-Brock, J. 1981. *Domesticated Animals.* London: Heinemann.

Cockrill, W.R. 1974. *The husbandry and health of the domestic buffalo.* Rome: FAO.

Coman, B.J. & H.Brunner 1972. Food habits of the feral house cat in Victoria. *J. Wildl. Manag.* 36: 848-853.

Corbett, L.K. 1979. Feeding ecology and social organization of wild cats (*Felis silvestris*) and domestic cats (*Felis catus*) in Scotland. Unpubl. Ph.D. thesis, Aberdeen University.

Corbett, L.K. 1985. Comparisons of Australian and Thai dingoes: a reappraisal of dingo status, distribution and ancestry. In M.G.Ridpath & L.K.Corbett (eds.), *Ecology of the Wet-Dry Tropics. Proc. Ecol. Soc. Aust.*: 13: 277-291.

Corner, L.A., R.H.Barrett, A.W.D.Lepper, V.Lewis & C.W.Pearsons 1981. A survey of Mycobacteriosis of feral pigs in the Northern Territory. *Aust. Vet. J.* 57: 537-542.

Crook, K.A.W. 1981. The break-up of the Australian-Antarctic segment of Gondwanaland. In A.Keast (ed.), *Ecological biogeography of Australia*: 1-14. The Hague: Junk.

Davies, S.J.J.F. 1988. The nest-building behaviour of the Magpie Goose, *Anseranas semipalmata. Ibis* 104: 147-157.

Digby, J. 1981. Buffalo in the National Diseases Eradication Programme. In F.G.Best (ed.), *Proc. Buffalo Industry Syposium*: 55-57. Darwin: NT Dept. Primary Production.

Dyce, A.L. & H.A.Standfast 1979. Distribution and Dynamics of suspected vectors of Bluetongue Virus Serotype 20 in Australia. In E.L.French & T.D.St George (eds.), *Arbovirus research in Australia*: 28-35. Brisbane: CSIRO and QIMR.

Eberhard, T. 1954. Food habits of Pennsylvania House Cats.*J. Wildl. Manage.* 18: 284-286.

Epstein, H. 1971. *The origin of the domestic animals of Africa.* New York: Africana Publ. Corp.

Flannery, K.V. 1969. Origins and ecological effects of early domestication in Iran and the Near East. In P.J.Ucko & G.W.Dimbleby (eds.), *The Domestication and Exploitation of plants and Animals*: 73-100. Chicago: Aldine.

Fogarty, P. 1982. A preliminary survey of environmental damage associated with activity of feral buffalo. *Tech. Report.* Darwin: Feral Animals Committee, Conservation Commission: NT.

Forbes, M.A. 1980. Preliminary assessment of Buffalo Impact on the aquatic environment in Kakadu National Park. Appendix I. In T.J.Bergin (ed.), A program for feral animal control in Kakadu National Park. Unpubl. report to Aust. Nat. Parks and Wildlife Service, Canberra.

Forbes, M.A. 1982. Management of feral Buffalo in Kakadu National Park. *Aust. Ranger Bull.* 2: 19.

Ford, B. 1977. Productivity of domesticated Buffalo. In B.D.Ford & D.G.Tulloch (eds.), *The Australian Buffalo − a collection of Papers. Tech. Bull.* 18: 73-80. Darwin: Dept. NT Anim. Industry and Agric. Branch.

Ford, B. 1981. Animal Production aspects. In F.G.Best (ed.), *Proc. Buffalo Industry Symposium*: 79-101. Darwin: NT Dept. Primary Production.

Ford, B. 1982. Productivity and Management of the Water Buffalo in Australia. *Tech. Bull.* 61. Darwin: NT Dept. Primary Production.

Fox, R.W., G.G.Kelleher & C.B.Kerr 1977. *Ranger Uranium Environmental Inquiry Second Report.* Canberra: Aust. Govt. Pub. Service.

Friend, G.R. & J.A.Taylor 1984. Ground surface features attributable to feral buffalo (*Bubalus bubalis* Lydekker): 2.

Their relationship to the abundance of small animal species. *Aust. Wild. Res.* 11: 311-323.

Frisch, J.E. 1981. Changes occurring in cattle as a consequence of selection for growth rate in a stressful environment. *J. Agric. Sci., Cambridge* 30: 127-133.

Frisch, J.E. & J.E.Vercoe 1982. Consideration of adaptive and productive components of productivity in breeding beef cattle for tropical Australia. *Proc. Second World Conference on Genetics Applied to Animal Breeding*: 307-321. Madrid: Neografis.

Frith, H.J. 1977. Band recoveries of Magpie Geese *Anseranas semipalmata. Aust. Wildl. Res.* 4: 81-84.

Frith, H.J. 1982. *Waterfowl in Australia*. Sydney: Angus & Robertson. .

Frith, H.J. & S.J.J.F.Davies 1961. Ecology of the Magpie Goose *Anseranas semipalmata* Latham (Anatidae). *CSIRO Wildl. Res.* 6: 91-141.

Gibb, J.A., C.P.Ward & G.D.Ward 1978. Natural control of a population of rabbits *Oryctolagus cuniculus* for ten years in the Daurarua enclosure. *DSIR Bull.* 223. Wellington:
‹ DSIR.

Gifford, G.F. & R.H.Hawkins 1978. Hydrologic impact of grazing on infiltration: a critical review. *Water Resources Res.* 14: 305-313.

Giles, J.R. 1980. The ecology of feral pigs in western New South Wales. Ph.D. thesis, University of Sydney.

Glazebrook, R. 1981. Melioidosis: a not so rare tropical disease. *Aust. Nurses J.* 10: 35.

Gooding, C.D. 1983. Donkey. In R.Strahan (ed.), *The Australian Museum Complete book of Australian Mammals*: 492-493. Sydney: Angus & Robertson.

Graham, A., R.Begg, P.Graham & S.Raskin 1982. An aerial survey of Buffalo in the Northern Territory. *Tech. Report.* Darwin: Conservation Commission, NT.

Griffiths, R.B. 1974. Parasites and parasitic diseases. In W.R.Cockrill (ed.), *The Husbandry and Health of the Domestic Buffalo*: 236-275. Rome: FAO.

Gwynne, M.D. & R.H.V.Bell 1968. Selection of vegetation components by grazing ungulates in the Serengeti National Park. *Nature* 220: 390-392.

Hafez, E.S.E. 1962. *The Behaviour of Domestic Animals*. London: Bailliere Tindall & Cox.

Hafez, E.S.E., A.L.Badredin & M.M.Shafei 1955. Skin structure of Egyptian Buffaloes and cattle with particular reference to sweat glands. *J. Agric. Sci., Cambridge* 46: 19-30.

Hill, B. 1981. Aerial Mustering and Cattle Control. *Tech. Bull.* 40. Darwin: NT Dept. Primary Production.

Hill, E. 1951. *The Territory.* Sydney: Angus & Robertson.

Hill, R. & G.Webb 1982. Floating grass mats of the Northern Territory floodplains – an endangered habitat? *Wetlands* 2: 45-50.

Hone, J. 1983. A short-term Evaluation of Feral Pig Eradication at Willandra in Western New South Wales. *Aust. Wildl. Res.* 10: 269-275.

Hone, J. & H.Pederson 1980. Changes in a feral pig population after poisoning. In J.P.Clark (ed.), *Proc. 9th Vert. Pest. Control Conf. Calif.*: 176-182. Davis: Univ. California.

Hone, J. & G.E.Robards 1980. Feral Pig Ecology and Control. *Wool Technol. and Sheep Breed.* 28: 7-11.

Hoogerwerf, A. 1970. *Udjung Kulon: the land of the last Javan rhinoceros*. Leiden: Brill.

Howard, F. & R.H.Edmunds 1969. McKinlays Expedition from Escape Cliffs. In C.C.Macknight (ed.), *The Farthest Coast*: 141-158. Carlton: Melbourne Univ. Press.

Howard, G.W. 1985. The Kafue flats of Zambia: a wetland ecosystem comparable with floodplain areas of northern Australia. In M.G.Ridpath & L.K.Corbett (eds.), *Ecology of the Wet-Dry Tropics, Proc. Ecol. Soc. Aust.* 13: 293-306.

Jarman, P.J. & A.R.E.Sinclair 1979. Feeding strategy and the pattern of resource partitioning in ungulates. In A.R.E.Sinclair & M.Norton-Griffiths (eds.), *Serengeti: Dynamics of an Ecosystem*. Chicago: University of Chicago Press.

Jennison, G. 1929. *Natural History – Animals*. London: Black.

Jewell, P.A. 1969. Wild mammals and their potential for new domestication. In P.J.Ucko & G.W.Dimbleby (eds.), *The Domestication and Exploitation of Plants and Animals*: 101-109. Chicago: Aldine.

Jones, E. 1983. Feral Cat. In R.Strahan (ed.), *The Australian Museum Complete book of Australian Mammals*: 489. Sydney: Angus & Robertson.

Jones, R. 1968. The geographical background to the arrival of man in Australia and Tasmania. *Arch. Phys. Anthropol. Oceania* 3: 186-215.

Kabay, E.D., A.S. George & K.F.Kenneally 1977. The Drysdale River National Park Environment. In *A Biological Survey of the Drysdale River National Park North Kimberley, Western Australia in August, 1975. Wildl. Res. Bull.* 6: 13-30. Perth: Dept. Fisheries and Wildlife.

Kelsey, D.E. 1975. *The shackle*. Blackwood SA: Lynton.

Kirby, G.W.M. 1979. Bali cattle in Australia. *FAO World Anim. Rev.* 3: 7.

Lack, D. 1954. *The Natural Regulation of Animal Numbers*. Oxford: Clarendon Press.

Lasry, M. 1983a. Fish River clean up – catching scrub cattle and buffaloes. *NT Rural News* 9: 4-6.

Lasry, M. 1983b. Mount Bundy Muster. *NT Rural News* 9: 14-19.

Leichardt, F.W.L. 1847. *Journal of an Overland Expedition in Australia 1844-1845*. London: Boone.

Lekagul, B. & J.A.McNeely 1977. *Mammals of Thailand*. Bangkok: Kurupsha.

Lemcke, B.G. 1981. Buffalo Industry Survey. In F.G.Best (ed.), *Proc. Buffalo Industry Symposium*: 88-94. Darwin: NT Dept. Primary Production.

Letts, G.A. 1977. History of Water Buffalo in Australia. In B.D.Ford & D.G.Tulloch (eds.), *The Australian Buffalo – a collection of papers*: 1-7. *Tech. Bull.* 18. Darwin: Dept. NT Anim. Industry and Agric. Branch.

Letts, G.A., A.Bassingthwaithe & W.E.L.de Vos 1979. *Feral Animals in the Northern Territory.* Darwin: Govt. Printer.

Lewis, J. 1922. *Fought and Won*. Adelaide: Thomas.

Lindsay, D. 1884. Mr D.Lindsay's Explorations through Arnhem's Land. *House of Assembly South Australia Parliamentary Paper* 239.

Loofs, H.H.E. 1974. Mainland south-east Asia. In A.L.Basham (ed.), *The Civilizations of Monsoon Asia*: 69-134. Sydney: Angus & Robertson.

McCool, C. 1981. Catching wild livestock and feral animals; some of the problems. *Tech. Bull.* 36. Darwin: NT Dept. Primary Production.

McCool, C., B.Radnuz, B.Fox, D.Stephens, B.Walsh, C.Watts & B.Hill 1981. Feral Donkeys in the Northern Territory. *Tech. Bull.* 81/39. Conservation Commission N.T.

MacDonald, D. 1981. The behaviour and ecology of farm

cats. *The ecology and control of feral cats*: 23-29. Univ. Fedn. for An. Welfare, Potters Bar UK.

McKnight, T. 1964. Feral livestock in Anglo-America. *Univ. Calif. Publ. Geog.* 16: 1-87.

McKnight, T. 1971. Australia's Buffalo Dilemma. *Ann. Assoc. Amer. Geog.* 61: 759-773.

McKnight, T. 1975. A Comparative view of feral hoofed livestock in Australia and the United States. In I. Douglas, J.E.Hobbs & J.J.Pigram (eds.), *Geographical essays in honour of Gilbert J.Butland*: 31-49. Armidale: Dept. Geography University of New England.

McKnight, T. 1976. Friendly Vermin: a Survey of Feral Livestock in Australia. *Univ. Calif. Publ. Geog.* 21.

Mahood, I.T. 1983. Feral Goat. In R.Strahan (ed.), *The Australian Museum complete book of Australian Mammals*: 516. Sydney: Angus & Robertson.

Mason, I.L. 1974a. Species, types and breeds. In W.R.Cockrill (ed.), *The Husbandry and health of the domestic buffalo*: 1-47. Rome: FAO.

Mason, I.L. 1974b. Environmental Physiology. In W.R.Cockrill (ed.), *The Husbandry and health of the domestic buffalo*: 88-104. Rome: FAO.

Merrilees, D. 1968. Man the destroyer: Late Quaternary changes in the Australian marsupial fauna. *J. Proc. R.Soc. West. Aust.* 51: 1-24.

Miller, I. 1983. Control of *Mimosa pigra*. *NT Rural News* 8 (2): 8-9.

Millham, P & P.Thompson 1976. Relative antiquity of human occupation and extinct fauna at Madura Cave, Southwestern Western Australia. *Mankind* 10: 175-180.

Minson, D.J. 1981. Nutritional differences between tropical and temperate pastures. In F.H.W.Morley (ed.), *Grazing Animals*: 143-157. Amsterdam: Elsevier.

Moran, J.B. 1973. Heat tolerance of Brahman cross, buffalo, Banteng and Shorthorn steers during exposure to sun and as a result of exercise. *Aust. J. Agric. Res.* 24: 775-782.

Morton, S.R., H.F.Recher, S.D.Thompson & R.W. Braithwaite 1982. Comments on the relative advantages of Marsupial and Eutherian reproduction. *Am. Nat.* 120: 128-134.

Newsome, A.E. & L.K.Corbett 1982. The identity of the dingo II. Hybridization with domestic dogs in captivity and in the wild. *Aust. J. Zool.* 30: 365-374.

Newsome, A.E., L.K.Corbett & S.M.Carpenter 1980. The identity of the dingo I. Morphological discriminants of dingo and dog skulls. *Aust. J. Zool.* 28: 615-625.

Norman, M.J.T. 1963. The pattern of dry matter and nutrient content changes in native pastures at Katherine, N.T. *Aust. J. Exp. Agric. Anim. Husb.* 3: 119-124.

Norman, M.J.T. 1966. Katherine Research Station 1956-64: A review of published work. Division of Land Research, *Tech. Paper* 28. Melbourne: CSIRO.

Ormeling, F.J. 1957. *The Timor Problem*. Groningen: Wolters.

Parker, P. 1977. An ecological comparison of marsupial and placental patterns of reproduction. In B.Stonehouse & D.Gilmore (eds.), *The Biology of Marsupials*: 273-286. London: Macmillan.

Parsons, J.J. 1980. Europeanization of the savanna lands of Northern South America. In D.R.Harris (ed.), *Human Ecology in Savanna Environments*: 267-289. London: Academic Press.

Parsons, P.A. 1983. *The evolutionary biology of colonizing species*. Cambridge: Cambridge Univ. Press.

Pavlov, P. 1983. Feral Pig. In R.Strahan (ed.), *The Australian Museum complete Book of Australian Mammals*: 495. Sydney: Angus & Robertson.

Phillip, A. 1789. *Voyage of Governor Phillip to Botany Bay*. London: Stockdale.

Pond, W.G. & K.A.Houpt 1978. *The biology of the pig*. Ithaca: Cornell Univ. Press.

Powell, C.McA., B.D.Johnson & J.J.Veevers 1981. The early Cretaceous break-up of Eastern Gondwanaland, the separation of Australia and India, and their interaction with Southeast Asia. In A.Keast (ed.), *Ecological biogeography of Australia*: 15-29. The Hague: Junk.

Protsch, R. & R.Berger 1973. Earliest radio-carbon dates for domesticated animals. *Science* 179: 235-239.

Pye, J. (n.d.). *The Tiwi Islands*. Darwin: Coleman.

Rees, W.A. 1978a. The ecology of the Kafue Lechwe: the food supply. *J. Appl. Ecol.* 15: 177-191.

Rees, W.A. 1978b. The ecology of the Kafue Lechwe: as affected by the Kafue Gorge Hydroelectric Scheme. *J. Appl. Ecol.* 15: 205-217.

Rich, T.H. 1982. Monotremes, Placentals, and Marsupials: Their record in Australia and its biasses. In P.V.Rich & E.M.Thompson (eds.), *The Fossil Vertebrate Record of Australasia*: 385-478. Melbourne: Monash University.

Rideout, B.L. 1977. The Buffalo Meat Industry. In B.D.Ford & D.G.Tulloch (eds.), *The Australian Buffalo – a collection of papers*: 50-56. *Tech. Bull.* 18, Dept. NT Anim. Industry and Agric. Branch.

Ridpath, M.G. 1981. The CSIRO wetlands research at Kapalga. In F.G.Best (ed.), *Proc. Buffalo Industry Symp.*: 69-73. Darwin: NT Dept. Primary Production.

Ridpath, M.G. 1985. Ecology in the Wet-Dry Tropics – How different? In M.G.Ridpath & L.K.Corbett (eds.), *The Ecology of the Wet-Dry Tropics, Proc. Ecol. Soc. Aust.* 13: 3-20.

Ridpath, M.G., R.J.Begg, M.L.Dudzinski, M.A.Forbes & A.Graham 1983. Counting the same populations of large tropical mammals from the ground and from the air. *Aust. Wildl. Res.* 10: 487-498.

Ridpath, M.G., J.A.Taylor & D.G.Tulloch 1985. Nature as a model? In R.C.Muchow (ed.), *Agroresearch for the semiarid Tropics: North West Australia*: 419-434. St Lucia: Univ. Queensland Press.

Ridpath, M.G. & J. Waithman 1988. Controlling feral Asian Water Buffalo in Australia. *Wildl. Soc. Bull.* 16: 385-390

Riney, T. 1982. *Study and Management of large Mammals*. New York: Wiley.

Rode, J.W. & D.D'A.Webling 1981. Melioidosis in the Northern Territory of Australia. *Med. J. Aust.*: 181-184.

Ruskin, F.R. 1981. *The Water Buffalo: New prospects for an underutilized animal*. Washington: Nat. Acad. Press.

Russell-Smith, J. 1985. A record of change: studies of Holocene vegetation history in the South Alligator River region, Northern Territory. In M.G.Ridpath & L.K. Corbett (eds.), *Ecology of the Wet-Dry Tropics, Proc. Ecol. Soc. Aust.* 13: 191-202.

Schmidt-Nielsen, K. 1979. *Animal Physiology*. Cambridge: Cambridge Univ. Press.

Silver, C. 1976. *Guide to horses of the World*. Oxford: Elsevier.

Simpson, T. 1981. Market Options and the future. In F.G.Best (ed.), *Proc. Buffalo Industry Symp.*: 45-47. Darwin: NT Dept. Primary Production.

Sinclair, A.R.E. 1977. *The African Buffalo*. Chicago: University of Chicago Press.

Sinclair, A.R.E. 1979. The Eruption of the Ruminants. In

A. R. E. Sinclair & M. Norton-Griffiths (eds.), *Serengeti Dynamics of an Ecosystem*: 82-103. Chicago: Chicago Univ. Press.

Snayden, R.W. 1981. The ecology of grazed pastures. In F. H. W. Morley (ed.), *Grazing Animals*: 13-31. Amsterdam:Elsevier.

Snowdon, W.A. 1979. Bluetongue Virus infection in Australia. In T. D. St George & E. L. French (eds.), *Arbovirus Research in Australia*: 16-19. Brisbane: CSIRO and QIMR.

Stanley, N. F. 1979. Problems related to the epidemiology of Murray Valley encephalitis and Kunjin Viruses created by development in North-West Australia. In T. D. St George & E. L. French (eds.), *Arbovirus Research in Australia*: 41-46. Brisbane: CSIRO and QIMR.

St George, T. D. & C. I. McCaughan 1979. The transmission of the CSIRO 19 Strain of Bluetongue virus type 20 to sheep and cattle. *Aust. Vet. J.* 55: 198-199.

St George, T.D., D. H. Cybinski, A. J. Della-Porta, D. A. McPhee, M. C. Wark & M. H. Bainbridge 1980. The isolation of two Bluetongue viruses from healthy cattle in Australia. *Aust. Vet. J.* 56: 562.

Stocker, G.C. 1971. The effects of Buffalo on paperbark forests in the Northern Territory. *Aust. For. Res.* 5: 29-34.

Stodart, E. 1965. A study of the biology of the wild rabbit in climatically different regions in eastern Australia III. Some data on the evolution of coat colour. *CSIRO Wildl. Res.* 10: 73-82.

Story, R. 1969. Grasses of the Adelaide-Alligator River area and their utilization. *Lands of the Adelaide-Alligator River Area, Northern Territory. Land Res. Ser.* 25: 130-139. Melbourne: CSIRO.

Taylor, J.A. & G. R. Friend 1984. Ground surface features attributable to feral buffalo (*Bubalus bubalis* Lydekker): I. their distribution relative to vegetation structure and plant lifeform. *Aust. Wildl. Res.* 11: 303-309.

Thompson, D. 1949. Arnhem Land: Exploration among an unknown people. *Geog. J.* 104: 53-67.

Tisdell, C.A. 1982. *Wild pigs: environmental pest or economic resource?* Sydney: Pergamon Press.

Tulloch, D.G. 1967. The distribution density, and social behaviour of the water buffalo in the Northern Territory. M.Sc. (Agric.) thesis, Univ. of Queensland, Brisbane.

Tulloch, D.G. 1969. Home range in feral water buffalo *Bubalus bubalis* Lydekker. *Aust. J. Zool.* 17: 143-152.

Tulloch, D.G. 1970. Seasonal movements, and distribution of the sexes in the water Buffalo *Bubalus bubalis* in the Northern Territory. *Aust. J. Zool.* 18: 399-414.

Tulloch, D.G. 1974. The feral swamp buffaloes of Australia's Northern Territory. In W. R. Cockrill (ed.), *The Husbandry and Health of the domestic buffalo*: 493-505. Rome: FAO.

Tulloch, D.G. 1977. Buffalo – His place in history of the North. *NT Rural Mag.* 2: 11-16.

Tulloch, D.G. 1978. The Water Buffalo, *Bubalus bubalis*, in Australia: Grouping and home range. *Aust. Wildl. Res.* 5: 327-354.

Tulloch, D.G. 1979a. Redomestication of Water buffaloes in the Northern Territory of Australia. *Anim. Regulation Studies* 2: 5-20.

Tulloch, D.G. 1979b. The Water Buffalo, *Bubalus bubalis*, in Australia: Reproductive and parent-offspring behaviour. *Aust. Wild. Res.* 6: 265-287.

Tulloch, D.G. 1981. Previous attempts at the development of a Buffalo Industry. In F.G. Best (ed.), *Proc. Buffalo Indus-*

try Symp.: 6-15. Darwin: NT Dept. Primary Production.

Tulloch, D.G. 1983. Water Buffalo. In R.Strahan (ed.), *The Australian Museum Complete Book of Australian Mammals: 514-515.* Sydney: Angus & Robertson.

Tulloch, D, K. M. Cellier & A. L. Hertog 1988. The distribution of the nests of the Magpie Goose (*Anseranas semipalmata* Latham) at Kapalga, N.T.: a four-year study. *Aust. Wildl. Res.* 15: 211-221.

Tulloch, D.G. & A.Grassia 1981. A study of reproduction in Water Buffalo in the Northern Territory of Australia. *Aust. Wildl. Res.* 8: 355-348.

Tulloch, D.G. & R.T.Litchfield 1981. Wallows for Buffalo. *Aust. Wildl. Res.* 8: 555-565.

Tulloch, D.G. & J.L.McKean 1983. Magpie Goose Populations on the Coastal Plains of the Northern Territory (1958-1980). *Corella* 7: 32-36.

Valdes, J. 1926. O Cavalo de Timor. *Boletin da agencia geral das colonias ano II* 17: 89-92.

Vercoe, J. E. & J. E. Frisch 1977. The importance of voluntary food intake and metabolic rate to production in different genotypes of cattle in different environments. *Proc. 3rd Int.Congress Soc. Advancement of Breeding Researches in Asia and Oceania (SABRAO):* 1(c) 42-45. Canberra.

Warburton, C. 1934. *Buffaloes.* Sydney: Angus & Robertson.

Webb, G.J.W., G.C.Sack, R.Buckworth & S.C.Manolis 1983. An Examination of *Crocodylus porosus* Nests in Two Northern Australian Freshwater Swamps, with an Analysis of Embryo Mortality. *Aust. Wildl. Res.* 10: 571-605.

Wesley-Smith, R., R. Piesse, B. Hill, B. Lemcke & B. Ford 1981. Fencing for Buffalo control. *Tech. Bull* 47. Darwin: NT Dept. Primary Production.

Wharton, C. H. 1968. Man, Fire and Wild Cattle in south east Asia. *Proc. Tall Timbers Fire Ecol. Conf.* 8: 107-167.

Whelan, P.I. 1981. The vulnerability and receptivity of the Northern Territory to mosquito borne disease. *Trans Menzies Foundn.* 2: 165-171.

Williams, A. R. 1979. Vegetation and stream patterns as indicators of water movement on the Magela floodplain, Northern Territory. *Aust. J. Ecol.* 4: 239-247.

Williams, C. K. & M. L. Dudzinski 1982. Ingestion rates, food utilization and turnover of water and sodium in grazing buffaloes *Bubalus bubalis* and cattle *Bos taurus* × *B. indicus* in monsoonal Northern Territory. *Aust. J. Agric.* 33: 743-754.

Williams, C. K. & B. Green 1982. Ingestion rates and aspects of water, sodium and energy in caged swamp buffalo *Bubalus bubalis*, from isotope dilution and materials balances. *Aust. J. Zool.* 30: 779-790.

Williams, C. K. & M. G. Ridpath 1982. Rates of Herbage Ingestion and Turnover of Water in Feral Swamp Buffalo *Bubalus bubalis* in relation to Primary Production in a cyperaceous swamp in monsoonal northern Australia. *Aust. Wildl. Res.* 9: 397-408.

Williams, M.A.J. 1976. Erosion in the Alligator Rivers area. *Lands of the Alligator Rivers area, Northern Territory. Land Res. Ser.* 38: 112-125. Melbourne: CSIRO.

Williams, O.J. & G.G.McPherson 1981. Brucellosis Eradication in the N.T. *Tech. Bull.* 44. Darwin: NT Dept. Primary Production.

Wraight, M.J. 1964. Modification of grasslands by grazing animals. *Proc. N.Z. Ecol. Soc.* 11: 27-32.

Zeuner, F. E. 1963. *A History of Domesticated Animals.* London: Hutchinson.

Part 4:
People, the environment and the future

Introduction

In November 1845 Dr. Ludwig Leichhardt made his way down the Arnhem Land escarpment and across the floodplains of the Alligator Rivers. As he did so he recorded in his diary some details of the life-style of the Aboriginal people he met, and their day to day use of the plants and animals in the area.

It is instructive to consider what changes have taken place in the seven score years since Leichhardt's journey, and this is one of the themes of Chapter 12, which derives from Dr. Betty Meehan's unrivalled knowledge of the Aboriginal people of northern Arnhem Land. Meehan draws an interesting comparison between the Aboriginal worlds of today and yesteryear, as well as between the contrasting present-day worlds of western and northern Arnhem Land. In so doing, she is offering us a glimpse of possible future social change.

The Anbarra people are a modern hunting community who live on the plains around the mouth of the Blyth River in northern Arnhem Land. Returing to their traditional homelands after an absence of fifteen years, they were initially hesitant about where and when they might find certain foods suitable for harvesting. With constant practice, they rapidly regained their former familiarity with the subtleties of seasonal change.

Depending upon the time of year, they culled meat, fish, eggs, fruit, shell-fish and vegetables from a variety of sources, including tidal estuaries, the sea, mangrove forests, freshwater swamps and billabongs, woodlands and grassy plains. The result was an ample and varied diet throughout the year, with every member of the community contributing to the quest for food, however young or old. Then came the buffalo and the seeds of change were sown.

Within a few short years the impact of the buffalo was becoming evident and the animals were resented and feared. Wells and springs upon which the people could previously rely for good water until late in the dry season became trampled quagmires and began to dry out prematurely. Once grassy meadows became denuded buffalo wallows, bare and rock-hard during the dry season. But the most insidious impact of all was upon the children and elderly members of the community. No longer able to contribute their share of foraging skills to the communal quest for food, they had perforce to rely upon the buffalo-hunting abilities of the younger men with their guns and vehicles who, torn between two utterly different cultures, are reliable in their own far less than in earlier times.

Much as in western Arnhem Land today, the Anbarra people are in danger of seeing a reliable, varied and nutritious diet superseded by a diet of buffalo meat, sweet tea and damper. The question now is whether the twin impact of European goods and feral buffaloes will mean that a broadly based foraging economy and a well integrated social fabric yield to the dubious delights of consumer goods, buffalo beef and flour.

In the final chapter to this book the three editors try to assess some of the biological and social reasons for the human successes and failures in adapting to the constraints imposed by the seasonally-wet tropical climate. Using a variety of criteria they conclude that the European occupation of the monsoonal lowlands of tropical Australia cannot be rated a great success.

Apart from the agricultural failures of the Daly River, Ord River, Humpty Doo, Tipperary and Willeroo schemes, the morbidity statistics also tell a sorry tale of a harsh environment taking its toll from those who lack the right adaptations or have ill-suited lifestyles, or both! Taking the Northern Territory population as a whole, the age-adjusted death rate among men and women is more than double that of the rest of Australia. Among adults, the death rates from cardiovascular disease and motor vehicle accidents are twice the national average. Yearly per capita alcohol consumption (expressed as equivalent pure alcohol consumption) is 1.5 times higher than the national average. A further cause for serious concern is the fact that the mortality rate among Aboriginal infants is four times the national average.

The authors of Chapter 13 conclude with a plea and a caution. The plea is for all of us to recognise that the climate, soils, plants and animals of monsoonal Australia are indeed different from those with which most Australians are familiar. The caution is that unless we try to understand the nature of those differences and take them into account in our plans to develop the North, we will again fail, and our failures will be detrimental to the land, costly to ourselves, but above all else, foreseeable and avoidable.

CHAPTER 12

Wetland hunters: Some reflections

BETTY MEEHAN

No part of the country we had passed, was so well-provided with game as this . . .

Leichhardt (1847: 496)

INTRODUCTION

Throughout this chapter the word 'hunters' denotes those societies which procure most of their food by various combinations of hunting, gathering and fishing. The human history of the present coastal plains of northern Australia is still being unravelled but it is clear that hunters have been exploiting the abundant resources of this area since the present sea level stabilised some 5,000-6,000 years ago (Schrire 1982). In historic times various people (explorers, missionaries, government servants, anthropologists) have described the lifestyle of the Aborigines, the most recent of these hunters. Their tales tell of a vigorous people thriving on the products which they harvested in abundance from their surroundings. For example, in November 1845, during the final stages of his epic journey from Jimbour Station in Queensland to Victoria, the short-lived settlement on the shores of Port Essington in northern Australia, Ludwig Leichhardt trekked across the plains which had been formed by the flooding of the Alligator Rivers (Fig. 65). His descriptions affirm their richness. It was the end of the dry season and water birds were everywhere:

Since the 23rd November, not a night had passed without long files and phalanxes of geese taking their flight up and down the river, and they often passed so low, that the heavy flapping of their wings was distinctly heard. Whistling ducks, in close flocks, flew generally much higher, and with great rapidity. No part of the country we had passed, was so well-provided with game as this; and of which we could have easily obtained an abundance, had not our shot been all expended.

(ibid., pp. 496-497)

Wallabies, fish and plant foods were also plentiful on the flood plains:

In skirting the bush, we came to a salt-water creek (the first seen by us on the north-west coast), when we immediately returned to the ridges, where we met with a well-beaten foot-path of the natives, which led us along brush, teeming with wallabies, and through the undulating scrubby forest ground to another large plain.

(ibid., p. 491)

On 8 December 1845 whilst following a tributary of the East Alligator River:

We met another foot-path at its northern bank, which led us between the river and ranges of rocky hills, over a country abounding with the scarlet Eugenia of which we made a rich harvest.

(Leichhardt: p. 518)

That night they:

. . . encamped at a fine lagoon, occupied as usual, with geese and ducks, and teeming with large fish, which were splashing about during the whole night.

(ibid., p. 518)

During this part of the journey members of the expedition developed a taste for the food from a small plant which the local Aborigines called 'Allamur', probably the delicious water chestnut *Eleocharis dulcis*, much of which still grows in the freshwater swamps of northern Australia:

The last had a sweet taste, was very mealy and nourishing, and the best article of food of the natives we had yet tasted . . . The plant grew in depressions of the plains, where the boys and young men were occupied the whole day digging for it.

(ibid., p. 504)

Of the Aboriginal inhabitants of the coastal plain Leichhardt was most complimentary, saying that:

There could not have been less than 200 of them present; they were all well made, active, generally well-looking, with an intelligent countenance: they had in fact all the characters of the coast blacks of a good country . . .

(ibid., p. 507)

A notable feature of this section of Leichhardt's journal is his reference to the presence of water buffa-

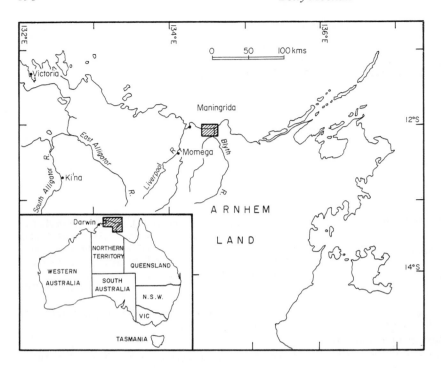

Figure 65. Location map of area covered in study area depicted in Figure 66.

loes in the area. The expedition constantly came upon single animals or groups of them and their tracks: 'Just before entering the forest, Brown observed the track of a buffalo on the rich grassy inlets, between the rocks' (ibid., p. 521). Later on the same day 'tracks of the buffaloes were again observed by Charley' (ibid., p. 521). Two days later, on 11 December 1845, evidence for the presence of the beasts was abundant:

When we approached the forest, several tracks of buffaloes were seen; and, upon the natives conducting us along a small creek which came into the plain from the N.N.E., we found a well-beaten path and several places where these animals were accustomed to camp.

(ibid., p. 524)

Having set up camp that evening, 'Charley' went off in search of a buffalo. He shot a 'young bull, about three years old, and in most excellent condition' (ibid., p. 524). This was a fortunate event for the expedition. It meant that they could replenish their own meagre meat supply and 'share freely with our black friends' who apparently 'had not the slightest objection to eat fresh meat, after baking it in their usual manner' (ibid., p. 524). The previous day Leichhardt had offered 'dried beef' to a group of Aborigines in exchange for plant food which they had given to him but they 'would not taste our dried beef, which they turned, broke, smelled and then with a feeling of pity and disgust returned to us' (ibid., pp. 522-523). The Ab-

origines called the buffalo 'Anaborro'. They said that the country ahead of the explorers was 'full of them' (ibid., pp. 524-525). Today Aborigines in this same region call them 'Nganaburru' (Gillespie 1979). (See Photographs 16 and 17.)

More than a century and a quarter after Leichhardt recorded his description of Aboriginal life on the Alligator River flood plains, 200 km to the east in Arnhem Land, groups of Aborigines still pursue a hunting lifestyle in a landscape into which Europeans and buffaloes have only just begun to penetrate.

Food species derived from the sea and from the estuarine rivers, creeks and mangrove forests play a major role in the diet of these hunters, but at certain times during the year so do species that thrive in the hinterland, on black soil plains, in monsoon thickets and open forests, and in the numerous fresh water swamps that dot the area. The richness of this area is reflected in its population density and in its social and cultural diversity (White 1979, White et al., in press).

THE ANBARRA

The Anbarra, one of these modern hunting communities, inhabit the coastal plains around the mouth of An-gatja Wana or Blyth River in Arnhem Land (Fig. 66). The 50 km^2 of coastal plain they own have been formed since the sea level ceased to fluctuate. Indeed, evidence accumulated so far suggests that no land

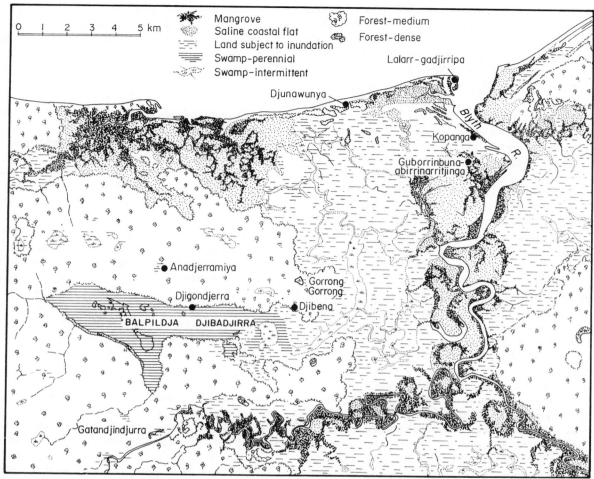

Figure 66. Major elements of the landscape and distribution of home bases and swamps, discussed in the text. Note that Djibena is an earth mound, and Guborrinbuna-abirrinarritjinga is a jungle patch.

presently owned by the Anbarra is older than about 2,000 years. In addition they have access to another 180 km² of land adjacent to their own which is owned by various communities with whom they have close social ties.

About 1970 the Anbarra, like many other groups living between the Liverpool and Blyth Rivers, returned to their own land where they established themselves in a series of what have become known as 'outstations' but which are in fact a select few of the home bases which they occupied before Europeans arrived there. Prior to 1970 and since 1957 many had lived more or less permanently at the government outpost on the eastern bank of the Liverpool River at Maningrida, which over a decade had developed into a substantial town. Before then some people, mostly men, had visited Darwin or worked at buffalo and timber camps between the Liverpool River and Dar-

win. From their outstations, the Anbarra hunt, gather and fish for products from their own territory and supplement this daily fare with food purchased from the supermarket at Maningrida – mainly tea, flour and sugar.

During a series of field trips over the past ten years, it has been possible to record in detail the foraging strategies employed by the Anbarra people and to monitor the success of these strategies in quantitative terms (Meehan 1982, Jones 1980). Whilst reading Leichhardt's Journals recently, I was struck by the similarity between some of the descriptions he had recorded in 1845 in western Arnhem Land and those that Rhys Jones and I have described in our journals over the past decade. The observations that follow are based on thirteen months of field work in 1972-1973, six weeks in 1974, five months in 1978, six weeks in 1979 and shorter visits every year since. I also spent two years at the newly established settlement of Maningrida during 1958 to 1960 as the wife of anthropologist Les Hiatt who was there to carry out social anthropological research. Rhys Jones, of the Department of Prehistory, Research School of Pacific Studies, Australian National University, Canberra, has collaborated with me on many of the expeditions carried out during the last decade.

Of course, the differences at one level are obvious. The Anbarra travel in motor vehicles, power boats and aeroplanes. They listen on their transistor radios to cyclone warnings and world news delivered in Pidgin from Papua New Guinea. They make tape recordings of their own song cycles which they play between the sounds of Charlie Pride and the Beatles. Some attend European-style schools and supplement their health care with that offered by the European health workers at Maningrida and in Darwin. These and other elements of Western European culture have been readily absorbed by the Anbarra but at the same time they have maintained a distinctively Aboriginal lifestyle. They inhabit an environment which allows them to satisfy their subsistence requirements with ease, so they find ample time to devote to religious and artistic pursuits.

By juxtaposing Leichhardt's snippets of information against a backdrop of Anbarra foraging life, it may be possible to interpret them as part of an Aboriginal subsistence system that presumably was functioning on the Alligator River flood plains when he was moving through them. His descriptions may be seen as a number of single frames surviving from a full length movie. If these frames are viewed against a complete movie of comparable content then perhaps it will be easier to grasp the full implications of the information contained in each of his frames.

'Red apples'

On 8 December 1972 the morning was 'still, overcast, sticky' – a typical (and unpleasant) beginning for a late dry season day. After breakfast Nancy Bandeiyama and her husband Frank Gurrmanamana decided to go to Djunawunya from their homebase Kopanga to collect some bukgrarrkgrarrk or 'red apples'. This fruit is probably Leichhardt's 'Eugenia'. On the way to Djunawunya, some four kilometres away, the foraging group consisting of men, women, children and dogs, captured some goannas, one blue-tongued lizard, a few small rats, and before quitting the inland plains for the Djunawunya dunes, they gathered some spike rush and lily corms from a dried out swamp.

At Djunawunya a 'dinnertime' camp was established under the shade of a fig tree where all the foragers sat protected from the hot sun and had access to freshwater which lay a few metres away in a deep well amidst a grove of pandanus. After lunch some children arrived from Kopanga, coming via the beach and not across the inland plains like the main party. On the way, those children had collected 'a great harvest of bukgrarrkgrarrk. They were in high spirits. It was difficult to keep track of how many fruits they had. They had been eating them all the way and still were'. I estimated that they must have accumulated about two hundred in all.

The senior man with the group, Gurrmanamana, began to crush these crisp astringent 'apples' by placing the flesh in a long, shallow wooden mortar and reducing it to pulp by crushing it with a pebble. People collected this crushed fruit with their fingers and ate it; it was 'all the same soft drink', they said.

When it was time to return to our home base the 'apples' which had not been eaten were broken in half, seeds removed and the resulting pieces threaded onto a supple stick whose ends were tied together with strips of pandanus frond. The attractive festoon was suspended from the forehead of a young girl so that it hung down her back (Photograph 18). Thus positioned the fruits were transported back to home base.

Swamp foods: 1972-1973

From swamps and annually inundated coastal plains the Anbarra harvest a suite of foods including the plants spike rush or water chestnut (*Eleocharis dulcis*) (Photograph 19) and various water lily products (e.g. *Nymphaea* sp.) (Photograph 20), ducks (e.g. *Dendrocygna arcuata, Anas superciliosa),* geese (e.g. *Anseranas semipalmata)* and their eggs; tortoises (probably *Chelodina rugosa*); several species of fish including *Oxyeleotris lineolatus,* the herring *(Megalops cyprinoides)* and at the beginning of the dry season barramundi *(Lates calcarifer)* when these large fish are feeding on the flooded plains; and some small crustaceans. In the past both freshwater and saltwater crocodiles were also culled from this habitat.

During the full year of dietary observation, 1972-1973, the Anbarra made little use of the largest freshwater swamp in the area, Balpildja, though they did harvest some foods from a series of small ephemeral billabongs that lay nearer to their coastal home bases (Photograph 21). Balpildja is situated just off the southern edge of Anbarra territory on Gunadba land but because these groups maintain good relations with each other the Anbarra are able to use the swamp at any time.

The Anbarra's failure to use the abundant resources contained in Balpildja during 1972-1973 was probably because they had not lived on their own land – and certainly not as part of a functioning foraging community – for some fifteen years. During this time they appear to have become temporarily uncertain of the

Photograph 18. The fruit *Syzygium suborbiculare* strung on a supple stick ready to be carried back to home base from Djunawunya, 1972 (Photograph: Meehan-Jones collection).

Photograph 19. Bandeiyama collecting water chestnuts from a freshwater swamp on Anbarra land. Middle of dry season, 1973 (Photograph: Meehan-Jones collection).

Photograph 20. Anbarra women examining a freshwater swamp early in the dry season, 1973. They are collecting edible water lily stems and corms (Photograph: Meehan-Jones collection).

Photograph 21. Anbarra women and children collecting the water chestnut *Eleocharis dulcis* from one of the ephemeral swamps close to Kopanga home base early in 1973 (Photograph: Meehan-Jones collection).

Photograph 22. Young Anbarra men and boys returning from Djibadjirra swamp with geese that they have just shot. Late dry season, 1978 (Photograph: Meehan-Jones collection).

Photograph 23. Anbarra people collecting fish from the freshwater swamp called Djigondjerra. Late dry season, 1978 (Photograph: Meehan-Jones collection).

Photograph 24. Aborigines in western Arnhem Land dining on buffalo meat in 1916 (Photograph: Ryko, by permission Institute of Aboriginal Studies).

Photograph 25. One of the mounds constructed from termite nests found in swampland areas on Anbarra territory. This one was visited after the abortive trip to Balpildja on 12 September 1972, and the whole area was burned. Goannas were captured from the mound (Photograph: Meehan-Jones collection).

Photograph 26. An Anbarra hunting expedition resting on the edge of the Djibadjirra swamp late in the 1978 dry season. They had been collecting geese and water chestnuts (Photograph: Meehan-Jones collection).

details of the way in which they had exploited their range before moving to Maningrida. They were no longer wholly familiar with the subtleties of seasonal change and consequently were unable to predict precisely when certain products would be suitable for harvesting.

The one trip the Anbarra did undertake to the big swamp during 1972 (12 September 1972) found it dry and parched. No geese were feeding on the swamp plants and the tasty spike rush was imprisoned in rock-hard lumps of clay. The foragers derived some compensation for the failure of their expedition by burning the swamp and its surrounding. This had not been done systematically for many years and everyone was pleased because that part of the terrain at last was 'clean' once again.

Swamp foods played an insignificant role in Anbarra diet during 1972-1973 providing a mere 2% of the gross weight of all food eaten during that time. About one quarter consisted of plant food, the rest was derived from five animal species (these calculations are based on data gathered during September 1972, and January, April and May 1973).

Despite the limited nature of these data, some predictable trends are recognisable. For example, more swamp foods were eaten during September 1972 than at any other time during the 1972-1973 year – about 1% of the total gross weight of all food consumed. Subsequent experience with the Anbarra has indicated that towards the end of the dry season, evaporating swamps are exploited heavily for tortoises, ducks, geese and spike rush. All these species were procured during September 1972 but only in small quantities. On the other hand, swamp foods played a more significant role in the total diet during January 1973 when they accounted for about 8% of the total gross weight of food procured during that wet month. January was a lean month all round during which any contribution to the diet, however small, was likely to have been significant (Meehan 1982: 152-153).

Swamp foods: post 1972-1973

Since 1973 the Anbarra have continued to live on their estates throughout the year, gradually regaining their mastery over the subtleties of the seasonal cycle. Freshwater swamps, including Balpildja, have again become an integral part of their annual round. During a five month period in 1978 freshwater swamps were being exploited heavily by the Anbarra and their neighbours. Between 11 September and 12 October seven expeditions were made to Djibadjirra, a specific location on the northeastern bank of Balpildja swamp. During that time 340 kg of geese were captured by the men and 180 kg of water chestnuts were harvested by the women.

Early on the morning of 1 October 1978 a group of women and children left Kopanga to spend several days at Djibadjirra where they intended to collect spike rush. By 4 October they had accumulated in excess of 100 kg of this corm. While this group was camped on the edge of the swamp some Nakara men from an outstation called Gorrong Gorrong arrived to shoot geese. I was able to count about 30 specimens but exact counting was a difficult exercise to record: 'It had obviously been a good haul – there were geese everywhere: alive and dead, cooked and uncooked. Legs and wings and heads were sticking out of dilly bags, baskets and billy cans, and everyone was eating them'.

The Anbarra women and children had been afraid while camped on the edge of Djibadjirra swamp. On the last night of their stay they had relocated their sleeping camp to a spot about one kilometre from the swamp's edge amidst the open sclerophyll forest which abuts it on its northern edge where they felt they would be more protected from the buffalo herds that were wandering about in the region.

On 10 October the Anbarra again visited Djibadjirra and again they were very successful. When they arrived early in the morning 'a huge number of geese (were) on the swamp' (Photograph 22). During their first foray into the grass-filled freshwater the hunters captured 54 geese. The second assault yielded another 29 making a total of 83 in all. While the men so successfully accumulated geese, the women gathered a mere 1.5 kg of spike rush and from ponds some fish which had become trapped as the water evaporated.

In 1979, while the Anbarra were participating in a ceremony at the Nakara site of Anadjerramiya, Balpildja was exploited extensively. On 12 October they travelled to Djigondjerra, a nearby location on the northern bank of Balpildja swamp. The stated purpose of the expedition was to procure fish from small waterholes within the swamp area (Photograph 23). In fact four food varieties were procured on that day: two species of fish, *Oxyeleotris lincolatus* and *Megalops cyprinoides* (36 kg), spike rush (2.0 kg), fifteen freshwater tortoises (21 kg) and two geese (5.0 kg), but fish contributed more than any other food – some 56% of the gross weight (see also Chapter 6 for Aboriginal use of fish poisons). On this day at Anadjerramiya, almost all the food consumed by its inhabitants came from swamp areas adjacent to the home base.

If this 520 kg of swamp food collected between 11 September and 12 October, approximately one month, is substituted for the amount collected in September 1972 (45 kg) then the contribution of swamp foods to the diet for that month would have been about 30% of the total gross weight. Based on my now considerable observation of Anbarra subsistence and the numerous discussions I have had with them about the importance of swamp foods during the late dry season, 30% seems to be a much more realistic figure than the tiny 1% recorded during September 1972.

OVERVIEW

There is no doubt that freshwater swamp foods play a significant role in the diet of the Anbarra people at certain times of the year. However, they form but one part of a complex system which allows the Anbarra to maintain such a high standard of living. The Anbarra characterise themselves as sedentary saltwater people who live chiefly on fish and shellfish. While this is an over-simplification of their actual diet, it is to some extent borne out by quantitative data. Seafood contributed about 60% of the gross weight of all food procured during 1972-1973, about one third of their energy (31.5%) and a highly significant two thirds (66%) of their protein intake. Of the total animal flesh consumed, seafood provided a substantial 86%, representing about 82% of the protein and 76% of the energy derived from this source (Meehan 1977: 1085-1095).

A summary of Anbarra diet over the four study months (Table 43 and Fig. 67) indicates that shellfish contributed the highest gross weight (31%) followed by fish (27%) and bought foods (18%). In terms of protein, fish provided most (51%), followed by European carbohydrates (18%) and shellfish (15%). Energy came from European foods (48%), fish (26%) and shellfish (5%). These percentages provide a general quantitative impression of the total diet, but they tend to disguise the fact that many food categories contri-

buted different amounts at different times of the year. For example, shellfish accounted for approximately 11% of the net weight of all food eaten by the Anbarra in September 1972. This rose to 20% in January 1973, dropped back to 11% in April and lower still to 6% in May (Meehan 1982: 152-155). We have already seen that the amount of swamp foods procured by the Anbarra varied markedly from month to month.

The content and quality of Anbarra diet varies from year to year as well as from month to month and day to day (as noted for the fauna in Part 3, and for the same climatic reasons). Variation over longer periods is well-illustrated by comparing the 1972-1973 year with the following one during which the marine shell beds were damaged by fresh water, and insignificant numbers of molluscan species, normally harvested from there in abundance, were collected. A similar dramatic change in foraging occurred between 1972-1973 (when only a small quantity of swamp food was procured), and 1978 and later, when items from that habitat have played a more substantial role in Anbarra diet.

The variation evident in Anbarra diet at daily, monthly, annual and longer intervals is reflected also in the number of species they collect. For example, during April 1973 well over 60 species of plants and animals were obtained. The wide range of products available to the Anbarra community ensures that they can enjoy an ample and varied diet throughout the year. It also means that almost everyone in that society can procure some food for themselves if the need arises.

ENTER THE BUFFALO

As yet, the buffalo does not play a significant role in the diet of the Anbarra people. A few individual animals only were noted during 1972-1973 and during that time buffalo flesh was included in the diet only a couple of times. Usually it came as a gift from Gunadba relatives at Gatandjindjurra. One beast which entered the wet season settlement at Lalarr-gadjirripa was chased some 12 km before it was killed. Its carcass was left unused because the successful hunter was not carrying a knife. The Anbarra community decided that a 24 km round trip was too far to travel even to collect such a large quantity of red meat.

Nowadays herds of buffalo containing up to thirty individuals can be seen grazing on the coastal plains near the Blyth River. Anbarra hunters still are a little wary of these large animals and only a few of the young men attempt to kill them. When they do, it is with guns from moving vehicles. Occasionally they are successful. Some of the middle-aged and older men relate stories about hunting buffalo with shovel-

Table 43. Approximate contribution made by various food categories to gross weight, energy and protein intake of Anbarra community during September 1972 and January, April and May of 1973.

	Percentage of gross weight	Percentage of energy	Percentage of protein
Animal flesh			
Shellfish	31	5	15
Crustacea	2	1	2
Fish	27	26	51
Reptiles	5	4	7
Birds	1	2	1
Mammals	4	8	4
Plant food			
Fruit and nuts	9	2	–
Roots and tubers	3	3	–
Honey	Present	Present	Present
European food			
Flour and sugar	18	48	18

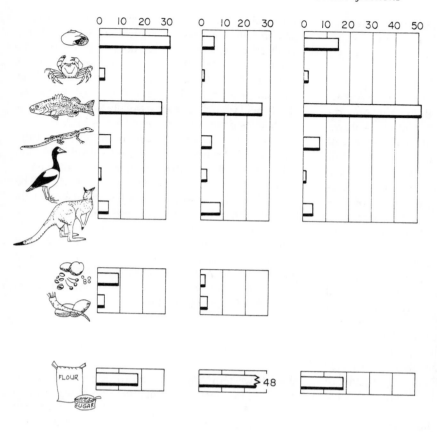

Figure 67. Food intake of Anbarra community during September 1972 and January, April and May 1973. Left column denotes percentage gross weight, centre column energy, right column protein (see also Table 43).

nosed spears when they were young and employed by European men in western Arnhem Land and near Darwin who were trading in hides before the Second World War.

During a short trip to the Blyth River in the middle of 1982, for the first time since I have known the Anbarra people, I was present when a buffalo was killed, butchered and eaten. This animal was sighted in the late afternoon of 18 June near a patch of monsoon forest called Guborrinbuna-abirrinarritjinga which lies about one kilometre from the home base Kopanga where we were camped. Three young men killed this animal using three cartridges fired from a shot gun. Some meat was removed and carried back to camp that night but early the next morning we all walked to the kill site where the rest of the carcass was butchered with tomahawks and butcher's knives. The flesh then was cooked in an oven composed of termite nest pieces. At the time of this catch the community was without a vehicle so it was fortunate that the animal was killed within an acceptable walking distance of the camp. Otherwise, as was the case during 1972-1973, the carcass may well have been left to rot. Nowadays the Anbarra community is in control of a two-way radio kept functional by a solar battery charger and it is relatively easy for them to ask kin who

own vehicles to come at short notice to transport buffalo meat. Even though the hunting of buffalo has not yet been perfected by the Anbarra hunters, their prowess has improved immensely since I lived with them for a year during 1972-1973.

Buffalo are much more common in the forested areas between the Liverpool and Blyth Rivers than they are on the coastal plains. The Gunadba, a Gidjingali-speaking community who inhabit this inland wooded zone south of Anbarra land have been culling them for well over a decade. This community contains several skilled hunters who learned their trade while employed in buffalo camps in western Arnhem Land. Nowadays, these men go hunting for buffalo regularly in order to acquire meat for the community at Gatandjindjurra. Though I have no figures for the relative success of these excursions, my impression is that they usually result in the slaughter of an animal.

For the Gunwinggu people living at Momega, an outstation situated to the south of Maningrida and southwest of Anbarra and Gunadba land, buffalo meat has become an important dietary item. According to Altman (1981: 280-281) 'about one buffalo a month has been shot regularly' and 'this meat is a significant source of bush calories and protein'. Altman (1981:

284) maintains that while 'this exploitation has required some modification in hunting techniques . . . The social organisation of production has, however, required no change'.

My assessment is that this would not be the case if the buffalo became a major dietary item for the Anbarra. If the buffalo is allowed to proliferate in Anbarra territory it is likely that the structure of the Anbarra diet will change radically and that this change will bring about considerable social upheaval within Anbarra society. All Anbarra land is coastal plain. It is a young and fragile landscape. Large numbers of buffalo grazing upon it over a prolonged period would be likely to damage it, especially those restricted areas of swamp, monsoon thicket and rainforest that it contains. Such damage would probably result in the disappearance of many plant and animal food species. Any reduction in the number and quantity of species available to the Anbarra would undermine the real strength of their subsistence – its broad base. It is unlikely at this stage that the entry of large numbers of buffalo onto Anbarra land would interfere with the productivity of the major contributor to their diet – the sea, but even this resource may be adversely affected in the long term – for example, by the degradation of the mangroves which constitute an important fish and shellfish breeding niche.

Already Anbarra foraging life has been influenced by the presence of a small number of buffalo on their territory. Wells and swamps once providing freshwater during the latter part of the dry season now dry up much earlier, become rock-hard and denuded of grass because of buffalo usage. The Anbarra have noted this, commented upon it and complain, but so far they have taken no steps to prevent its happening. Furthermore, because the Anbarra are still learning about buffalo behaviour they remain somewhat afraid of them and frequently alter their hinterland foraging strategies to avoid them. Children, in the past a normal part of foraging expeditions, are nowadays more frequently left at home base because parents fear for their safety. Adult foragers, who would normally fan out over the plains en route to a chosen hunting spot, now tend to travel along wooded tracts of land which they suppose will afford them some protection from any hostile buffalo in the area. The immediate effect of this adaptation is that people are unable to use these occasions to inspect the plains for resources which can be harvested at some later date. They are in danger of becoming less knowledgeable about what is happening in the landscape.

In 1983 buffalo were hunted by young men in vehicles using guns. Vehicles are driven by young men as older males and women tend not to acquire this skill. Young men are frequently away from their community, having driven the vehicles to Darwin or other communities, so that buffalo hunting tends to be a random rather than a systematic element of Anbarra subsistence strategy. Contemporary Anbarra subsistence is contributed to in some way by most members of that community – men, women and children of all ages. This proliferation of providers, like the wide variety of species available to them, is yet another insurance against want that is built into the Anbarra dietary system especially in the case of the socially and physically weak women with children and old women. If the buffalo population is allowed to multiply, the diversity of available food species will be reduced and the community may become dangerously dependent upon unreliable youthful male hunters. One optimistic element in this rather gloomy scenario is that the Anbarra will still have access to their coastal resources. Perhaps the richness of these will make it possible for the Anbarra to avoid the seemingly inevitable progression towards a diet made up almost entirely of buffalo beef and flour.

RETURN TO WESTERN ARNHEM LAND

We left western Arnhem Land for resource-rich Anbarra territory just as Leichhardt was moving through it in 1845 when the lives of the Aborigines located there had been little altered by the arrival of Europeans but where the buffalo already had become a familiar figure in the landscape. Since that time considerable environmental and social changes have taken place.

The water buffalo was introduced to Melville Island from Timor in 1828, only seventeen years prior to Leichhardt's journey, as a source of meat (Gillespie 1979: 5, Letts et al. 1979: 12). Subsequently some were released on the mainland where they soon increased in number. However, at the time of Leichhardt's expedition the effect of their expansion upon the environment appears to have been minimal. The Alligator Rivers wetlands were still in prime condition and Aborigines living on them were probably enjoying a high standard of living.

In the 135 years that have elapsed since Leichhardt's observations were recorded, substantial environmental changes have occurred in this region. Some of these have been the result of buffalo herds living there. Gillespie (1979) argues that this animal has caused a degradation of the wetlands and, by implication, of Aboriginal lifestyle. He concludes that:

What is certain is that *Bubalus bubalis* has changed the landscape very often to the detriment of the species upon which Aboriginal people depended and the species (the buffalo) was a potential competitor for man's water resources in the region.

(ibid., p. 69)

He also believes that Aboriginal social life in the Alligator Rivers region has been adversely affected by stress induced because of the economic and environmental results of buffalo presence (Gillespie 1979: 65) (Photograph 24). The explosion of their population attracted to the area European people who were anxious to exploit the animals for their hides. Some Aborigines were drawn to these camps where they became acquainted with an alien way of life which included alcohol, European disease and prostitution as well as a wide range of new foods and material objects.

A diminished number of Aboriginal people still inhabit the Alligator Rivers flood plains today – some 800 as opposed to about 2000 prior to European settlement (Keen 1980), while the buffalo population has increased dramatically to somewhere in the vicinity of between 15,000 and 20,000 (Letts et al. 1979: 12). Only recently has any attempt been made to arrest the destruction of valuable wetland resources.

How do the Aborigines living on the Alligator Rivers floodplain satisfy their subsistence requirements today? Although no long term quantitative work has been carried out in that area, qualitative data accumulated by a number of people with extensive experience in the area (McLaughlin 1982, Gillespie, Lindner & Morris, pers. comm.) indicate that the bulk of Aboriginal diet consists of buffalo meat and European foods such as bread, tea and sugar. Indigenous foods still are included in their diet especially those culled seasonally from the wetlands such as ducks, geese and their eggs, tortoises, file snakes and fish, spike rush and water lily. Yams and fruits are collected from the forests adjacent to the wetland areas. But these foods, although at certain times of the year harvested in abundance, are merely an adjunct to the staple diet of beef and carbohydrate. There is some indication that with the eradication of buffalo herds, and the regeneration of sensitive areas of the countryside, wetland foods could again take on renewed importance in Aboriginal diet.

THE WETLANDS IN THE PAST

There is archaeological evidence for the past use of the wetlands by Aborigines at the Blyth River and in the area of the Alligator Rivers. In both cases the evidence suggests that the wetland areas were used somewhat differently in the past. At the Blyth River, archaeological evidence indicates that in the recent past Balpildja swamp was somewhat more important than it is at present. Dotted around the banks of this swamp lie a series of humanly constructed mounds some 20 m long, 10 m wide and 2 m high (Photograph 25). They are sited where the highest level of the swamp meets the surrounding woodland. These mounds are composed largely of nodules of termite nest, which are still used as a cooking agent by the Anbarra, and various remains of human activity such as bone, shellfish remains, egg shell and stone tools. Preliminary results from a proposed long term archaeological project indicate that the construction of these mounds began round about 1500 years ago. Some were being occupied as recently as the mid 1950s by living Anbarra people (Sweeney 1955). The number and size of these mounds suggest that in the past large numbers of Aborigines spent considerable time camped around the edges of the Balpildja swamp complex.

Nowadays the Anbarra and neighbouring peoples visit the swamp mainly during the late dry season when the ground surface is dry and bare. If they camp there for more than a day they do so not on the mounds or even in the area of the mounds, but on flat ground closer on the water's edge (Photograph 26). They continue to use nodules of termite nest in ovens in which they cook goose, duck, tortoise and wallaby. On one occasion during 1979 Frank Gurrmanamana, his wife and oldest son constructed an oven from 50 kg of termite nest in which they cooked two wallabies. This earth was carried to a spot a few metres from Djibena, a mound which Rhys Jones and I had been excavating a few days before. The 1500 year old date came from the bottom of a two metre deep pit excavated into this mound.

Although considerable archaeological research has been carried out in western Arnhem Land (Schrire 1982, Kamminga & Allen 1973) little has been focussed on the wetland sites. However, a short survey of wetland sites carried out recently by a team from the Department of Prehistory at the Australian National University (Jones 1985) revealed evidence for dense human occupation of the edges of the wetlands in the Alligator Rivers area. These sites differ from those on the Blyth River. Most consist of dense surface scatters of stone tools sometimes covering large areas. One mound only was recorded and it was small compared with those found at Balpildja Swamp. The bottom of this mound, Ki'ina, is about 700 years old. Brockwell (1983) believes that the size and density of the Alligator Rivers wetland sites suggests that Aborigines living in the area in the past occupied sites for most of the year, merely moving a kilometre or so onto higher ground during the wet season.

What is to be learnt about Aboriginal use of wetlands by comparing the Blyth River region with western Arnhem Land? To the Anbarra, the coast and its resources are the mainstay of their subsistence. Wetland foods add seasonal variety to their diet. Their exploitation of the wetlands was more intense in the past, as shown by the large mounds present around the edge of Balpildja swamp. Nowadays people have vehicles, and so can exploit the swamps during the daytime and return to their preferred coastal home bases at night. So far the buffalo has only marginally affected the Blyth River wetland environment but ominous signs of destruction can already be observed.

Today, in the Alligator Rivers area, wetlands are of minor importance to the local Aborigines for whom buffalo meat and European foods take precedence. Swamp foods are exploited seasonally as they are at the Blyth River, but no precise data are available as to the magnitude of this exploitation. Whatever quantity it turns out to be, it does provide variety in a restricted diet. It seems likely that before Aboriginal subsistence was markedly modified by the entry of Europeans and buffalo into the area, the wetlands may have been the major focus for some Aboriginal groups inhabiting that region. The wetlands may have been to those people what the coast is to the Anbarra – the under-pinning of their subsistence: an inland coast. The archaeological evidence supports this view.

In western Arnhem Land Aborigines have moved away from a subsistence focussed on wetland products. This shift seems to have been encouraged by the arrival of Europeans in the area and by the impact that buffaloes have had on the environment. If buffalo herds were now eliminated, the Aborigines might face a chronic shortage of meat. And perhaps after so long the contemporary Aborigines do not wish to return to a pre-European type subsistence. The rub appears to be: the buffalo and Europeans have caused a diminution of wetland foods while providing an alternative food source in the form of buffalo meat. To rejuvenate the wetlands the buffalo have to be removed. Do some Aborigines wish to revert to the wetland subsistence and if they do is it too late to reverse a process begun so long ago by the releasing of a few *Bubalus bubalis* into the countryside with so little thought to its conse-quences?

Are the Anbarra heading in the same direction as the people of western Arnhem Land? Will the combined influence of European settlement and the incursion of more and more buffalo into their terrain mean that they too will become dependent upon this animal for their food? Or will the difference in the history of these two peoples plus the fact that the Anbarra continue to own a relatively undamaged coastal foraging area mean that changes in their subsistence will go in a different direction? Unlike the Aborigines in western Arnhem Land, the Anbarra have time to contemplate the op-tions. It will be interesting to see which path they will choose. Will they opt for buffalo beef and bread or for the maintenance of the abundant and varied sub-sistence they enjoy today? Or will the decision be made for them by Government authorities in Darwin and Canberra who at present are committed to a policy of eradication of wild buffalo in order to protect the Australian beef herd from brucellosis and tubercu-losis?

ACKNOWLEDGEMENTS

I thank the Anbarra people for what they have taught me during the last twenty years; CSIRO and ANPWS for inviting me to the workshop; Winifred Mumford and Betsy Jane Osborne for drawing the figures; Dragi Markovic for preparing the plates; Jill Johnston and Janine Mummery for typing the manuscript; and Danny Gillespie for introducing me to Ryko's photos.

REFERENCES

Altman, J.C. 1981. Hunting buffalo in north-central Arnhem Land: a case of rapid adaptation among Aborigines. *Oceania* 52: 274-285.

Brockwell, C.J. 1983. Wetlands in focus. MA qualifying essay, Department of Prehistory and Anthropology, Aus-tralian National University, Canberra.

Gillespie, D.A. 1979. Nganaburru gumwam (then the buffalo came). Unpubl. manuscript.

Jones, R. 1980. Hunters in the Australian coastal savanna. In D.R. Harris (ed.), *Human ecology in savanna environ-ments*. London: Academic Press.

Jones, R. (ed.) 1985. *Archaeological research in Kakadu National Park*. Canberra: Australian National Parks and Wildlife Service and the Department of Prehistory, Research School of Pacific Studies, Australian Nat. Univ.

Kamminga, J. & H. Allen 1973. Report of the archaeological survey. Alligator Rivers region environmental fact-finding study. Darwin: Govt. Printer.

Keen, I. 1980. The Alligator Rivers Aborigines: retrospect and prospect. In R. Jones (ed.), *Northern Australia: options and implications:* 171-186. Canberra: Research School Pacific Studies, Aust. Nat. Univ.

Leichhardt, L. 1847. *Journal of an overland expedition in Australia*. London: Boone.

Letts, G.A., A. Bassingthwaighte & W.E.L. de Vos 1979. *Feral animals in the Northern Territory.* Report of the Board of Enquiry. Darwin: Government Printer.

McLaughlin, D. 1982. A checklist of biotic resources utilized by Aboriginal people in the Alligator Rivers region. Northern Land Council in co-operation with the Austra-lian National Parks and Wildlife Service. Unpubl. report.

Meehan, B. 1977. The role of seafood in the economy of a contemporary Aboriginal society in coastal Arnhem Land. *Official Hansard Report*, 3 May. Canberra, Australian Government Publishing Service: 1085-1095.

Meehan, B. 1982. *Shell Bed to Shell Midden*. Canberra: Australian Institute of Aboriginal Studies.

Schrire, C. 1982. *The Alligator Rivers. Prehistory and eco-logy in western Arnhem Land.* Terra Australis 7. Canberra: Department of Prehistory, Research School of Pacific Studies, Australian National University.

Sweeney, G. 1955. Report of survey in the Liverpool-Blyth River region, August 1955. Darwin: Northern Territory Administration. Unpubl. report.

White, N.G. 1979. Tribes, genes and habitats: genetic divers-ity among Aboriginal populations in the Northern Terri-tory of Australia. PhD thesis, Latrobe University.

White, N.G., B. Meehan, R. Jones & L.R. Hiatt (in press). Diltjingara and madjuwa – back country and beach: Aboriginal population diversity and environmental corre-lates in Arnhem Land.

CHAPTER 13

A torrid land

M.G.RIDPATH, M.A.J.WILLIAMS & C.D.HAYNES

In the final chapter some of the key themes that run through this volume are drawn together in an attempt to evaluate the influence of the monsoonal climate upon the northern environment and its biota, and to take note of some of the human successes and failures in accommodating – culturally as well as physiologically – to this very special environment. We begin by considering the region in the context of its past.

A DYNAMIC LANDSCAPE – THE LAST TWO MILLION YEARS

In strong contrast to Indonesia, Papua New Guinea or the Philippines, much of tropical Australia is geologically old and stable, with the last major earth movements in the region discussed in this book dating back to 1,850 million years ago. The next significant tectonic events to influence this region were associated with the final breakup of Gondwanaland. India (or more precisely the lithospheric plate comprising Greater India) became separated from Australia 125 million years ago, and the Australian plate began to split from Antarctica and move north about 90 million years ago (Veevers 1984). As the Southern Ocean separating a drifting Australian plate from a stationary Antarctic plate grew wider and cooler, ice began to accumulate on Antarctica, first as mountain glaciers and eventually as major ice caps. The global effect of significant ice accumulation in Antarctica, apart from causing a fall in sea-level, was to increase the temperature and pressure contrasts between equator and south pole, thereby accentuating the intensity of atmospheric circulation in the southern hemisphere. In addition, as Australia moved closer to the equator, at a mean rate of about 5 cm a year during the last 30 million years, it came increasingly under the climatic influence of the zonal high pressure system located athwart the tropic of Capricorn, so that its hitherto humid and equable climate became more and more arid (Bowler 1982, Williams 1984).

A third but much more recent geological event of far reaching significance as far as world climates are concerned was the closure of the Panama Isthmus slightly over three million years ago. This caused diversion of a warm ocean current – the Gulf Stream – into the North Atlantic, and triggered a rapid growth of the North American ice caps, particularly from two and a half million years ago onwards. The waxing and waning of these great continental ice sheets involved a corresponding rise and fall in world sea-level, with the oceans up to 150 m lower than today during times of maximum ice extent. During each glacial-interglacial cycle – there were perhaps 25 such cycles during the last two and a half million years, each of roughly 100,000 years duration – there were associated changes in temperature, evaporation and precipitation. At a very broad level of generalisation, times of maximum northern hemisphere glaciation (the most recent of which culminated towards 18,000 years ago) were characterised in Australia by a climate that was colder, drier and windier than today. Interglacial climates, on the other hand, were comparable to the present-day climate of Australia, or may even have been somewhat wetter and perhaps slightly warmer than today (Williams 1985).

A clearer idea as to how the alternation from interglacial to glacial and back again to interglacial might have influenced the tropical northern environment may be gained from Figure 68 which summarises a great deal of palaeoclimatic information relating to the eastern half of Australia during the past 140,000 years.

The top curve (A) shows how sea-level has fluctuated from a level equal to or slightly above present level towards 120,000-130,000 years ago to over 140 m below present level at the height of the last glaciation, 20,000-18,000 years ago. The curve is based on dated flights of terraces of uplifted coral reefs on the Huon Peninsula of Papua New Guinea (see Fig. 68F for location) but is probably reasonably representative for northern Australia (Chappell 1983: Fig. 102). Also shown are the sea-level depths below which the Gulf of Carpentaria would have been exposed as land, and Torres Strait would have formed a land bridge linking Australia to New Guinea, thereby

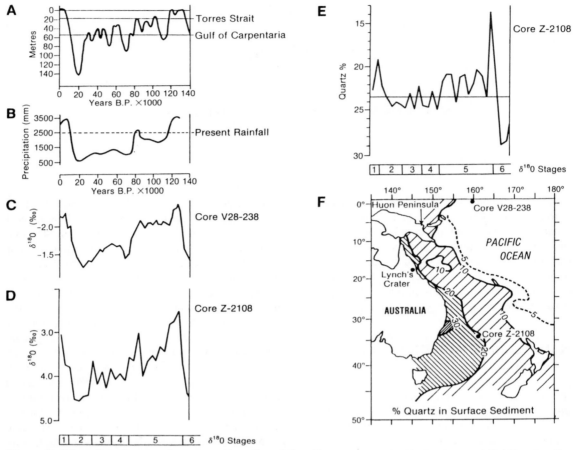

Figure 68. Palaeoclimatic curves for eastern Australia and the adjacent western Pacific for the past 140,000 years. Curve (A) refers to sea level; (B) to rainfall; (C) and (D) to ice volume and sea surface temperatures based on the oxygen isotopic composition of foraminifera in two deep-sea cores; (E) to the percentage of quartz (desert dust) in deep-sea cores, shown as a distribution map in (F). Note that in curve (E) the quartz content increases down the vertical scale. For sources of data, see text.

diverting the warm ocean current that flows from the Pacific past the Gulf and into the Arafura Sea. Two other conclusions of considerable palaeoenvironmental significance may be drawn from this curve. One is the oft-forgotten fact that interglacial times occupied but a small fraction of the full glacial-interglacial cycle – possibly a tenth or less, so that the present-day climate and the present familiar disposition of land and sea are in fact quite atypical of the past 100,000 years, and equally so of much of the preceding two million years. In fact, for nearly 80% of the time there was a fluctuating but inexorable build-up towards an ever lower sea-level associated with growth of the northern ice. Second, and very evident in the last section of the curve from 30,000 years BP onwards, is the great rapidity with which the final change can take place into and out of full glacial conditions. A sea-level rise of 140 m or so in the 12,000 years from 18,000 to 6,000 years BP implies a

phenomenally fast rate of flooding of the shallow Arafura Shelf, and an equally rapid change in the climate along the changing coastal margins.

Curve B is based on Kershaw's reconstruction of the vegetation history of the Atherton Tablelands of northeastern Queensland deduced from the types and abundance of pollen grains preserved in deep peats within Lynch's Crater (Kershaw 1978). The dotted horizontal line depicts the present-day mean annual rainfall at this site - 2,500 mm of rain a year. Rainforest species were even more abundant in this general locality during the last interglacial some 125,000 years ago and again at the start of the Holocene some 9,000-11,000 years BP, indicating that rainfall may have been up to 140% of present values at those times. However, except for a short-lived moist interval towards 80,000-85,000 years ago, inferred precipitation at all other times was less than today, and from about 80,000-20,000 years ago significantly so, with the 18,000 BP precipitation amounting to barely 20% of the present yearly average. The pollen data from Lynch's Crater therefore confirm that there was a long lead-up time from last interglacial until glacial maximum, during which the climate of northeastern Queensland was very much drier than it is today.

The next two curves (C and D) provide information about the ocean surface temperatures north and east of Australia. Both curves show the stable oxygen isotopic composition of the calcareous tests of marine foraminifera collected from two deep-sea cores in the Pacific. Core V28-238 is from the equatorial Pacific, at longitude 160°, slightly north of the equator (Shackleton & Opdyke 1976), and core Z-2108 is from latitude 35° and about the same longitude as the equatorial core (Thiede 1979) (see Fig. 68F for locations). The curves are expressed as the relative deviation (per thousand) of the ^{18}O content when compared to a known standard, and reflect changes in global ice volume as well as in sea temperature and salinity. Once again, the two curves show a saw-toothed progression from high ocean temperatures and low global ice volumes 125,000 years ago to low ocean temperatures and high ice volumes towards 25,000-20,000 years BP.

One consequence of the colder and drier climates which prevailed in eastern Australia during the long interval between the end of the last interglacial some 120,000 years ago and the start of the Holocene 10,000 years ago was an increase in the amount of quartz dust blown out to sea from the much expanded arid and semi-arid zones of Australia. Curve E shows that the quartz particle content of core Z-2108 is low during those times when curves A and B indicate high rainfall and high sea-levels, and very high during the cold episode preceding the last interglacial (Thiede 1979). That the quartz input is relatively high today may reflect the combined influence of modern devegetation and a general reduction in precipitation since the early Holocene, but it would be misleading to believe that all of this quartz was desert dust, since some may have been washed in from rivers and reworked by deep ocean currents. Nevertheless, as the percentage values on map F show, the source was predominantly from the Australian mainland, the amount of quartz diminishing rapidly with distance from the land.

To sum up thus far, during the last two million years the now submerged shelf north of Australia was frequently exposed as dry land as a result of glacially lowered sea-levels. The climate was in general both drier and cooler than today, with a much expanded arid zone. Plants and animals adapted to aridity would have been selectively favoured, although periodic returns to conditions at least as warm and wet as today would have enabled those more mesic elements of the coastal flora and fauna to move south once more as the seas rose and flooded their former habitats.

THE INFLUENCE OF THE MONSOONAL CLIMATE

Landscape and climate

The monsoonal lowlands of tropical northern Australia constitute a geographically unique environment. Failure to acknowledge and adapt to the physical and biological constraints imposed by this environment – constraints that are sometimes brutal, sometimes subtle – has periodically generated considerable economic hardship and social distress among the white Australian settlers of this region (Bauer 1964, Mollah 1980). The traditional Aboriginal communities have had their share of seasonal want, but today their troubles flow rather from the presence in their midst of sizeable urban settlements like Darwin than from any specific biological or social maladaptations to the tropical climate.

During the dry half of the year, from April until October, the land effectively functions as a semi-desert. Rivers and swamps dry out, weathering and leaching slow down and virtually cease in many areas, and annual plants die off. Dry season fires sweep through the woodlands and grasslands, burning off the brittle straw of spear-grass or wild sorghum, leaving the stony granite and sandstone hillocks and ridges exposed to the full erosional impact of the convectional downpours of late November and early December which herald the advent of the summer monsoon. Soil loss at this time of the year may amount to a net downslope movement of several hundred cubic metres per square kilometre from slopes of only a few per cent, but the rate falls off dramatically to perhaps a fiftieth of that rate with rapid growth of the wet season grass cover (Williams 1969, 1976). Needless to say, any disturbance of that plant cover during the height of the monsoon, whether by pigs, buffaloes or humans, can lead to vastly accelerated slopewash and gully erosion, with sediment yields locally exceeding 100,000 m³/km²/yr (Williams 1976). In this landscape there is a virtual absence of anything but skeletal soils on slopes steeper than 5%, and deeper soils are generally protected from rainsplash erosion by a layer of lag gravel – observable facts which the would-be sorghum farmers of Tipperary chose to neglect, to the detriment of both land and farmer.

During certain years, as a number of contributors to this volume have been at pains to emphasise (see e.g. Chapters 7, 8, 10 and 11), the harsh distinction between wet and dry seasons becomes blurred, and sporadic winter rains occasionally may disrupt the customary dry season drought, or, more commonly, a delayed start to the wet season may herald a season of intermittent rainstorms separated by long dry spells. Since annual rainfall in the seasonally-wet tropics of northern Australia is less variable than anywhere else on the continent (Jennings 1967), it may at first blush appear surprising that there are years when the wet season is intermittently dry and the dry season intermittently wet. Is there a discernible pattern to these years of extreme flooding or of prolonged drought – years which can cost the nation as a whole billions of dollars (Pickup & Minor 1980: 7)? And can we detect a common cause or set of causes?

One likely cause, which was referred to fleetingly in Chapter 2, involves the Southern Oscillation. Earlier this century a number of perceptive meteorologists came to the conclusion that Australia, India and much of Africa were part of a coherent climatic zone linked by the Indian Ocean (Williams et al. 1986). Rainfall within this zone is strongly influenced by year to year variations in the atmospheric pressure over the central equatorial Pacific, and that over the western Pacific/eastern Indian Ocean, including Indonesia and Darwin. The Southern Oscillation Index (there are several such indices, but all are variations on the same theme) is a measure of these relative pressure differences: positive during years of very low pressure off Darwin, when abundant summer rains fall over northern and eastern Australia, India and northeastern Africa; and negative during years of prolonged and anomalously high pressure, when droughts are widespread throughout the region.

The effects of the monsoonal climate upon the landscape are thus not to be assessed solely in terms of whether erosion will skim off one, ten or a hundred millimetres of soil from the surface in any one wet season, significant though that might be. Equally important is a knowledge of whether a rainfall surplus or deficit is purely local, or whether it is part of a wider pattern of global weather controlled by the Southern Oscillation, in which case the rivers and lakes, plants, animals and humans throughout northern Australia will be affected, as they will in countries as seemingly remote from the monsoonal lowlands of Australia as China, India, Ethiopia and the Sudan.

Climate and biota

The aquatic environment undergoes two very important annual changes as a result of the heavy but seasonal rainfall. In the first place, the area covered by water up to two metres deep expands enormously during the wet seasons: for example from a little over 10 km^2 on the Magela floodplains to cover 150-200 km^2 in most years, and occasionally 300 km^2 (Chapter 6). Thus, for a few months, there is a huge increase in the amount of habitat available to aquatic vegetation, aquatic invertebrates, fishes, tadpoles, aquatic reptiles such as crocodiles, and waterbirds. The great majority of these species reproduce then, when the extent of their habitat, and the resources with which it provides

them, are at their maximum. However, the effect on any non-amphibious species, such as wallabies and most small mammals, is to force them off the floodplains at this time. In the second place the very high tropical evaporation rates, due to constantly high temperatures, mean that most of this aquatic habitat disappears during the dry season, while some of the remaining patches of water become saline. Hence, in the dry season most aquatic plants and invertebrates enter a dormant phase and most of the aquatic vertebrates endure in or near deep freshwater refuges. However, a few species of aquatic vertebrates, such as the saltwater crocodile and some fishes, are adapted to withstand salinity and are able to use the estuaries during the dry season when they become salty.

The terrestrial environment, although it is never flooded, is as greatly affected by the highly seasonal rainfall as is the aquatic environment. The amount of moisture in its soils alternates annually from that found in a hot humid rainforest to that in a hot dry desert. The severity of this annual oscillation is shown most strikingly in Figure 55 of Chapter 10. Indeed the severity of the effects of the annual tropical aridity on the biota is such that much of the monsoonal lowlands could practically be regarded as a kind of semi-desert, ameliorated by a heavy annual rainfall! The annual desiccation of most soils explains the absence of rainforest or monsoon forest in all but a few tiny pockets, where ground-fed moisture persists near the surface all year round (Braithwaite et al. 1984). There is thus a marked paucity of characteristically rainforest species in the flora and fauna of the monsoonal lowlands as a whole. This also explains why the vertebrate animals on the whole are more closely related to those of the adjacent but rather drier Kimberley region than to those of more humid northeast Queensland or Papua New Guinea.

Inevitably, some degree of adaptation to greater aridity has occurred among a number of species with rainforest origins, which today are found in less mesic habitats in the monsoonal lowlands, for instance, species of *Terminalia* and *Buchanania* among the plants. By and large, however, it appears that any long-term swing to aridity has been accompanied by the replacement of species of plants and animals adapted to moist conditions by dry-adapted species already in existence. Thus, for instance, reptiles which as a group withstand aridity well, form a particularly significant component of the fauna of the monsoonal lowlands.

In keeping with the seasonality of reproduction enforced by the climatic regime, primary and secondary production of the aquatic and terrestrial ecosystems are seasonally pulsed. In broad terms, the pulse starts during the wet season and peters out in the early part of the dry season, although some species have a different pattern (Chapters 8, 9 and 10). However the timing, duration and amplitude of the pulse varies significantly between years because of the variable pattern of the annual wet season (Ridpath 1985, and as discussed earlier). The erratic character of the rainfall differs only in timescale from that in much drier regions of Australia and likewise gives rise to very flexible patterns of breeding by some species of animals (Chapter 10). Thus we see in the monsoonal lowlands the response of opportunistic breeding and the occurrence of plaguing, both of which are usually considered characteristic of much drier climates.

Local mobility allows many species of birds, mammals, fishes and some reptiles to escape the rigours of the dry season in the monsoonal lowlands by moving to refuges. This is another parallel with drier regions. Denied the mobility of animals, perennial plants largely cease growth in the dry season, some relying on storage of food to tide them over till next year's wet season. This phenomenon is important to some herbivorous animals, which have adaptations to make use of it. For instance the underground food-storage corms of the prolific rushes *Eleocharis* spp. on the floodplains are the major resource upon which the very abundant Magpie Geese and Dusky Rats depend for their survival in the dry season. They were also an important source of food for Aborigines (Leichhardt 1847, Jones 1980, Meehan Chapter 12).

Much of the annual vegetation of the monsoonal lowlands, including that of the floodplains, senesces and dries out over the dry season. Although the breakdown of litter is rapid in the monsoonal lowlands (P.A.Werner, pers. comm.), much of this very dry annual material remains during this period, and a significant part of the perennial vegetation also becomes dry. All this flammable matter creates a relatively large fuel load each dry season. As a result fire, whether set by humans or lightning, forms one of the most significant seasonal influences on the environment (Chapters 4, 5 and 10). In fact, fires are more frequent, though less intense, in the grassy-floored forests and woodlands of the wet-dry tropics than they are in any other part of Australia (Walker 1981). We now have a good idea as to how Aborigines have used fire in this environment (Chapters 5 and 12). Nonetheless, the immediate effects of present-day European and Aboriginal fire regimes in the monsoonal lowlands are not yet fully described. The dynamic consequences for the biota appear complex and they comprise one of the major gaps in our understanding of tropical ecosystems.

HUMANS IN THE MONSOONAL LOWLANDS

Humans have been in Australia for at least 40,000 years, but probably for less than 120,000-130,000 years (Jones & Bowler 1980, Singh et al. 1981). Present evidence shows they have occupied the monsoonal lowlands for a minimum of 23,000 years (White 1971). By any count they are a late arrival in a land which hardly any other eutherian land mammals even reached. We consider here the impact humans have had upon the monsoonal lowlands, and then the influence this tropical environment has had upon them, their way of life and their prospects.

Human impact on the monsoonal lowlands

Early humans on every continent used fire as their principal means of local manipulation of their habitat to their advantage. There seems little doubt that Australia's first people were already using fire when they arrived (Clark 1983). It is equally certain that natural fires, normally caused by lightning, occurred regularly long before humans arrived (see Chapter 5). Sclerophyllous vegetation burns more readily and recovers more readily than that associated with lower fire frequencies, and also with colder and/or wetter conditions. Thus increases and decreases in the frequency of fire (as indicated by charcoal in deposits) during at least the last 350,000 years in Australia have been associated with corresponding increases and decreases in the amount of sclerophyllous vegetation (Singh et al. 1981). Until the advent of humans in Australia, it must be presumed that climatic oscillations determined the extent of this flammable vegetation. We may conclude that the frequency of fire before humans arrived would have depended indirectly on temperature and annual rainfall, and directly on the frequency of lightning.

What happened in consequence of the arrival of humans is still much in dispute (Merrilees 1968, Latz & Griffin 1978, Stocker & Mott 1981, Horton 1982). Compared with the period beforehand, it is clear that after about 120,000 BP the extent of *Eucalyptus* and certain other notably fire-tolerant sclerophyllous vegetation, and the frequency of fires, all began to increase markedly (Singh et al. 1981). That change in the frequency of fire started c. 80,000 years before the earliest authenticated human presence in Australia. We do not know whether this change was due to a change in climate and associated vegetational changes, or was in fact due to a much earlier advent of fire-using humans than presently recognised, or to some combination thereof (Singh et al. 1981). Whatever the date of the arrival of humans in Australia, what we know about their traditional burning regimes in the monsoonal lowlands today (Chapters 5 and 12, Jones 1980) suggests that the prevailing pattern and frequency of fire would have changed once humans were present. Whether this change was sufficient to cause major shifts in vegetation is still hard to judge. The most we have to go on is some depositional evidence from a site within Kakadu National Park (Jones 1985). This indicates a sudden major increase in the erosion from higher land early in the known human occupancy of the monsoonal lowlands. The deposits would be consistent with some change in vegetation that could be argued to have been caused by a perturbation such as a change in fire regime or climate.

Burning aside, it seems unlikely that any other activities of these early hunter-gatherers would have had a major impact on the Australian biota (Calaby 1971), including that of the monsoonal lowlands. The only other obvious ecological consequence of the arrival of hunter-gatherers was their importation of dogs c. 3,500 BP (Millham & Thompson 1976). These gave rise to the wild predatory dingo, although we do not know what effects it had on other animals. It may be significant that its arrival more or less coincided with the disappearance from the monsoonal lowlands of the indigenous predatory Thylacine *Thylacinus cynocephalus* of about the same size and possibly a competitor (Calaby 1980). Many Aboriginal rock-paintings of the Thylacine still exist in Arnhem Land today.

European people only settled in the monsoonal lowlands c. 160 years ago, yet have been responsible already for certain major changes. Firstly, after their arrival the Aboriginal population disappeared almost altogether from a wide area while the pattern of occupation and use of the remainder of their range changed drastically (Keen 1980). This has resulted in considerable changes in fire regimes, with burning occurring more erratically and mostly later in the dry season than was the case under traditional Aboriginal regimes (Chapters 5 and 12, Braithwaite & Estbergs 1985). The consequences for the biota have yet to be investigated. Secondly, European people brought with them a number of plants and domesticated animals from other parts of the world. Some of these have escaped and now live wild, certain species causing important changes in the ecosystem (Chapters 6 and 11). In the case of feral buffaloes a major campaign to reduce numbers has partially reversed these changes. However feral pigs and many exotic weeds, such as *Hyptis suaveolans*, are much less amenable to control. The spread of the South American prickly bush *Mimosa pigra* continues unabated across the wetlands (Miller et al. 1981). Its effects on the wetlands ecosystem may well turn out to be even more drastic than those of buffaloes (e.g. Braithwaite et al. 1989). It is highly probable that other introduced species, such as the Cane Toad *Bufo marinus* and Mission Grass *Pennisetum polystachyon* will invade the monsoonal lowlands in the foreseeable future. Indeed the latter species has already begun to do so.

It is clear that the introduction of buffaloes has affected both the resources available to Aboriginal communities still living traditionally, and their lifestyle (Chapter 12). All in all, it can be seen that the advent of Europeans has had some important repercussions on humans and the biota of the monsoonal lowlands, and that more are in train. Nevertheless, compared with the more densely settled parts of southern Australia, the monsoonal lowlands remain significantly less affected by European colonisation, so far.

The influence of the monsoonal lowland environment on humans

Human utilisation of any environment depends ultimately on water. People need it to drink and, equally vitally, to drive the system on which they depend for their food, whatever that may be. In the wet-dry tropics the presence of a large and reliable supply of water sufficient to meet these needs year-round is a much more acute constraint on human life than in cooler climates. That is because the permanently high temperatures mean that there is constant heavy loss by evaporation both from the surface of the soil and of free water (Chapter 2).

The geomorphic and hydrological processes described in Chapters 1, 2 and 3 gave rise to vast, flat, low-lying areas of impoundment. This has resulted in very extensive and fertile wetlands and freshwater swamps in the monsoonal lowlands, fed by heavy annual rainfall and a constantly high input of solar radiation (Chapter 2). Thus humans had around them there just the sources of moisture, free water, nutrients and energy to generate, in abundance, the food resources that they needed. These tropical sub-coastal wetlands were, and still are, a remarkably rich habitat for humans in terms of the quantity and variety of their natural foods that are provided, particularly aquatic ones (Photographs 27a, b, 28, Plate 30; Chapter 12 and Jones 1980). This is in line with an ecological situation in which, for instance, the net annual primary production of swamps is up to five times greater than that of adjacent savanna or forest in the tropics generally (Golley & Misra 1972). Indeed, when Europeans first arrived in Australia the density of Aboriginal tribes in the monsoonal lowlands was equalled only by that in some parts of tropical coastal northeast Queensland (from recorded boundaries in Tindale 1974, see also Keen 1980). The territories of hunter-gatherer social groupings can be related essentially to the availability of food (Peterson 1975).

Early evolution of humans probably occurred in the tropical savannas of Africa (Harris 1980, Jones 1980). There the climate must have been torrid and, indeed, would have much resembled that of the Australian monsoonal lowlands today, apart from being somewhat drier. This would explain possession by humans of an outstandingly efficient cooling system. Humans keep cool by excretion of copious sweat from very numerous glands and subsequent evaporation from the very large surface area of their hairless skin. Though not unique, such external evaporative cooling is much more highly developed than in other mammals which rely largely on internal evaporation via respiration.

The human cooling system is particularly well-adapted to the environment of the tropical savannas (Harris 1980). Nevertheless, from their tropical origins humans later extended their range into subtropical, temperate and even sub-arctic environments. In view of this dispersion it is not surprising that human representatives remaining in the tropics have retained more marked tropical adaptations than those from higher latitudes. Thus Aborigines in tropical Australia sweat nearly twice as much and drink twice as much as people of temperate provenance on arrival in the tropics (Macfarlane 1981a). However, after arrival immigrants to the tropics go through a long process of acclimatisation at the end of which they actually sweat twice as much as indigenous tropical inhabitants (Fox et al. 1974, Edholm 1978). This suggests that sweating in immigrants from temperate regions has to compensate for some other, less adaptable aspect of the physiology of their regulation of body temperature. Furthermore, in the tropics people from temperate regions take five times longer to adjust their sweating rates according to changes in the ambient relative humidity than do Aborigines (Macfarlane 1981b). Such relative inflexibility of response is also indicative of poorer European adaptation to tropical conditions.

The proper functioning of this vital human cooling system is at risk in the tropics because of the potentially damaging effects on skin and sweat glands of medium wavelength ultraviolet solar radiation, which is at its maximum intensity near the equator (Hunter 1981). However, Aborigines, in common with all other indigenous tropical peoples, have melanin dispersed widely in their skin and this protects their sweat glands by blocking such radiation (Harris 1980). By contrast, white-skinned people living in tropical Australia lack this protection and have the highest incidence of skin cancer in the world (Gordon et al. 1972, Hunter 1981, Macfarlane 1981b). This practically never occurs among Aborigines. Most such skin cancer is curable, but one rare form tends to be lethal.

The dispersion of humans to cooler climates during their evolutionary history entailed a need to produce more bodily heat than was necessary in their original tropical environment. Thus basal metabolic rate (rate of production of heat) is c. 10% higher among people of European descent than it is among Aborigines (Macfarlane 1981b). In the Indian tropics, present evidence suggests that only around half those coming from cooler latitudes are able to reduce their rate of heat production to accommodate to a hotter climate (Mason & Jacob 1972). Again we see a significant lack of physiological adaptation among immigrants from cooler latitudes to the tropics. In fact, in the monsoonal tropics of Australia, the values of ambient temperature modified by relative humidity often exceeds the maximum recommended for sustained effort by workers of temperate regions (Thrift 1982).

The evidence discussed in this section has emphasised two important ways in which the environment of the monsoonal lowlands has influenced humans. These are the richness of the aquatic and wetland systems as human habitats, and the physiological demands of the torrid climate. Aborigines exploited these natural riches very effectively, and possess the various necessary physiological adaptations to the tropical climate. It is pertinent that full-scale settlement of lowlands within 12° from the equator by people originally derived largely from populations from northern Europe (latitudes 50-60°) has occurred nowhere else in the world in recent times, except along the climatically more benign coast of northeast Queensland (see below). It constitutes a gigantic experiment. So meagre remains our knowledge of the basic ecology of humans in the tropics – we know more about cattle! – that it is impossible to predict the outcome. We venture only that the results are virtually certain to be different from the human pattern in the rest of Australia.

Environmental conditions and demography

One of the outstanding features of our species is its adaptability. *Homo sapiens* has spread to the four corners of the earth and has developed and prospered everywhere. Nonetheless historians, geographers, social scientists, doctors and biologists remain divided upon whether groups from one environment can manage equally well in all others; and whether or not human success in colonisation is purely a matter of tradition and culture. At the end of the last section we put the view that only time will tell what weight any genetic differences in human adaptations actually have in populating the monsoonal lowlands. We shall not speculate on the issues involved. Instead we summarise now some of the outstanding features of human ecology in this environment.

Widespread use of various cooling procedures, nowadays especially of air-conditioning, highlights the pressure of the climate on temperate humans in tropical environments. Relative heat-strain on humans can be expressed by an index which takes account of physiological parameters, relative humidity, ambient temperature, solar radiation and clothing (Hounam 1970). This index was established relative to people with temperate origins. In the monsoonal lowlands, on the basis of the index, there are on average 180-225 days each year when Europeans experience strain from heat. This compares with only 50-75 such days in the coastal regions of tropical northeast Queensland, where the climate is ameliorated by the influence of the Pacific Ocean. Coastal northeast Queensland is more intensively settled by Europeans than the monsoonal lowlands, although the causes of this difference almost certainly also include economic factors.

Another striking aspect of human ecology in tropical Australia is the very high death-rate. Age-adjusted annual rates of death from all causes, for men and women, are more than twice as high in the Northern Territory as they are in the other Australian states (Fig. 69). Whereas almost all the Northern Territory lies within the tropics and two-thirds of its population lives within 15° of the equator, the populations of the other states live entirely or almost entirely outside the tropics.

Among the possible factors contributing to this difference three can be identified at present. Rates of death from accidents in motor vehicles are at least

twice the national average (Wigglesworth 1981, Stack, n.d.), as is the death rate from cardiovascular disease (Fig. 69). Aboriginal infant mortality stands at four times the national average (Burnley 1984). It may also be significant that annual consumption of alcohol in the Northern Territory, computed as litres of pure alcohol per capita of population (in 1982-1983), was c. 1.5 times higher than the national average (Australian Bureau of Census and Statistics in 1983 and figures provided by the Drug and Alcohol Bureau, Northern Territory Government). This rate of alcohol consumption may be implicated in the higher mortality rate from cardiovascular disease in the Northern Territory. However it is open to question whether alcohol abuse alone is responsible for the fact that mortality from cardiovascular disease stands at double the rate of anywhere else in Australia. Unlike heart disease, the death rate from all forms of cancer (most of the skin cancer is not lethal) does not differ from elsewhere in Australia (Fig. 69). The higher death rate from motor vehicle accidents seems due mainly to the much higher proportion of long distance travel on roads outside major urban centres than in other states (Hewitt & Sanders 1984).

Tropical infectious diseases do not appear to be an important cause of human mortality in the monsoonal lowlands today. Malaria and Dengue Fever have been eliminated during the last forty years (Whelan 1981, Stanley 1984). Aborigines exhibit evidence of high exposure to the arboviruses responsible for Australian Encephalitis and Epidemic Polyarthritis, but seem to possess substantial immunity to these diseases (Whelan 1981). Both diseases can be more serious for Europeans and, though fairly sporadic at present, their incidence could increase with any more extensive settlement in the monsoonal lowlands (Stanley 1981, Whelan 1981). Even after taking into account motor vehicle accidents, alcohol consumption, cardiovascular disease and Aboriginal infant mortality, the very high annual death rate in the Northern Territory remains hard to explain.

A third notable feature of human ecology in tropical Australia has always been the transience of the European population, especially in the monsoonal lowlands (Young 1983, Bauer 1984). Thus, the proportion of the population over the age of five in the Northern Territory which had arrived from outside between 1976 and 1981 was five times greater than the proportion in southern states (Heatley 1983). By contrast Aborigines in the Northern Territory are very sedentary. Furthermore, Aborigines make up a quarter of the total population (Black 1983), which means that the rate of European transience must actually be higher than that indicated by the figures for arrivals just quoted which lump both groups. Transience is also strongly indicated by the turnover of the workforce,

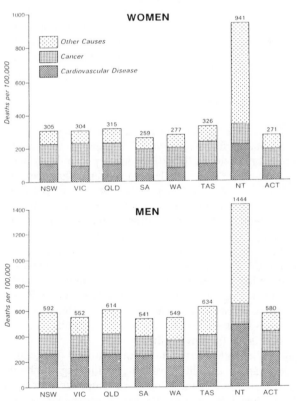

Figure 69. Age-adjusted death rates per 100,000, aged 30-64, in Australia (by courtesy National Heart Foundation). CVD = Cardiovascular disease. The upper portions of the histograms represent all causes of death other than cancer and cardiovascular disease.

almost a third of which in the Northern Territory comprises government employees of one kind or another and are mostly European (from data in Cameron 1982). Thus a recent analysis by the Public Service commissioner estimated that 25-30% of the staff leave Northern Territory Government employment each year (Pope 1985). Of those departing, 87% of the locally recruited and 97% of the externally recruited left within five years of joining. Such a degree of transience after more than 100 years of settlement is unusual; even though it is understandable in the smaller, recently established mining communities in the North (Neil et al. 1984). An incidental but significant consequence of European transience, largely overlooked, is what can be termed the short community memory of the European population. Thus a European tradition embodying past ecological experience in the monsoonal lowlands is notably lacking. As a result past mistakes are often repeated, and possibly genuine promise fails to be realised through European ignorance, for want of continuity from one generation to the next.

In spite of European transience there has been a steady increase in the total population of the Northern Territory, normally at a rate of 3-4% p.a. in recent years (Black 1983). Thus the total population grew from c. 98,000 to c. 134,000 between 1976 and 1983. In the case of Europeans the major component of the increase was due to migration, mainly from elsewhere in Australia, whereas among Aborigines natural increase was the only factor involved (Mitchell 1981). Compared with the entire Australian population, that of the Northern Territory comprises an appreciably higher proportion of people below the age of forty (Heatley 1983). The Aboriginal component has an even younger age structure than that of the Northern Territory population as a whole (Black 1983). This is largely due to the Aboriginal rate of natural increase which is double the Australian average (Burnley 1984). This high rate of natural increase occurs in spite of the fact that Aboriginal infant mortality, now falling, is still four times the Australian average.

We can sum up this section as follows. By temperate human standards, ambient environmental heatloads are distinctly stressful in the monsoonal lowlands. An unusually youthful and remarkably transient European population is growing fast, mostly because of high net inward migration. An even younger, wholly sedentary, Aboriginal population of about a third the size, is growing about as fast, but entirely because of natural increase. Its rate of natural increase has risen as mortality, especially among infants, has been reduced (as has occurred widely among the populations of developing countries as medical services have improved). The death rate of the whole Northern Territory population is double that of all non-tropical populations in Australia. This can be attributed in part to much higher death rates from cardiovascular disease and motor vehicle accidents, and among Aboriginal infants, and probably also to the significantly higher consumption of alcohol. There could well be more to it than that.

The study of human ecology, especially environmental physiology and demographic patterns and processes, has only just begun in tropical Australia (see Parkes 1984). Our discussion suggests it offers unique and exciting opportunities for objective scientific research. The results could also be of benefit to the community as a whole. They should provide a clearer understanding of human aspects of what is involved in attempts to develop the Australian tropics. In this connection the recent establishment of the Menzies School of Health Research in Darwin is a promising step within the medical arena.

GETTING A LIVING

Hunter-gatherers were very successful in exploiting the seasonal and variable resources of the monsoonal lowlands, as indicated by the high original density of tribal groupings in this environment (Tindale 1974). They achieved this by a strategy of local mobility, in conjunction with technologically well-developed sets of implements, traps and other procedures such as the use of fire (Chapters 5 and 12, Jones 1980, Schrire 1982). Their cultural adaptations had been acquired over many generations of living under these tropical conditions. By contrast, the temperate context and culture of European explorers and colonists of Australia had little application to the conditions of the tropics. No doubt this fact influenced their first perception of the monsoonal lowlands.

The tropical north coast may well have been the first part of Australia ever reconnoitred by Europeans, early in the 16th century (Wallis 1982). The region had to wait a hundred years to be seen by Europeans again, this time by the Dutch in the early 17th century (Bauer 1964). In view of this and other such knowledge of the north, it can hardly have been a mere historical accident that the first Europeans chose to colonise the temperate south instead. Europeans had established tropical trading outposts in Timor and other parts of Indonesia as early as the beginning of the 16th century. Nevertheless, it took another 350 years before they even attempted the establishment of a self-sufficient community, lacking the cushion of indigenous labour, in the nearby tropics of northern Australia. In fact the patterns of settlement of Australia shows how unequivocally most of the early colonists chose the south; even though, from the outset, the government made great efforts to promote settlement of the monsoonal lowlands (Jones & Bowler 1980). This was even in spite of the fact that the colonists included some with tropical experience in Asia and Africa. Nonetheless, it is not surprising that they chose as they did.

We contend that the first colonists favoured the south because of their perception of the basic natural differences between temperate and tropical Australia. Thus they recognised in particular that their own agricultural technology was based upon roughly similar natural foundations to the former (though there were some surprises in store!), but were quite different from those of the tropics. Only later, hesitantly, and with limited success so far, has a temperate culture spread into the monsoonal lowlands. Indeed, attempts to use the monsoonal lowlands still often suffer from assumptions imported from temperate experience. Nowhere has this been more obvious than in attempts to introduce intensive agriculture.

Sustained, high-yielding agricultural systems, other than intensive peasant agriculture, were developed in temperate regions of the world, and only in comparatively recent times (Williams 1981). Development of such systems for the tropical world has lagged far behind. Thus five major attempts this century to develop cropping in the Australian monsoonal lowlands (the Daly River, Ord River, Humpty Doo, Tipperary and Willeroo schemes) have all failed (Christian 1977, Fisher et al. 1977, MacKenzie 1980, Davidson 1980, Bauer 1985). A sixth attempt, the Douglas-Daly scheme, started in 1982 (Cameron & Hooper 1985). It is still too early to assess its future, but first results were disappointing (Anon. 1983). Apart from the economic disadvantages of a remote and sparsely populated region, these failures could be attributed to characteristic features of all ecosystems in the wet-dry tropics (Ridpath 1985, Ridpath et al. 1985). Pre-eminent among these factors was the variability of the onset and duration of the annual rains, speedy desiccation of soil resulting from the permanently high tropical evaporation rate, and the restricted growing season. Leaving aside the seasonally flooded but relatively fertile sub-coastal plains, the great majority of the soils are too infertile for dryland cropping. Even the best are still poor in nutrients by temperate standards (Isbell 1985, Williams et al. 1985). Other special difficulties experienced include frequent occurrence of erosion and waterlogging due to the intensity of tropical storms, lethal soil surface temperatures, hard soil crusts, and the variety, abundance and persistence of insect pests and plant diseases (see also Muchow 1985). Some of these difficulties might perhaps be overcome by minimum-tillage, leguminous, ley-farming systems, which are more appropriate to tropical conditions. These are now being investigated (McCown et al. 1985).

Attempts to develop plantation forestry in the monsoonal lowlands have been no more successful than attempts to develop cropping (Lacey 1979, Cameron 1985). The main tropical impediments have been the general infertility of the soils, the curtailment of growth due to the stringency of the long annual dry season and frequent attack by abundant termites. All of these factors can be seen at work in the native forests too. An equally important reason for failure, as in other fields, has been that attempts were often made to use species relatively ill-suited to the rigorous tropical conditions of the monsoonal lowlands. This aspect is now becoming better understood. Thus more recent trials with a conifer *Pinus caribaea* from provenances with a similar environment to that of the monsoonal lowlands, have shown some limited promise (Cameron 1985).

Horticulture is limited by the same constraints in the monsoonal lowlands as are cropping and plantation forestry (Scholefield & Blackburn 1985). Nonetheless certain species of tropical fruits, particularly mangoes, are well adapted to this environment. They also have certain economic advantages because their fruits ripen well before those in other tropical regions of Australia or overseas. Thus production of mangoes and some other tropical fruits and vegetables, including cashew nuts, in a few favourable areas in the monsoonal lowlands appears to have some potential. This will also be stimulated as local demand for fruit and vegetables increases with any growth in population. The limited areas where fertility and soil moisture in the dry season are good enough for horticulture usually adjoin the wetlands and their varied and abundant wildlife. In that situation crops are bound to suffer some damage from native rodents, fruit-bats, wallabies and feral pigs, an issue which will need attention.

Beef production in the monsoonal lowlands is limited by the low nutritional quality of tropical native pastures and the severity and length of the annual dry season, when cattle may lose up to one fifth of their weight (Chapter 11). The value of the industry is less than that of tourism and is declining (Winter et al. 1985b). There has been considerable degradation of native pastures. Introduction of tropical breeds of cattle has occurred slowly in recent times to replace temperate breeds ill-adapted to the poor forage, high temperatures and parasite stresses in the monsoonal lowlands. Nevertheless there is a trade-off between superior tropical adaptations on the one hand and, on the other, reduced animal productivity and increased pressure on the pasture (Ridpath et al. 1985, Winter et al. 1985b). This is a biological factor of which developers need to be aware. Research on the intensification of beef production has concentrated on mineral supplements (which somehow allow cattle to utilise low quality pastures better), and on the introduction of better quality pasture plants (Winter et al. 1985a, b). Buffaloes are a tropically adapted species. As feral

Photograph 27a. A catch of aquatic file snakes, Mary River 1916.

Photograph 27b. A catch of catfish, Mary River 1916 (Photographs: Ryko, by permission Institute of Aboriginal Studies).

Photograph 28. A catch of Saratoga *Scleropages jardini* at Meeguinea in north central Arnhem Land, October 1979. Note the same basic method of carrying big hauls of aquatic food is used as 60 years previously, shown in Photographs 27a and b (Photograph: G.J.W.Webb).

Plate 30. Aboriginal rock-painting of a Silver Barramundi *Lates calcarifer*, one of the major items of their diet (Photograph: G. Miles).

animals, they have colonised the monsoonal lowlands up to 100 km from the coast far more effectively than cattle, which originated further from the equator in drier habitats (Chapter 11). Buffaloes clearly have the biological potential for intensive utilisation, and already support a small but growing industry.

It can be seen from this account that so far all attempts at large-scale, high-yielding, monocultural exploitation of the monsoonal lowlands have failed. Elsewhere in the tropics the only intensive agricultural exploitation to succeed and persist has been peasant agriculture. It is characterised by its heterogeneity of crops and livestock, their close adaptation to the environment, and the marked flexibility and fine-tuning of management to meet the demands of a situation in which most environmental fluxes are of pronounced amplitude. It may seem to be crying in the wilderness to suggest that those trying to develop agriculture and other exploitation of the monsoonal lowlands should pay more attention to such strategies. It is an approach which would involve a gradualism of development and limitation of potential reward hitherto unfashionable. It might, nevertheless, be safer and ultimately more rewarding.

The wetland environment, in which Aboriginal hunter-gatherers were so successful, is the part of the monsoonal lowlands to which European colonists have paid least attention. It is biologically highly productive and may well have interesting potential for human use, as in other parts of the tropics (Ridpath et al. 1985). Artificial and semi-natural aquaculture is widely practised in tropical Asia, sometimes involving harvesting of more than one food (De Silva 1980, Fernando 1980). Considerable and sophisticated technical expertise exists there in the use of such resources. Some wetland and aquatic habitats in the monsoonal lowlands lend themselves to forms of semi-intensive farming of large edible fish, such as Barramundi, and of crocodiles, the skins of which command high prices. Enterprises such as these might also be attractive to Aboriginal communities seeking self-sufficiency in the emerging social order in which they find themselves now. These, or similar enterprises, have been initiated previously at the hand of non-Aborigines, and have failed (Rowley 1972). Lack of success has largely been the result of the entrepreneur's failure to acknowledge the very strong cultural base held by traditional Aboriginal societies. Such past failure need not indicate that appropriate schemes, preferably initiated by Aborigines themselves, cannot succeed in the future. The life-style of the Aborigines has been in the process of gradual change and this is likely to continue. Use of these traditional resources in a modified way seems to link past ways of life with today's new vistas. There are signs that such intensification of use of this environment is beginning

among Aboriginal communities. The sustainability and ecological impact of such increased use would no doubt call for careful exploration and assessment, as do other more conventional options such as the farming of buffaloes in different ways.

We conclude this section by noting that two of the present top three industries in the monsoonal lowlands are either not biologically based or only indirectly so. In terms of the value of its products, mining has been by far the most successful industry established in the monsoonal lowlands (Table 44). However it is an industry the profits of which tend to flow outside the local community, and even beyond Australia. By contrast the profits from tourism, equal in value to those from crops and livestock combined (Table 44), remain largely within the community. Tourism has grown rapidly, coming second only to mining by the late 1980s. It is an industry which depends upon the very features which have been the subject of most of this book.

CONSERVATION OF A TROPICAL HERITAGE

Some of the biggest increases in world population have taken place in tropical regions in recent times. One outcome is the growing disturbance and destruction of tropical ecosystems and the threat of extinction to an ever increasing number of plant and animal species. This has caused most concern in the humid tropics, for instance in the basin of the Amazon and the rainforests of southeast Asia.

Less concern has been shown about human impact on the wet-dry tropics, to which the monsoonal lowlands of northern Australia belong. That may well be because we are aware that humans have long been inhabitants of that environment. They have been moulded by this habitat, as we saw earlier in this chapter and, whatever their effects on it, they are of long standing. This habitat appears to represent their original home. Prevailing evidence suggests that humans evolved in the wet-dry tropics of Africa during the Pliocene and arrived in Australia only during the

Table 44. Gross values in 1981 of the top four industries in the wet-dry tropics of north-west Australia (north of 19°S in the Kimberley region of Western Australia, and the Top End of the Northern Territory). Sources: Cameron (1982) and Winter et al. (1985a).

Industry	Value of production	
	(10^6)	(%)
Mining	530	77
Livestock and crops	64	9
Tourism	63	9
Fishing and pearling	35	5
Total	692	100

late Pleistocene. Nevertheless the Australian monsoonal lowlands still provide us with some insights into the environmental conditions under which human evolution must have taken place, different though the fauna and flora of the two continents actually are. The Australian monsoonal lowlands, unlike equivalent regions of Africa, Asia or South America, have been much less altered by the increases of human populations of recent times. For this reason, conservation of these lowlands is of particular human as well as biological interest worldwide. This landscape preserves, still relatively intact, a significant habitat from our heritage. The unique galleries of rock-art in the Kakadu National Park depict the use of that habitat and provide a priceless window on the abiding concern humans have with the view of the natural world, based on what nowadays we know as ecology.

The monsoonal lowlands are also the home of a unique flora and fauna, as described in earlier chapters. The huge and varied population of waterbirds, the rich array of marsupials and insects, the unusual vegetation and spectacular scenery make the region of outstanding attraction to the general public, students and scientists alike. Yet its biota is still insufficiently described or understood. Much effort and resources are still needed to protect it from the effects of introduced plants and animals, mining development, and even tourists. There are several species such as the Oenpelli Python *Morelia oenpelliensis*, the Hooded Parrot *Psephotus chrysopterygius*, the Gouldian Finch *Erythrura gouldiae*, the Little Rock Wallaby *Peradorcas concinna*, and the Northern Nail-tailed Wallaby *Onychogalea unguifera*, to name but a few, which are rare and may be in danger of disappearance. The way in which fire is used as a management tool is so crucial, yet its effects on the biota remain to be properly assessed (see Chapter 5). All these issues need management based on knowledge of the complex manner in which nature works in the tropics. The managers of National Parks and conservation authorities can now call on a slowly growing body of research on the biota of the monsoonal lowlands to make decisions about conservation. As in the Kakadu National Park, they can also draw on the traditional knowledge of those to whom the region has always been home. We must manage our tropical resources wisely, outside as well as within National Parks and reserves. Thus we need much more knowledge from science and also about traditional Aboriginal management practices, based as they are on the empiricism of survival (Ridpath & Corbett 1985). We must know realistically what are our options.

Harry Frith devoted his career as a professional ecologist to trying to answer such problems with objective, insightful and pragmatic research. He had a particular enthusiasm for the monsoonal lowlands and was responsible for substantial research on its wildlife, notably its waterfowl. It is fitting that we end with a quotation about the region from his classic *Wildlife Conservation* (1979), revised three years before his death: '. . . it has been obvious, in my own work in the north, that nearly all of the present problems could have been forecast by a naturalist working in the area in the planning period. It is entirely possible that many of the major disasters to crops and to animal species could be prevented by appropriate modification of the development plan'.

REFERENCES

Anon. 1983. Farm Report. *Northern Territory Agricultural Newsletter*, April.

Australian Bureau of Census and Statistics 1983. *Apparent consumption of foodstuffs and nutrients.* Canberra: Australian Government Printer.

Bauer, F.H. 1964. *Historical geography of white settlement in part of northern Australia Part 2. The Katherine-Darwin region.* Canberra: CSIRO Division of Land Research and regional survey Report 64/1.

Bauer, F.H. 1984. What Man hath wrought: Geography and Change in northern Australia.In D.Parkes (ed.), *Northern Australia. The Arenas of Life and Ecosystems on Half a Continent*: 1-20. Sydney: Academic Press.

Bauer, F.H. 1985. A Brief History of Agriculture in North-West Australia. In R.C.Muchow (ed.), *Agro-research for the semi-arid Tropics: North-West Australia:* 12-28. St Lucia: University of Queensland Press.

Black, D.W. 1983. *Northern Territory Statistical Summary 1983.* Darwin: Australian Bureau of Statistics.

Bowler, J.M. 1982. Aridity in the late Tertiary and Quaternary of Australia. In W.R.Barker & P.J.M.Greenslade (eds.), *Evolution of the Flora and Fauna of Arid Australia*: 35-45. Frewville, South Australia: Peacock.

Braithwaite, R.W., M.L.Dudzinski, M.G.Ridpath & B.S.Parker 1984. The impact of water buffalo on the monsoon forest ecosystems in Kakadu National Park. *Aust. J. Ecol.* 9: 309-322.

Braithwaite, R.W. & J.A.Estbergs 1985. Fire patterns and woody vegetation trends in the Alligator Rivers region of northern Australia. In J.C.Tothill & J.J.Mott (eds.), *Ecology and Management of the World's Savannas:* 359-364. Canberra: Aust. Acad. Sci.

Braithwaite, R.W., W.M. Lonsdale & J.A. Estbergs 1989. Alien vegetation and native biota in tropical Australia: The impact of *Mimosa pigra. Biol. Conser.* 48: 189-210.

Burnley, I.H. 1984. Aborigines in northern Australia. In D.Parkes (ed.), *Northern Australia. The Arenas of Life and Ecosystems on Half a Continent*: 115-123. Sydney: Academic Press.

Calaby, J. 1971. Man, Fauna and Climate. In D.J.Mulvaney & J.Golson (eds.), *Aboriginal Man and Environment in Australia*: 80-93. Canberra: Australian National University Press.

Calaby, J. 1980. Ecology and Human Use of the Australian Savanna Environment. In D.R.Harris (ed.), *Human Ecology in Savanna Environments*: 321-337. London: Academic Press.

Cameron, B.J. & A.D.L.Hooper 1985. The Agricultural Development and Marketing Authority Scheme in the Northern Territory. In R.C.Muchow (ed.), *Agro-research for the semi-arid tropics: North-West Australia*: 488-503. St. Lucia: University of Queensland Press.

Cameron, D.M. 1985. Forest crops. In R.C.Muchow (ed.), *Agro-research for the semi-arid Tropics: North-West Australia*: 165-178. St Lucia: University of Queensland Press.

Cameron, R.J. 1982. *Northern Territory Statistical Summary 1982*. Darwin: Australian Bureau of Statistics.

Chappell, J. 1983. *CLIMANZ, Proceedings of first CLIMANZ 1981, Vol. 2.*: Figure 102. Canberra: Department of Biogeography and Geomorphology, R.S.Pac.S., Australian National University.

Christian, C.S. 1977. Agricultural cropping in northern Australia. In F.H.Bauer (ed.), *Cropping in northern Australia: Anatomy of Success and Failure*: 11-28. Canberra: Australian National University Press.

Clark, R.L. 1983. Pollen and charcoal evidence of the effects of aboriginal burning on the vegetation of Australia. *Archaeol. Oceania* 18: 32-37.

Davidson, B. 1980. The economics of pastoral and agricultural development in northern Australia. In R.Jones (ed.), *Northern Australia: Options and Implications*. Canberra: Australian National University Press.

De Silva, S.S. 1980. Aquaculture: A synthesis. In J.I.Furtado (ed.), *Tropical Ecology and Development, Proc. Int. Soc. Trop. Ecol.* 16: 1207-1208.

Edholm, O.G. 1978. *Man – Hot and Cold*. London: Arnold.

Fernando, C.H. 1980. Ricefield ecosystems: a synthesis. In J.I.Furtado (ed.), *Tropical Ecology and Development, Proc. Int. Soc. Trop. Ecol.* 16: 939-942.

Fisher, M.J., A.L.Garside, P.J.Skerman, A.L.Chapman, R.W.Strickland, R.J.K.Myers, D.F.Beech & E.F.Henzell 1977. The role of technical and related problems in the failure of some agricultural development schemes in northern Australia. In F.H.Bauer (ed.), *Cropping in northern Australia: Anatomy of Success and Failure*: 35-82. Canberra: Australian National University Press.

Fox, R.H., G.M.Bud, P.M.Woodward, A.J.Hackett & A.L.Hendrie 1974. A study of temperature regulation in New Guinea people. *Phil. Trans. R. Soc. London B.* 268: 375-391.

Frith, H.J. 1979. *Wildlife Conservation*. Sydney: Angus & Robertson.

Golley, F.B. & R.Misra 1972. Organic production in tropical ecosystems. *Bioscience* 22: 735-736.

Gordon, D., H.Silvertone & B.A.Smithurst 1972. The Epidemiology of skin cancer in Australia. In W.B.McCarthy (ed.), *Melanoma and Skin Cancer*. Sydney: NSW Government Printer.

Harris, D.R. 1980. Commentary: Human Occupation and Exploitation of Savanna Environments. In D.R.Harris (ed.), *Human Ecology in Savanna Environments*: 31-39. London: Academic Press.

Heatley, A. 1983. Socio-economic and political context. In D.Jaensch & P.Loveday (eds.), *Territorians or Mobile Australians:* 7-19. Darwin: Australian National University, North Australia Research Unit.

Hewitt, J.V. & B.Sanders 1984. Urban and non-urban aspects of road safety. *National Road Safety Symposium Canberra, 29-31 October 1984:* 45-57. Canberra: Australian Government Publishing Service.

Horton, D.R. 1982. The burning question: Aborigines, fire and Australian ecosystems. *Mankind* 13: 237-251.

Hounam, C.E. 1970. Climate and air-conditioning requirements in sparsely occupied areas of Australia. *World Meteorological Organization Technical Notes 109 WMO* 225: 175-183.

Hunter, I.J. 1981. The long-term effects of sunlight. *Trans. Menzies Foundation* 2: 121-123.

Isbell, R.F. 1985. Natural Environment – Soils. *Rural Research for northern Australia*: 12-17. Commonwealth Council for Rural Research and Extension. Canberra: Australian Government Publishing Service.

Jennings, J.N. 1967. Two maps of rainfall intensity in Australia. *Australian Geographer* 10: 256-262.

Jones, R. 1980. Hunters in the Australian Coastal Savanna. In D.R.Harris (ed.), *Human Ecology in Savanna Environments*: 107-146. London: Academic Press.

Jones, R. (ed.) 1985. Archaeological Research in Kakadu National Park. *Special Publication* 13. Canberra: Australian National Parks and Wildlife Service.

Jones, R. & J.Bowler 1980. Struggle for the savanna: northern Australia in ecological and prehistoric perspective. In R.Jones (ed.), *Northern Australia: Options and Implications*: 3-31. Canberra: Australian National University Press.

Keen, I. 1980. The Alligator Rivers' Aborigines – Retrospect and prospect. In R.Jones (ed.), *Northern Australia: Options and Implications*: 171-186. Canberra: Australian National University Press.

Kershaw, A.P. 1978. Record of last interglacial-glacial cycle from northeastern Queensland. *Nature* 272: 159-161.

Lacey, C.J. 1979. Forestry in the Top End of the Northern Territory – Part of the Northern Myth. *Search* 10: 174-180.

Latz, P.K. & G.F.Griffin 1978. Aboriginal land management in relation to fire and to food plants in central Australia. In B.S.Hetzel & H.J.Frith (eds.), *The nutrition of Aborigines in relation to the ecosystem of Central Australia*: 77-85. Melbourne: CSIRO.

Leichhardt, R.W.L. 1847. *Journal of an overland Expedition in Australia from Moreton Bay to Port Essington, 1844-1845*. London: Boone.

Macfarlane, W.V. 1981a. Water, salt and food for tropical survival. *Trans. Menzies Foundation* 2: 113-119.

Macfarlane, W.V. 1981b. Living and working in the heat. *Trans. Menzies Foundation* 2: 185-195.

MacKenzie, I. 1980. European incursions and failures in northern Australia. In R.Jones (ed.), *Northern Australia: Options and Implications*: 43-72. Canberra: Australian National University Press.

Mason, E.D. & M.Jacob 1972. Variations in basal metabolic rate responses to changes between tropical and temperate climates. *Human Biol.* 44: 141-172.

McCown, R.L., R.K.Jones & D.C.I.Peake 1985. Evaluation of a No-Till, Tropical Legume Ley-Farming Strategy. In R.C.Muchow (ed.), *Agro-research for the semi-arid tropics: North-West Australia*: 450-469. St Lucia: University of Queensland Press.

Merrilees, D. 1968. Man the destroyer: late Quaternary changes in the Australian marsupial fauna. *J. Roy. Soc. West. Aust.* 51: 1-24.

Miller, I.L., L.Nemestothy & S.E.Pickering 1981. *Mimosa pigra* in the Northern Territory. *Tech. Bull.* 51. Darwin: Northern Territory Department Primary Production.

Millham, P. & P.Thompson 1976. Relative antiquity of human occupation and extinct fauna of Madura Cave, South-western Western Australia. *Mankind* 10: 175-180.

Mitchell, W.H. 1981. Movement North into the Northern Territory. *Trans. Menzies Foundation* 2: 51-61.

Mollah, W.S. 1980. The Tipperary story: an attempt at large-scale grain sorghum development in the Northern Territory. *Northern Australia Research Bulletin* 7: 59-183.

Muchow, R.C. (ed.) 1985. *Agro-research for the semi-arid Tropics: North-West Australia*. St Lucia: University of Queensland Press.

Neil, C.C., J.A.Jones, T.B.Brealey & P.W.Newton 1984. Population stability in Northern Australian Resource Towns: Endogenous versus Exogenous Influences. In D.Parkes (ed.), *Northern Australia. The Arenas of Life and Ecosystems on Half a Continent*: 363-393. Sydney: Academic Press.

Parkes, D. 1984. The human population and some dimensions of ecological structure in northern Australia. In D. Parkes (ed.), *Northern Australia. The Arenas of Life and Eco-systems on Half a continent*: 89-113. Sydney: Academic Press.

Peterson, N. 1975. Hunter-Gatherer territoriality: the perspective from Australia. *Amer. Anthropologist* 77: 53-68.

Pickup, G. & J.E.Minor 1980. Assessment of research and practice in Australian natural hazards management. *Northern Australia Research Bulletin* 6: 1-207.

Pope, K.S. 1985. Study on departures from the Northern Territory Public Service. In P.Loveday & D.Wade-Marshall (eds.), *Economy and People in the North*: 202-211. Darwin: Australian National University, North Australia Research Unit.

Ridpath, M.G. 1985. Ecology in the Wet-Dry Tropics – how different? In M.G.Ridpath & L.K.Corbett (eds.), *Ecology of the Wet-Dry Tropics, Proc. Ecol. Soc. Aust.* 13: 3-20.

Ridpath, M.G. & L.K.Corbett 1985. VII Basis of Manage-ment – Introduction. In M.G.Ridpath & L.K.Corbett (eds.), *Ecology of the Wet-Dry Tropics, Proc. Ecol. Soc. Aust.* 13: 307-308.

Ridpath, M.G., J.A.Taylor & D.G.Tulloch 1985. Nature as a Model? In R.C.Muchow (ed.), *Agro-research for the semi-arid Tropics: North-West Australia*: 419-434. St Lucia: University of Queensland Press.

Rowley, C.D. 1972. *The Remote Aborigines*. Ringwood, Victoria: Penguin.

Scholefield, P.B. & K.J.Blackburn 1985. Horticulture. In R.C.Muchow (ed.), *Agro-research for the semi-arid tropics: North-West Australia*: 192-208. St Lucia: Univer-sity of Queensland Press.

Schrire, C. 1982. The Alligator Rivers. Prehistory and eco-logy in Western Arnhem Land. *Terra Australis* 7. Canberra: Australian National University Press.

Shackleton, N.J. & N.D.Opdyke 1976. Oxygen-isotope and palaeomagnetic stratigraphy of Pacific Core V28-239. Late Pliocene to Latest Pleistocene. *Geol. Soc. Amer. Mem.* 145: 449-464.

Singh, G., A.P.Kershaw & R.Clark 1981. Quaternary vegeta-tion and fire history in Australia. In A.M.Gill, R.H.Groves & I.R.Noble (eds.), *Fire and the Australian Biota*: 23-54. Canberra: Aust. Acad. Sci.

Stack, E. (n.d.). Human Health and Disease. In D.Giese (ed.), *Present Indicative and Future Conditional. A Selection of papers on Northern Development*: 125-136. Darwin: Northern Territory University Planning Authority.

Stanley, N.F. 1981. Changing patterns of infectious disease and human behaviour – Australia's North. *Trans. Menzies Foundation* 2: 147-159.

Stanley, N.F. 1984. Infectious disease: Human Ecosystems and Health in northern Australia. In D. Parkes (ed.), *North-ern Australia. The Arenas of Life and Ecosystems on Half a Continent*: 395-409. Sydney: Academic Press.

Stocker, G.C. & J.J.Mott 1981. Fire in the tropical forests and woodlands of northern Australia. In A.M.Gill, R.H.Groves & I.R.Noble (eds.), *Fire and the Australian Biota*: 425-439. Canberra: Aust. Acad. Sci.

Thiede, J. 1979. Wind regimes over the late Quaternary southwest Pacific Ocean. *Geology* 7: 259-262.

Thrift, K.J. 1982. Working in Heat and Cold. *CSIRO Occupa-tional Safety and Health Circular* 82/1.

Tindale, N.B. 1974. *Aboriginal Tribes of Australia*. Canberra: Australian National University Press.

Veevers, J.J. (ed.) 1984. *Phanerozoic earth history of Austra-lia*. Oxford: Clarendon Press.

Walker, J. 1981. Fuel dynamics in Australian vegetation. In A.M.Gill, R.H.Groves & I.R.Noble (eds.), *Fire and the Australian Biota*: 101-127. Canberra: Aust. Acad. Sci.

Wallis, H. 1982. The Rotz Atlas. A Royal Presentation. *The Map Collector* 20: 40-42.

Whelan, P.I. 1981. The vulnerability and receptivity of the Northern Territory to mosquito-borne disease. *Trans. Menzies Foundation* 2: 165-171.

White, C. 1971. Man and Environment in northwest Arnhem Land. In D.J.Mulvaney & J.Golson (eds.), *Aboriginal Man and Environment in Australia*: 141-157. Canberra: Australian National University Press.

Wigglesworth, E.C. 1981. Death on the roads: Road traffic mortality in the Northern Territory 1969-1980. *Trans. Menzies Foundation* 2: 173-178.

Williams, J., K.J.Day, R.F.Isbell & S.J.Reddy 1985. Constraints to Agricultural Development – Soils and Cli-mate. In R.C.Muchow (ed.), *Agro-research for the semi-arid Tropics: North-West Australia*: 31-92. St Lucia: University of Queensland Press.

Williams, M.A.J. 1969. Prediction of rainsplash erosion in the seasonally wet tropics. *Nature* 22: 763-765.

Williams, M.A.J. 1976. Erosion in the Alligator Rivers area. In R.Story et al. (eds.), *Lands of the Alligator Rivers area, Northern Territory*: 112-125. Melbourne: CSIRO, Land Research Series 38.

Williams, M.A.J. 1984. Cenozoic evolution of arid Australia. In H.G.Cogger & E.E.Cameron (eds.), *Arid Australia*: 59-78. Sydney: Australian Museum.

Williams, M.A.J. 1985. Pleistocene aridity in tropical Africa, Australia and Asia. In I.Douglas & T.Spencer (eds.), *Environmental Change and Tropical Geomorphology*: 219-233. London: Allen & Unwin.

Williams, M.A.J., D.A.Adamson & J.T.Baxter 1986. Late Quaternary environments in the Nile and Darling basins. *Australian Geographical Studies* 24: 128-144.

Williams, R.T. 1981. Potential. *Rural Research in northern Australia*: 122-132. Commonwealth Council for Rural Research and Extension. Canberra: Australian Govern-ment Printing Service.

Winter, W.H., A.C.Cameron, R.Reid, T.G.Stockwell & M.C.Page 1985a. Improved Pasture Plants. In R.C. Muchow (ed.), *Agro-research for the semi-arid tropics: North-West Australia*: 165-178. St Lucia: University of Queensland Press.

Winter, W.H., T.H.McCosker, D.Pratchett & J.D.A.Austin 1985b. Intensification of beef production. In R.C. Muchow (ed.), *Agro-research for the semi-arid tropics: North-West Australia*: 395-418. St Lucia: University of Queensland Press.

Young, E. 1983. Demography: Implications for the electorate. In D.Jaensch & P.Loveday (eds.), *Territorians or Mobile Australians*: 32-51. Darwin: Australian National University, North Australia Research Unit.

Glossary of scientific terms

Words of a scientific nature which are defined in most dictionaries have been omitted.

Abiotic – Non-living

Aestivate – To lie dormant during the dry season (summer).

Aggradation – Building up of the land through deposition of river alluvium.

Alkali earth – Soils which contain abundant calcium, sodium or potassium carbonates.

Allopatric – Occupying a different area to that of (an)other species.

Alluvial fan – Fan-shaped deposit of alluvial sand or gravel or mud, laid down as rivers emerged from confined valleys onto open plains.

Anabranch – A watercourse which leaves a river and later rejoins it lower down.

Anoestrous – Non-breeding (in mammals).

Anoxia – Absence of molecular oxygen (e.g. from water) (See *Hypolimnion*).

Aquifer – Water-bearing rock formation.

Archaean – That span of *Precambrian* geological time older than *Proterozoic* (See Table 1).

Autochtohonous – Original, native.

Base – Alkaline constitutents of soil, especially calcium, magnesium, potassium and sodium.

Benthos, Benthic – Floor of a water-body (e.g. of lagoon, lake or sea).

Billabong – Any lagoon or waterbody in northern Australia (derived from European use of an Aboriginal word of Southern Australia).

Bioenergetics – Concerning the flow of energy through a living system.

Biomass – Weight of living material.

Biota – The entirety of living organisms in a region.

Biotic – Having to do with life.

Blastocyst – An early stage in the development of a mammal, comprising a hollow ball of cells.

BP – Before present.

Caecum – Blind extension of the intestine (in which cellulose is often digested by micro-organisms).

Caprock – Resistant summit rock layer which protects underlying less-resistant rocks.

Catadromous – Refers to fish which migrate from freshwater to the sea (e.g., to spawn).

Chenier – Shelly beach ridge, overlying intertidal mud flat or freshwater swamp deposits.

Cleistogamous – Closed.

Cliseral shift – Successional development of vegetation from one climax to another.

Craton – Region of ancient rocks that have long been stable.

Crust – Outer relatively solid layer of the earth, between about 5 and 50 km thick.

Denudation – Weathering, erosion and lowering of the land surface.

Detritivor(s) – Animal(s) which eat(s) dead matter.

Diadromous – Refers to fish which undertake extensive migrations into freshwater from estuaries.

Ecological 'wides' – Plant species found on a variety of soil types.

Ecosystem – A biotic community and its abiotic environment.

Ecotone – Transition zone between two different biotic communities.

Ectothermic – (*Poikilothermic*). Refers to animals the body temperatures of which fluctuate according to that of their external environment (popularly referred to as 'cold-blooded').

Edaphic – Pertaining to or influenced by conditions of soil or substratum.

Endemism, Endemic – Presence within a localized area. Usually refers to species restricted to such an area.

Endoheic – Hydrologically closed (system); internally drained.

Epeirogenic uplift – Broad, continental-scale uplift.

Eutrophic, Eutrophication – With a high concentration of nutrients (refers to soils, lakes and other water-bodies) (See *oligotrophic*).

Fault – Rock fracture and displacement caused by compression or extension.

Feral – Refers to animals of domestic origin which have gone wild, living and breeding successfully outside human control.

Field capacity – The maximum amount of water stored by a soil and available for plant growth.

Fold – A bend or flexure of crustal rocks produced by compression.

Fossorial – Living underground.

Frugivory, Frugivorous – The dietary habit of eating fruit.

Geomorphic – Related to landforms and landscape.

Geomorphology – The study of the origin and distribution of landforms.

Geosyncline – Major downfolded basin in the earth's crust which receives sediment from a region of sub-continental extent.

Glacial (period) – A cold phase during an ice age characterised by lower world temperatures and lower sea levels (See *Quaternary*).

Gonad maturity stage index – A seven stage system in which gonad maturity stages are subjectively assigned according to the appearance and size of the gonads within the body cavity.

Gondwana(land) – An ancient southern land mass, comprising Antarctica, Australasia, India, South America and Africa, which began to separate some 160 million years ago.

Granivorous – Seed-eating.

Greywacke – Poorly sorted sandstone or gritstone composed mainly of angular to sub-angular rock fragments in a matrix of clay or silt-sized particles. A rock type common in *Geosynclines*.

Hadley cell – Part of world atmospheric circulation characterised by warm, rising air above the equator and by cool, subsiding air over latitudes 30°N and S (See *Intertropical Convergence Zone*).

Halophytic – Pertaining to plants of saline habitats.

Herpetofauna – That part of the fauna composed of amphibia and reptiles.

Holocene – The last 10,000 years of geological time till the present day. Also called Recent (See *Quaternary*).

Holomixis – The situation in which a waterbody has complete free circulation throughout the water column at the time of surface cooling (winter).

Homeotherm(ic) – An animal which maintains a constant body temperature (See *poikilothermic* and *ectothermic*).

Hypereutrophic – Strongly *eutrophic* (q.v.).

Hypolimnion, Hypolimentic – The zone in a waterbody comprising water of relatively uniform (and low) temperature which extends from the bottom to the area in which there is a marked change of temperature.

Hypotonic – Used to describe a concentration of dissolved substance lower than that of a specified concentration.

Icthyocidal – Fish-killing.

Igneous rocks – Rocks formed through crystallization and solidification of molten rock (or magma). Includes granitic and volcanic rocks, the former having large crystals as a result of very slow cooling.

Inlier – Outcrop of rocks wholly surrounded by yonger rocks (See *outlier*).

Intercalated – Rock layer sandwiched (or interbedded) between two other layers of rock.

Interglacial (period) – A warm phase during an ice age characterised by warmer world climates and higher sea levels (See *Quaternary*).

Intertropical convergence zone (I.T.C.Z.) – That part of the *Hadley Cell* in which moist equatorial air is rising and rainfall is high. The I.T.C.Z. moves N and S across the equator with the seasonal migration of the overhead sun.

Intrusive rocks – *Igneous* rocks which have been injected with molten into the existing solid rocks. Granites are common intrusive rocks which have cooled and solidified some 20-30 km below the earth's surface.

Ion(ic) – An electrically charged atom or group of atoms.

Irian – refers to the species of birds occupying lowland in northeastern Queensland and New Guinea and small pockets in the Top End and Kimberley.

Jet streams – A narrow band of fast, high-altitude winds. Wind speeds range from 100-400 km/hr; they are mainly found more than 12 km above sea level.

Joint – A fracture within rocks which otherwise show no relative displacement, unlike a *fault*.

Laterite – Highly weathered rock, depleted of bases and silica, but with high concentrations of iron or aluminium hydroxides.

Limnology, Limnological – The study of fresh waters and their biota.

Lithological – Relating to rock characteristics especially rock particle size.

Lithosphere – The earth's *crust*.

Macroinvertebrates – Large invertebrates.

Macrophytes – Aquatic plants, other than algae.

Mesic – Humid or moist.

Mesoeutrophic – *Eutrophic* (q.v.) to an average extent.

Mesotrophic – Intermediate between *oligotrophic* and *eutrophic* (q.v.).

Metamorphic rock – Rock altered by heat and pressure (See *metamorphism*). Metamorphism – Process(es) by which rocks are changed by heat and pressure so that they acquire a different physical and chemical composition.

Metasediments – Metamorphosed rocks that were originally sediments or sedimentary rocks.

Monsoon Forest – A form of depauperate rainforest found under monsoonal conditions, where there is a marked dry season.

Mutualism – Interaction between two species in which both benefit and are unable to persist separately. They have co-evolved.

Nectarivorous – Nectar-eating.

Neolithic – Cultural period, characterized by the use of polished stone tools, primitive cropping and animal husbandry. The period lasted in S.W. Asia from roughly 12,000 BP to 8,000 BP.

Oestrous – Refers to the cycle of reproductive events in female mammals in the absence of pregnancy.

Oligotrophic – With a low concentration of nutrients (refers to soils, lakes and other waterbodies) (See *eutrophic*).

Omnivore(s), Omnivorous – Refers to animals with a wide dietary range from plant to animal matter.

Oocyte – A cell which forms the ovum, as a result of the process of meiosis, after fertilization.

Ordovician – Geological period, 500-435 million years ago (See Table 1).

Orogenic uplift – Uplift during a major phase of mountain building (or orogenesis), often accompanied by folding and *faulting* of former *geosynclinal* sediments, and by *igneous intrusions*.

Orographic – Concerned with the physical relief of the land.

Osmosis, Osmotic – The movement of a solvent, such as water, from a solution of lower to a solution of higher concentration, through a semi-permeable membrane through which the solute cannot pass.

Outlier – Sedimentary rock outcrop separated by erosion from main body of rock and completely surrounded by older rocks (See *inlier*).

Oviparity, Oviparous – The laying of shelled eggs.

Oxygen-18 – Heavy isotope of oxygen. When ice caps melt they help to dilute the amount of oxygen-18 in sea water, so that the ratio of heavy to light oxygen in sea water can be used as an index of global ice.

Palaeocene – Geological period, 65-54 million years ago (See Table 1).

Palaeomagnetism – Traces of the earth's former magnetic field preserved in certain igneous and sedimentary rocks.

Palynological – Pertaining to the study of fossil spores, pollen (See *pollen spectra*).

Pandanus – A widely dispersed genus of small medium-sized palm-like plants which occurs throughout the tropical region except the neotropics.

Paperbark – *Melaleuca*. A genus of trees and shrubs, many of which possess a brown or pale-coloured, paper-like bark; often occurring in humid places.

Pentadactyle – Having five digits (i.e. toes, fingers) on a limb.

Periphyton – Biotic community on the surface of submerged objects.

pH – The chemical measure of acidity/alkalinity on a scale of 1 to 14.

Physiognomy – Physical characteristics or structure (of plants or plant communities).

Physiography – The study of landforms and landscape evolution (See *geomorphology*).

Phytoplankton – Plant community which floats free in water, mainly consisting of algae and diatoms.

Piscivorous – Fish eating.

Pisolithic – Composed of spherical, 3-6 mm diameter (pea-size) particles.

Placental – (Eutherian). Refers to mammals which nourish their young before birth via a membrane (placenta). cf. Marsupials.

Pleistocene – Geological epoch which began about two million years ago, lasting until the *Holocene* period (See *Quaternary*).

Pliocene – Geological epoch which began 5 million years ago, lasting until the *Pleistocene* (See Table 1).

Pneumatophore – Specialised root of mangrove and some other swamp plants that grows upwards and exchanges gases directly with the atmosphere.

Poikilotherm(ic) – Refers to animals, the body temperatures of which fluctuate according to that of their external environment (popularly called 'cold-blooded') (See *Ectothermic*).

Pollen spectra – Diagrams in which the varying amounts of different pollen grains are recorded as a guide to past and present regional and local plant cover (See *Palynological*).

Polygenic surface – A surface fashioned by different combinations of erosional and depositional processes at various stages in its development.

Precambrian – That span of geological time that includes the *Archaean* and the *Proterozoic* (See Table 1).

Prograding coast – A coast that is advancing seawards as a result of nearshore deposition.

Pyric succession – Succession of plants after fire.

Quaternary – The last 2 million years of geological time, including the *Pleistocene* and the *Holocene* (or *Recent*) (See Table 1).

Radiometric dating – Any method of dating rocks based upon measuring rates of decay of their radioactive elements.

Recent – See *Holocene*.

Rift(ing) – Formation of one or more linear troughs as a result of subsidence of a block of rocks bounded on each side by *faults*.

Riparian – Associated with rivers.

Ruminant – Mammals with a complex stomach in which plant food is retained for long periods, regurgitated for further chewing, and where cellulose is chemically broken down by micro-organisms. The species are often horned, lack upper incisor teeth and include deer, goats and cattle.

Sclerophyllous – Hard-leafed; an adaptation to a dry environment.

Serotiny – Retention of a mature fruit on a plant in an unopened condition.

Speciation – The process of evolution of a species.

Stratabound uranium – Concentration of uranium within rock layers (= strata) of a particular age or level.

Stratigraphy – The study of the characteristics, distribution and age of layered rocks.

Strike-ridge – Ridge aligned parallel to the long axis (or strike) of a major *fold*. Differential erosion removes the less resistant rocks leaving the more resistant layers as hills or ridges.

Suffrutescent – Refers to a plant many of the branches of which die after flowering, to leave a persistent woody base.

Sympatric – Living within the same area.

Taxon (plural Taxa) – A *taxonomic* (q.v.) group.

Taxonomy (ic) – The study of the classification of organisms mainly based on the degree of similarity of their structure and origins.

Tectonic – Related to internal forces of crustal deformation (See *crust*).

Tensional joint – A joint developed as a result of extensional stresses within the rock, for example, on the outside of a *fold*.

Thermoluminescence dating – Method of determining the age of a sediment by measuring the light emitted by particular minerals during heating (See *radiometric dating*).

Torresian – Refers to the biogeographic province comprising the monsoonal tropics of northern Australia, bordered inland by the arid Eyrean province.

Transgression – Advance of the sea across former land as a result of a rise in sea level.

Triptonic turbidity – Refers to turbidity caused by non-living particles in suspension in water.

Trophic – To do with feeding. *Trophic level* is a stage of feeding representing one of the links in the food chain; all organisms belonging to that trophic level will obtain their energy and food in broadly comparable ways.

Tropics – Between the latitudes of Capricorn and Cancer (i.e., up to $23\frac{1}{2}°$ from the equator).

Tumbunan – Refers to the species of birds occupying temperate, montane rainforests in the ranges of Eastern Australia and the central highlands of New Guinea.

Unconformity – Break in a geological sequence caused either by erosion or by non-deposition. The break may be abrupt, and may represent the lapse of considerable time.

Ungulate – Mammals with hooves, most of which belong to the orders Artiodactyla (even-toed hooves including pigs, deer and cattle) and the Perrisodactyla (odd-toed hooves including horses).

Vagile – Denotes ability to disperse effectively and establish.

Variance – Statistical term; a measure of the dispersion of the probability of occurrence of numerical values about their highest point; more precisely the square of the standard deviation.

Vertebrate – Animal(s) with backbones.

Vicariant – Closely related (in regard to species).

Viviparity – Producing live young.

Walker Circulation – A quasi-periodic circulation over the equatorial Pacific reflected in anomalous temperatures and pressures over northern Australia and the eastern Pacific. Often called the Southern Oscillation, it is associated with extreme floods and droughts in Australia, and many other regions.

Xeric – Dry.

Xeromorphic – Pertaining to parts of a plant which protect it against aridity.

Zoogeography – The study of the distribution of animals.

Zooplankton – The free-floating animal community which is independent of the shore and bottom of water bodies.

Contributors

Bishop, K.A., Fisheries Research Institute, Dept. of Agriculture, Cronulla, New South Wales. (Previously Alligator Rivers Region Research Institute, Jabiru, Northern Territory)

Braithwaite, Dr. R.W., CSIRO Division of Wildlife and Ecology, Darwin

Brennan, K.G., Alligator Rivers Region Research Institute, Jabiru, Northern Territory

Dunlop, C.R., Herbarium of the Conservation Commission of the Northern Territory, Darwin

Forbes, M.A., Australian National Parks and Wildlife Service, Canberra

Friend, Dr. G.R., Department of Conservation and Land Management, Perth. (Previously CSIRO Division of Wildlife and Ecology, Darwin)

Haynes, C.D., Department of Conservation and Land Management, Perth. (Previously Australian National Parks and Wildlife Service, Darwin)

Kingston, D., Groundwater consultant, Brisbane. (Previously Water Division, Darwin)

MacDonald, Dr. N.S., Department of the Environment, Canberra

McAlpine, J., CSIRO Division of Water and Land Resources, Canberra

Meehan, Dr. Betty, Australian Museum, Sydney

Morton, Dr. S.R., CSIRO Division of Wildlife and Ecology, Alice Springs. (Previously Alligator Rivers Region Research Institute, Jabiru, Northern Territory)

Newsome, Dr. A.E., CSIRO Division of Wildlife and Ecology, Canberra

Ridpath, Dr. M.G., CSIRO Tropical Ecosystems Research Centre, Darwin

Webb, Dr. G.J.W., Biological Consultant, Conservation Commission of the Northern Territory, Darwin

Webb, Prof. L.J., School of Australian Environmental Studies, Griffith University, Brisbane

Williams, Dr. C.K., CSIRO Division of Wildlife and Ecology, Canberra

Williams, Prof. M.A.J., Department of Geography and Environmental Science, Monash University, Melbourne

Wombey, J.C., CSIRO Division of Wildlife and Ecology, Canberra

Index